SPECTRAL THEORY OF LINEAR OPERATORS

Volume I

ABRAM I. PLESNER

SPECTRAL THEORY OF LINEAR OPERATORS

Volume I

Translated from the Russian
by
MERLYND K. NESTELL *and* ALAN G. GIBBS

FREDERICK UNGAR PUBLISHING CO.
NEW YORK

Title of the Russian original:
Spektral'naia Teoriia Lineinykh Operatorov

Volume 1: Chapters 1–5

Volume II: Chapters 6–10

Standard Book Number: 8044-4766-7

Library of Congress Catalog Number: 68-20524

AMS 1969 Subject Classifications: 4730

Printed in the United States of America

PREFACE

The name of Abram Iezekiilovich Plesner, and his lectures on the theory of operators,* are widely known, so this book needs no recommendation. My task is to summarize briefly its history.

The book was conceived by the author long ago, even before the war. He thought at first that he would not modify his lectures greatly, but after later reflection on the way that the spectral theory of linear operators should be presented, he reconsidered and decided to write everything anew. The work was apparently begun in 1948, and it progressed very quickly when it was not interrupted by illness. During the following ten years, whenever his health permitted he returned to the work, but he did not succeed in finishing it. On April 18, 1961, he passed away. Not long before that he entrusted to me the task of completing the work.

According to the author's plan the book was to consist of eleven chapters. He had time to complete more or less eight chapters and to sketch the ninth. With respect to the contents of Chapter 10, he left only a few verbal wishes which did not affect the manner of exposition; concerning Chapter 11, he said only that he intended to devote it to applications: differential operators, dynamical systems, and perhaps quantum mechanics. Not a single line of Chapters 10 or 11 had been worked out. Under these circumstances, it was decided to write Chapters 9 and 10 following the suggestions of the author, using his lectures, and to publish the book without Chapter 11.

The work on Chapters 9 and 10, and the two final, unfinished sections of Chapter 8, was taken on by L. M. Abramov, and the difficult responsibility of editing the book was assumed by B. M. Makarov. They did everything to retain the style of A. I. Plesner. Now, finally, the book to which this remarkable man and mathematician gave so much of himself, goes out into the world.

V. A. Roklin

June 1965

* Uspekhi Matematicheskikh Nauk, Vol. IX (old series) 1941, and part 1, Vol. 1, 1946.

TRANSLATORS' PREFACE

This book is intended for both mathematicians and workers in the applied sciences. The reader should have some acquaintance with measure theory and elementary topology. Occasionally the author used subtle facts from analysis without explanation, for example, the existence almost everywhere of boundary values for an analytic function of the class H_2.

Unfortunately, the author left neither historical background nor a bibliography. In fact, there are very few references in his papers to the literature. It is with some reservations that we have not included some references; however, the reader is referred to other well-known works on spectral theory for references to the general literature.

We have corrected errors that we found and would like to be informed about any other errors the reader may come across.

For citations in the text, the following scheme is used: §10.2 means Chapter 10, Section 2; and 8.2.3 means article 2.3 of §8.2. Each section has an independent indexing of equations. In each section, only the number of the equation is cited; for references to equations in other sections, this is indicated by using the number of the section; for example, Eq. (3.2.1) means Eq. (1) of §3.2. A triple number is also used for theorems and lemmas; for example, Theorem 8.5.1. refers to the first theorem in §8.5.

We would like to thank Professor James Brown of Oregon State University for critically reading parts of the translation. We also thank Mrs. Ramona Nestell for her expert preparation of the typescript.

M.K.N.

A.G.G.

January 1969
Battelle Memorial Institute
Richland, Washington

CONTENTS

SPECTRAL THEORY OF LINEAR OPERATORS

Volume I

CHAPTER 1

Vector Systems and Order Relations

§ 1. Ordered Sets

1.1. Equivalence and order relations. Let the arguments ξ and η of a function $\theta(\xi, \eta)$, which takes on the values 0 and 1, range independently over a set Z. Then by definition * *the function $\theta(\xi, \eta)$ defines a relation θ in Z. Elements ξ and η in Z are said to be in the relation θ if*

* Editor's Note: By means of the words "by definition" the author always *introduces* definitions. These words will not refer to definitions given earlier, or in the literature.

$$\theta(\xi, \eta) = 1. \tag{1}$$

The function

$$\bar{\theta}(\xi, \eta) = 1 - \theta(\xi, \eta)$$

defines a *dual relation* $\bar{\theta}$ in Z: *if ξ and η in Z are not in the relation θ, then they are in the relation $\bar{\theta}$.* Sometimes $\theta[\xi; \eta]$ or $\xi\theta\eta$ is written in place of (1). Correspondingly, if ξ and η are not in the relation θ, then we write $\xi\bar{\theta}\eta$.

By definition, a relation θ in Z is:

(1) *reflexive*, if $\xi\theta\xi$ for any $\xi \in Z$;

(1′) *irreflexive*, if $\xi\bar{\theta}\xi$ for any $\xi \in Z$;

(2) *symmetric*, if $\xi\theta\eta$ implies $\eta\theta\xi$;

(2′) *asymmetric*, if for a pair, ξ, η, it is impossible that $\xi\theta\eta$ and $\eta\theta\xi$ simultaneously;

(3) *transitive*, if $\xi\theta\eta$ and $\eta\theta\zeta$ imply that $\xi\theta\zeta$.

1°. A reflexive, symmetric and transitive relation is called an *equivalence relation*, or simply an *equivalence*.

In the case of an equivalence relation θ we will write

$$\xi \equiv \eta\,(\theta) \tag{2}$$

instead of $\xi\theta\eta$, and we will call (2) a *congruence*. The elements ξ and η are said to be *congruent* (*equivalent*) *with respect to* θ, or *modulo* θ.

If elements ξ and η in Z are congruent modulo θ, then by definition they have the same type $\mathring{\xi}$ *with respect to* θ, and are *representatives* of this type.

Two types (with respect to θ) are equal if and only if they have the same representatives. An equivalence θ decomposes the set Z into disjoint classes of elements which are equivalent with respect to θ; each class consists of precisely all representatives of the same type. Conversely, any decomposition of the set Z into disjoint classes of elements defines an equivalence relation for which the congruence of two elements means that they belong to the same class.

The mapping which takes an element $\xi \in Z$ into $\mathring{\xi}$, its type with respect to the equivalence θ, will also be denoted by θ, and we will write

$$\mathring{\xi} = \theta(\xi), \qquad \mathring{Z} = \theta(Z), \tag{3}$$

where $\mathring{Z}$ is the set of all types with respect to θ.

Two trivial equivalences can be found in every set Z:

(1) $\xi \equiv \eta\ (\theta)$ for any ξ and η in Z. In this case, the set $\mathring{Z}$ consists of exactly one element whose representatives are all the elements of Z.

(2) $\xi \equiv \eta\ (\theta)$ if and only if $\xi = \eta$. In this case equivalence coincides with equality and the mapping θ is one-to-one.

REMARK. An equivalence relation possesses the basic properties of equality, and the transition from a set Z to the set $\mathring{Z}$ of types with respect to the equivalence (2) may be understood in the following sense: we introduce a new definition of equality in the set Z, namely, two elements ξ and η in Z are said to be equal if (2) holds. The set $\mathring{Z}$ now denotes the set Z equipped with a new concept of equality, and elements, equal (in the new sense) to ξ, are representatives of the type $\mathring{\xi}$.

2°. A reflexive and transitive relation is called an *order relation*.

An order relation θ is said to be *proper* if $\xi\theta\eta$ and $\eta\theta\xi$ always imply that $\xi = \eta$. Otherwise the relation is called *improper*.

An example of a proper order relation is: let Z be the set of real numbers with inequality $\xi \geqslant \eta$ as $\xi\theta\eta$. An example of an improper order relation is the relation $Z_1\theta Z_2$ defined in the set of complex numbers by the condition:

$$Z_1\theta Z_2, \qquad \text{if} \qquad \operatorname{Re} Z_1 \leqslant \operatorname{Re} Z_2.$$

According to our definition, equivalence is an order relation, but a proper equivalence relation θ is synonymous, in view of its symmetry, with equality.

An improper order relation θ generates a proper order relation in the following manner. Let $\xi\theta_0\eta$ denote that $\xi\theta\eta$ and $\eta\theta\xi$ hold simultaneously. Then:

(1) θ_0 is an equivalence relation;

(2) the relation θ induces a proper order relation in the set Z_0 of types with respect to θ_0.

It is evident that the relation θ_0 is reflexive and symmetric since it follows from $\xi\theta_0\eta$ and $\eta\theta_0\zeta$, i.e. from $\xi\theta\eta$, $\eta\theta\xi$, $\eta\theta\zeta$, $\zeta\theta\eta$ that $\xi\theta\zeta$ (since $\xi\theta\eta$ and $\eta\theta\zeta$)

and $\zeta\theta\xi$ (since $\zeta\theta\eta$ and $\eta\theta\xi$). Thus, $\xi\theta_0\eta$ and $\eta\theta_0\zeta$ imply that $\xi\theta_0\zeta$, i.e. θ_0 is transitive.

We now let $\mathring{\xi}\mathring{\theta}\mathring{\eta}$ for types $\mathring{\xi}$, $\mathring{\eta} \in Z_0$, if $\xi\theta\eta$ for their representatives ξ and η, and we will show that the definition of $\mathring{\theta}$ is independent of the choice of the representatives ξ and η.

Suppose ξ', η' are other representatives of the types $\mathring{\xi}$, $\mathring{\eta}$. Then together with $\xi\theta\eta$, we have $\xi'\theta\xi$ and $\eta\theta\eta'$ (since θ is reflexive) and this means that $\xi'\theta\eta'$ (by the relations $\xi'\theta\zeta$, $\xi\theta\eta$, $\eta\theta\eta'$ and the transivity of θ). The relation $\mathring{\theta}$ induced by θ is also an order relation, and moreover, is proper: $\mathring{\xi}\mathring{\theta}\mathring{\eta}$ and $\mathring{\eta}\mathring{\theta}\mathring{\xi}$ imply that $\xi\theta\eta$ and $\eta\theta\xi$ for representatives ξ and η of the types $\mathring{\xi}$ and $\mathring{\eta}$, i.e. ξ and η are congruent with respect to θ_0, hence $\mathring{\xi} = \mathring{\eta}$.

If, instead of $\theta\,(\xi, \eta)$, we consider a function $\theta\,(\xi^{(1)}, \xi^{(2)}, \dots, \xi^{(n)})$ $(n \geqslant 1)$ which has the values 1 and 0, and whose arguments $\xi^{(k)}$ range independently over a set Z, then we have a generalization of the concept of relation. The function $\theta(\xi^{(1)}, \xi^{(2)}), \dots, \xi^{(n)})$ gives by definition an *n-term relation* θ: the elements $\xi^{(1)}, \xi^{(2)}, \dots, \xi^{(n)}$ in Z (in the given order) stand in the relation θ if

$$\theta\,(\xi^{(1)}, \xi^{(2)}, \dots, \xi^{(n)}) = 1. \tag{1'}$$

The function

$$\bar{\theta}\,(\xi^{(1)}, \xi^{(2)}, \dots, \xi^{(n)}) = 1 - \theta\,(\xi^{(1)}, \xi^{(2)}, \dots, \xi^{(n)})$$

defines the dual relation $\bar{\theta}$ in Z.

A two-term relation will be called simply a relation.

A one-term relation defines a property of elements in Z, namely the property that an element $\xi \in Z$ satisfies the equation $\theta(\xi) = 1$, i.e. belongs to a set whose characteristic function is θ; an n-term relation $(n > 1)$ can be considered as a property of points of the sets

$$Z_n = Z \times Z \times \dots \times Z \quad (n \text{ factors})$$

of all systems $\{\xi^{(1)}, \xi^{(2)}, \dots, \xi^{(n)}\}$ of n elements of the set Z, i.e. as one-term relation in Z_n.

We will write (1′) in the form $\theta[\xi^{(1)}, \xi^{(2)}, \dots, \xi^{(n)}]$.

The relation θ is by definition the identity in the set $Z' \subset Z$, if $\theta[\xi^{(1)}, \xi^{(2)}, \dots, \xi^{(n)}]$ holds when $\xi^{(k)}$ $(k = 1, 2, \dots, n)$ range independently through Z', and in particular it is the identity in Z if $\theta[\xi^{(1)}, \xi^{(2)}, \dots, \xi^{(n)}]$ always holds.

1.2. Majorants and minorants. Chains. A set Z in which a proper order relation is defined is said to be an *ordered set*.

For a fixed order relation θ defined in the set Z, we introduce a special notation, namely we will write $\xi \geqslant \eta$ or $\eta \leqslant \xi$ if $\xi\theta\eta$. Furthermore, we will write $\xi > \eta$ or $\eta < \xi$, if $\xi\theta\eta$ and $\xi \neq \eta$.

The relation $>$ is: (1) irreflexive ($\xi > \xi$ is impossible), (2) asymmetric ($\xi > \eta$ and $\eta > \xi$ are inconsistent), and (3) transitive. We note that if a relation is irreflexive and transitive, then it is asymmetric. Indeed, if a pair of elements ξ, η could be found for which $\xi \theta \eta$ and $\eta \theta \xi$ simultaneously, then by the transitivity we would have $\xi \theta \xi$ for the element ξ, which contradicts the irreflexibility.

Conversely, if a relation θ is irreflexive and transitive, then it defines a proper order relation in the following way: $\xi \geqslant \eta$ if and only if $\xi \theta \eta$ or $\xi = \eta$. Indeed, the relation $\geqslant$ is reflexive and transitive, and since θ is irreflexive the order relation $\geqslant$ is proper. It is obvious that the corresponding relation $>$ coincides with the relation θ.

An element η in an ordered set Z is said to be *subordinate* to ξ if $\xi \geqslant \eta$. It is *strictly subordinate* if $\xi > \eta$.

The relation $\geqslant$ induces a proper order relation in every subset of Z, and therefore any subset of an ordered set is also an ordered set.

An order relation is introduced in a natural manner in the system $Z = \mathfrak{P}(T)$ of subsets of the set T, namely: $M \geqslant N (M \in \mathfrak{P}(T),\ N \in \mathfrak{P}(T))$ if $M \supset N$, i.e. all elements of the set N belong to M. The relation $\geqslant$ is a proper order relation (by inclusion) and it converts the system $\mathfrak{P}(T)$ and each of its subsets into ordered sets.

Other important examples of ordered sets are families of functions. We consider the collection $\mathfrak{G}_r(T)$ of all bounded real functions $F(\tau)$, defined on the set T. We introduce the proper order relation $\geqslant$ in $\mathfrak{G}_r(T)$ by letting $F \geqslant G$ for functions $F = F(\tau)$ and $G = G(\tau)$ in $\mathfrak{G}_r(T)$, if $F(\tau) \geqslant G(\tau)$ for all $\tau \in T$. The definition $F \geqslant G$ can be extended to functions of the family $\tilde{\mathfrak{G}}_r(T)$ of all finite real functions on T. In both cases the proper order relation $\geqslant$ converts $\mathfrak{G}_r(T)$ and $\tilde{\mathfrak{G}}_r(T)$ into ordered sets. Every subset of $\tilde{\mathfrak{G}}_r(T)$ (with the relation $\geqslant$ induced from $\tilde{\mathfrak{G}}_r(T)$) is itself an ordered set.

Let Z be an ordered set and let N be a subset of Z. An element $\zeta \in Z$ is called a *majorant* of N, if all elements in N are subordinate to ζ, and a *minorant* of N if ζ is subordinate to each of the elements of N.

If a set $N \subset Z$ has a majorant (minorant), then it is called *bounded above* (*below*). The set N is by definition *bounded* (in Z) if it is bounded both above and below.

If the majorant (minorant) of a set $N \subset Z$ belongs to N, then it is the greatest (least) element of the set N.

An ordered set Z is bounded (in itself) above (below) if and only if it has a greatest (least) element. For example, the ordered set $\mathfrak{P}(T)$ is bounded: T, and the empty set are its greatest and least elements.

It is necessary to distinguish so-called *maximal* and *minimal* elements from greatest and least elements. An element η is called a *maximal element* of the set N if N contains no elements to which η is strictly subordinate, and a *minimal element* if there is no element in N which is strictly subordinate to η.

A subset N of an ordered set Z is said to be a *chain* in Z if for any two elements of the set N, one is subordinate to the other. For a chain N, the maximal (minimal) element is the greatest (least) and, hence, is unique.

A chain N is said to be *upper maximal* in Z if there is no element in Z to which all elements of the chain are strictly subordinate, and correspondingly *lower maximal* in Z if there is no element in Z which is strictly subordinate to every element of the chain.

A chain which is upper maximal is either not bounded in Z, or it has a greatest element which is a maximal element of the set Z. A similar remark holds for chains which are lower maximal.

If all elements of a chain N belong to a chain N', then N' is called a continuation of N. A chain N which is not upper maximal can always be continued "above" by adding to it an element $\eta \in Z$ to which all elements of N are strictly subordinate. If the chain obtained in this manner is not upper maximal, then the process of continuation can be repeated. By using the theory of transfinite numbers and transfinite induction, it is possible to continue this process until the chain is upper maximal.

By such a procedure it is possible to prove:

LEMMA 1.1.1. *Every chain in an ordered set can be continued to a chain which is upper maximal.*

Lemma 1.1.1. is equivalent to the axiom of choice. In view of the intuitive clarity of Lemma 1.1.1, we will consider it to be a postulate. In what follows, the use of this lemma permits us to avoid recourse to the unwieldly, occasional application of the theory of transfinite numbers. Essentially, Lemma 1.1.1 is one of the variants of the well-known Zorn's lemma.

1.3. Dual sets. A dual relation θ_d corresponds to a proper order relation θ: $\xi\theta_d\eta$, if $\eta\theta\xi$. The relation θ_d is also a proper order relation. In particular, if θ is an order relation in an ordered set Z, then the set Z with θ replaced by its dual θ_d is called the dual to the set Z (with the relation θ).*

Under the transition from the ordered set Z to its dual the following roles are interchanged: (1) majorants and minorants, (2) greatest and least elements, (3) maximal and minimal elements, (4) boundedness above and boundedness below; the concepts of boundedness and of a chain are preserved (upper maximal chains become lower maximal chains and vice versa). In the statement of Lemma 1.1.1 the words "chain which is upper maximal", would thus be replaced by the words "chain which is lower maximal".

* Translator's Note: The word "dual" is used in two different senses in this sentence. When "dual" occurs, it should be clear from the context in what sense this word is being used.

§ 2. Vector Systems

2.1. Vector operations. A set $\mathfrak{V}$ is said to be a *vector system over a field K*, if

(1) $\mathfrak{V}$ is closed with respect to the linear operations of addition and scalar multiplication, namely: the sum $f+g$ of elements f and g in $\mathfrak{V}$ belongs to $\mathfrak{V}$, and the product αf of an element $f\in\mathfrak{V}$ and an element α of the field K belongs to $\mathfrak{V}$;

(2) $\mathfrak{V}$ is an abelian (commutative) group with respect to addition;

(3) for $f\in\mathfrak{V}$, $g\in\mathfrak{V}$ and $\alpha, \beta\in K$,

(a) $$\alpha(f+g) = \alpha f+\alpha g,$$

(b) $$(\alpha+\beta)f = \alpha f+\beta f,$$

(c) $$\alpha(\beta f) = (\alpha\beta)f,$$

(d) $$1\cdot f = f.^*$$

Elements of a vector system $\mathfrak{V}$ are also called *vectors*, and the linear operations are called *vector operations*.

Since $\mathfrak{V}$ is a group with respect to addition it has a zero element which, as in the case of the zero of the field K, we will denote by 0.

It is not difficult to see that

$$0\cdot f = 0, \quad \alpha\cdot 0 = 0$$

for $f\in\mathfrak{V}$ and $\alpha\in K$, and that the elements f and $(-1)f$ are inverses in the group $\mathfrak{V}$, i.e.

$$(-1)f = -f.$$

The zero of the group $\mathfrak{V}$ is also called the *zero vector*, and if it is the only element of the vector system then the latter is called *null*.

A subset $\mathfrak{V}'$ of a vector system $\mathfrak{V}$ is called a *vector subsystem* if whenever vectors f and g belong to $\mathfrak{V}'$, the elements $\alpha f (\alpha\in K)$ and $f+g$ belong to $\mathfrak{V}'$.

For fixed $f\in\mathfrak{V}$, and arbitrary $\alpha\in K$, the elements αf form a vector subsystem which we denote by (f). In particular, when $f=0$, (0) is the null vector system.

A vector subsystem $\mathfrak{V}'$ is called *proper* if it is neither $\mathfrak{V}$ nor (0).

Let $\mathfrak{V}_\tau$ $(\tau\in T)$ be any collection of vector subsystems of a system $\mathfrak{V}$. Elements which belong simultaneously to all the $\mathfrak{V}_\tau$ form a vector subsystem $\mathfrak{V}'$ which is called the intersection of the subsystems $\mathfrak{V}_\tau$. For the intersection $\mathfrak{V}'$ we write

$$\mathfrak{V}' = \bigcap_\tau \mathfrak{V}_\tau.$$

For each subset M of a vector system $\mathfrak{V}$ there exists a smallest vector subsystem $\mathfrak{V}' = \mathscr{L}(M)$ which contains M, namely the intersection of all vector

* 1 denotes the identity of the field K.

subsystems of $\mathfrak{B}$ which contain the set M. The vector system $\mathscr{L}(M)$ is called the *vector envelope* or *linear hull* of the set M, and it can be obtained in the following, more effective manner.

Every expression of the form

$$\alpha_1 f_1 + \alpha_2 f_2 + \dots + \alpha_n f_n = \sum_{k=1}^{n} \alpha_k f_k$$

with coefficients α_k belonging to the field K will be called a *linear combination* of the elements $f_1, f_2, \dots, f_n$ in $\mathfrak{B}$. If all of the coefficients α_k equal zero, then the linear combination is called *trivial.*

An element $h \in \mathfrak{B}$ is by definition *linearly dependent on the set M*, if it can be represented as a linear combination of elements of this set, i.e. if

$$h = \sum_{k=1}^{m} \alpha_k f_k, \tag{1}$$

where

$$f_k \in M, \quad \alpha_k \in K, \quad k = 1, 2, \dots, m.$$

The collection of elements h which are linearly dependent on M is the linear hull $\mathscr{L}(M)$ of the set M. Indeed, elements h which are dependent on M obviously form a vector subsystem of $\mathfrak{B}$ and every vector subsystem of $\mathfrak{B}$ which contains the set M also contains all linear combinations of its elements.

Let the set M be the union (the set-theoretic sum) of vector subsystems $\mathfrak{B}_\tau$ $(\tau \in T)$ of the system $\mathfrak{B}$. The linear hull $\mathscr{L}(M)$ then consists of those elements h that can be represented in the form

$$h = \sum_{k=1}^{n} f_{\tau_k} \qquad (n \text{ arbitrary}) \tag{2}$$

where $f_{\tau_k} \in \mathfrak{B}_{\tau_k}$. Indeed, elements h of the form (2) form a vector system and belong to $\mathscr{L}(M)$. We will usually write (2) in the form of a sum $(f_\tau \in \mathfrak{B}_\tau)$

$$h = \sum_{\tau} f_\tau, \tag{2'}$$

which has been extended to all τ in T, keeping in mind that only a finite number of vectors f_τ in (2′) are different from zero.

The vector subsystem $\mathfrak{B}'$ of elements h which can be represented in the form (2′), where $f_\tau \in \mathfrak{B}_\tau$ and only a finite number of vectors f_τ are different from zero, will be called the *vector sum of the vector subsystems* $\mathfrak{B}_\tau$ $(\tau \in T)$ of the system $\mathfrak{B}$.

The sum $\mathfrak{B}'$ is the linear hull of the union of sets $\mathfrak{B}_\tau$, and we will write

$$\mathfrak{B}' = \sum_{\tau} \mathfrak{B}_\tau .$$

2.2. Examples of vector systems.

1°. The field K. The field K itself, with its operations of addition and multiplication, is a vector system $\mathfrak{V}_1$ over K.

2°. The system $\mathfrak{S}(T; K)$. Let T be an arbitrary set. We denote by $\mathfrak{S}(T; K)$ the family of all functions $f = f(\tau)$ with domain $T(\tau \in T)$ and range in the field K. We convert $\mathfrak{S}(T; K)$ into a vector system by defining vector operations in $\mathfrak{S}(T; K)$, namely

$$h = f + g, \qquad f' = \alpha f \qquad (\alpha \in K), \tag{3}$$

if

$$h(\tau) = f(\tau) + g(\tau), \qquad f'(\tau) = \alpha f(\tau) \tag{3'}$$

for any $\tau \in T$ (ordinary addition of functions and multiplication of functions by constants). The vector system $\mathfrak{S}(T; K)$, along with its vector subsystems, gives a large supply of vector systems which we will call *vector systems of functions.*

Among the vector subsystems of $\mathfrak{S}(T; K)$ we note:

(a) the system $\mathfrak{S}^{(0)}(T; K)$ of functions in $\mathfrak{S}(T; K)$ which differ from zero at not more than a finite number of points;

(b) the system $\mathfrak{S}^{(1)}(T; K)$ of functions in $\mathfrak{S}(T; K)$ which differ from zero perhaps only on a countable set of points of T.

If the set T is finite and contains n elements, for which it is always possible to take the natural numbers $1, 2, \ldots, n$, then the vector system $\mathfrak{S}(T;K) = \mathfrak{S}^{(0)}(T; K)$ consists precisely of all finite sequences of the form $\{c_1, c_2, \ldots, c_n\}$ in K $(c_k = f(k))$. If T is a countable set, then $\mathfrak{S}(T; K) = \mathfrak{S}^{(1)}(T; K)$ and it is possible to choose T to be the sequence of all natural numbers. Then the vector system $\mathfrak{S}(T; K)$ consists of precisely all (infinite) sequences $\{c_n\}$ of elements of the field K $(c_n = f(n))$.

If K is the field of complex numbers K_c, then we will write $\mathfrak{S}(T; K_c) = \tilde{\mathfrak{S}}(T)$, $\mathfrak{S}^{(0)}(T; K_c) = \mathfrak{S}^{(0)}(T)$, $\mathfrak{S}^{(1)}(T; K_c) = \mathfrak{S}^{(1)}(T)$, and $\mathfrak{S}(T)$ for the systems of bounded functions in $\tilde{\mathfrak{S}}(T)$. In the case when K is the field of real numbers K_r, then we will write $\mathfrak{S}(T; K_r) = \tilde{\mathfrak{S}}_r(T)$ and the collection of all bounded functions in $\tilde{\mathfrak{S}}_r(T)$ will be denoted, as previously, by $\mathfrak{S}_r(T)$.

3°. The system $\mathfrak{S}(T; \mathfrak{V}_\tau)$. To each value $\tau \in T$ let there correspond a vector system $\mathfrak{V}_\tau$ over the field K. The collection of functions $f(\tau)$ with domain T and range $f(\tau) \in V_\tau$ forms the vector system $\mathfrak{S}(T; \mathfrak{V}_\tau)$ if the definitions (3)—(3′) are preserved for vector operations. Among the vector subsystems of $\mathfrak{S}(T; \mathfrak{V}_\tau)$ we again note:

(a) the system $\mathfrak{S}^{(0)}(T; \mathfrak{V}_\tau)$ of functions in $\mathfrak{S}(T; \mathfrak{V}_\tau)$ which differ from zero perhaps only at a finite number of points;

(b) the system $\mathfrak{S}^{(1)}(T; \mathfrak{V}_\tau)$ of functions in $\mathfrak{S}(T; \mathfrak{V}_\tau)$ which differ from zero on a countable set of points from T.

If $\mathfrak{V}_\tau = \mathfrak{V}$ for all $\tau \in T$, then we will write respectively $\mathfrak{S}(T; \mathfrak{V})$ $\mathfrak{S}^{(0)}(T; \mathfrak{V})$, and $\mathfrak{S}^{(1)}(T; \mathfrak{V})$ instead of $\mathfrak{S}(T; \mathfrak{V}_\tau)$, $\mathfrak{S}^{(0)}(T; \mathfrak{V}_\tau)$, and

$\mathfrak{G}^{(1)}(T; \mathfrak{B}_\tau)$. In particular, $\mathfrak{G}(N_0; \mathfrak{B}) = \mathfrak{G}^{(1)}(N_0; \mathfrak{B})$, where N_0 is the set of natural numbers, is the family of all sequences $\{f_n\}$ which consist of elements of $\mathfrak{B}$.

2.3. Basis and rank of a vector system. The zero element of the vector system $\mathfrak{B}$ is dependent on any (non-empty) set since it can be represented by a trivial linear combination.

A set M is said to be *linearly independent* if the zero vector can be represented only by a trival linear combination of elements of M. A set N is said to be a basis of its linear hull $\mathscr{L}(N)$ if it is linearly independent.

A set $N \subset \mathfrak{B}$ is a *basis of the vector system* $\mathfrak{B}$ if it is linearly independent and its linear hull $\mathscr{L}(A)$ coincides with $\mathfrak{B}$.

Elements of the linear hull $\mathscr{L}(N)$ can thus be uniquely represented as linear combinations of elements of its basis N. In particular, this is true for the vector system $\mathfrak{B}$ and its basis N. Otherwise an element $h \in \mathscr{L}(N)$ could be represented by different linear combinations of elements of the basis

$$h = \sum_1^n \alpha_k f_k, \qquad h = \sum_1^{n'} \beta_k g_k \qquad (f_k, g_k \in N).$$

Then

$$0 = h - h = \sum_1^n \alpha_k f_k - \sum_1^{n'} \beta_k g_k,$$

and we would obtain a representation of zero as a nontrivial linear combination of elements of the basis N, which is impossible.

If a vector system $\mathfrak{B}$ has a basis with a finite number of elements, then it is called a system of finite rank, and the number n of elements of the basis of the system $\mathfrak{B}$ which, as is known from linear algebra, is independent of the choice of the basis, is by definition the *rank* of the system $\mathfrak{B}$.

It is easy to find a basis for the system $\mathfrak{G}^{(0)}(T; K)$. Let the function $e_\rho(\tau)$, where $\rho \in T$, be given by:

$$e_\rho(\tau) = 0, \qquad \tau \neq \rho, \qquad e_\rho(\rho) = 1.$$

The functions $e_\rho(\tau)$, which belong to the system, form a basis. This follows since:

(1) The set $\{e_\rho(\tau)\}$ is linearly independent. In fact, if

$$0 = \sum_{k=1}^n \alpha_k e_{\rho_k}(\tau) \tag{4}$$

is a representation of the zero function, then substituting $\tau = \rho_k$ in (4) we obtain $\alpha_k = 0$, i.e. (4) is the trivial combination.

(2) The linear envelope of the set $\{e_\rho(\tau)\}$ coincides with $\mathfrak{G}^{(0)}(T, K)$. Indeed, let $\rho_1, \rho_2, \ldots, \rho_n$ be those values of the variable τ for which the function $f(\tau) \in G^{(0)}(T; K)$ is different from zero, and let $f(\rho_k) = c_k$. Then

$$f(\tau) = \sum_{k=1}^{n} c_k e_{\rho_k}(\tau).$$

The system $\mathfrak{G}^{(0)}(T; K)$ has finite rank if and only if the set T is finite; in that case the number of elements T is the same as the rank of $\mathfrak{G}^{(0)}(T; K)$.

If the vector system $\mathfrak{V}$ does not have finite rank, but is the linear hull of a countable set of vectors $\{f_n\}$, then we will obtain a basis of $\mathfrak{V}$ if we successively remove from $\{f_n\}$ those vectors which linearly depend on the preceding elements of $\{f_n\}$. By making use of the theory of transfinite numbers and transfinite induction, it would be possible to apply a similar process to construct a basis in the case of an arbitrary vector system. Instead of this, we make use of an equivalent method here, as in similar questions in what follows; namely Lemma 1.1.1. With its help, a somewhat more general theorem will be proved.

THEOREM 1.2.1. *Every linearly independent set N_1 of a vector system $\mathfrak{V}$ can be extended to a basis of the system $\mathfrak{V}$.*

PROOF. We establish a proper order relation in the set $\mathfrak{N}$ of all linearly independent sets N in $\mathfrak{V}$, which converts $\mathfrak{V}$ into an ordered set; namely, we let $N' \geqslant N$ for N and N' in $\mathfrak{N}$ if N is contained in N'. By Lemma 1.1.1 there exists an upper maximal chain $\mathfrak{Z}$ in $\mathfrak{N}$ which is an extension of a chain of the one given linearly independent set N_1.

The chain $\mathfrak{Z}$ has a greatest element which is a basis of the system $\mathfrak{V}$ (which contains the set N_1). Indeed, the union $\bar{N}$ of the sets N of the chain $\mathfrak{Z}$ is linearly independent, i.e. $\bar{N} \in \mathfrak{N}$, and since $N \leqslant \bar{N}$ for each $N \in \mathfrak{Z}$, and since $\mathfrak{Z}$ is upper maximal, the majorant $\bar{N}$ belongs to the chain $\mathfrak{Z}$ and is its greatest element. Thus the set $\bar{N}$ (which contains N_1) is a maximal element of $\mathfrak{N}$ and hence $\bar{N}$ cannot be supplemented by a vector which is linearly independent of $\bar{N}$, i.e. $\bar{N}$ is a basis of $\mathfrak{V}$.

REMARK. Since any non-zero vector can be taken for N_1, Theorem 1.2.1 asserts, in particular, the existence of a basis for every non-null vector system.

THEOREM 1.2.2. *Two bases of a vector system are equivalent.*

Let $M = \{f_\sigma\}$ and $N = \{g_\tau\}$ be two bases of the vector system $\mathfrak{V}$ with respective cardinalities $\mathfrak{m}$ and $\mathfrak{n}$. If one of the cardinalities is finite, then the theorem, as was already mentioned, is well known from algebra. We consider the case when both cardinalities are infinite. Each element $f_\sigma \in M$ has a representation

$$f_\sigma = \sum_\tau c_{\sigma\tau} g_\tau,$$

where (for fixed σ) only a finite number of the coefficients $c_{\sigma\tau}$ are different from zero. We let

$$g_\tau = \varphi(f_\sigma) \tag{5}$$

if $c_{\sigma\tau} \neq 0$.

The mapping φ is a finite-valued mapping of the system M onto the system N. Indeed, for every vector $g_{\tau_0} \in N$ there exists at least one vector $f_\sigma \in M$ with $c_{\sigma\tau_0} \neq 0$, since otherwise all of the vectors f_σ would be dependent on the system $N \setminus \{g_{\tau_0}\}$, and thus any vector in $\mathfrak{V}$, including g_{τ_0}, would be dependent on $N \setminus \{g_{\tau_0}\}$, which would contradict the linear independence of the set N. The mapping φ can be considered as a single-valued mapping from M into N', where N' is the set of all finite subsets of N. In addition, it is obvious that the mapping φ takes only a finite number of elements of the system M into each element of the set N'. It follows that $\mathfrak{m} \leqslant \mathfrak{n}$.

Since the roles of the bases are interchangeable in the above arguments, we also have $\mathfrak{n} \leqslant \mathfrak{m}$, so that in fact $\mathfrak{n} = \mathfrak{m}$.

The *rank* of a vector system is by definition the (uniquely determined) *cardinality of its basis.* Vector systems of any rank exist over an arbitrary field K. Indeed, let $\mathfrak{n}$ be a given cardinality and let T be a set of cardinality $\mathfrak{n}$. Then the vector system $\mathfrak{G}^{(0)}(T; K)$ has rank $\mathfrak{n}$: the functions $e_\rho(\tau)$, where ρ ranges over the set T (of cardinality $\mathfrak{n}$), form a basis for $\mathfrak{G}^{(0)}(T; K)$.

2.4. The direct sum of vector systems. The collection $\{\mathscr{L}(f_\tau)\}$ of vector subsystems of rank 1 of the system $\mathfrak{V}$ has been given with the set $N = \{f_\tau\} \subset \mathfrak{V}$, where $f_\tau \neq 0$, $\tau \in T$. The property of linear independence (or dependence) of the set N can be formulated in a similar manner for the collection $\{\mathscr{L}(f_\tau)\}$, and is generalized in this form to any set $\mathfrak{V}_\tau$ of vector subsystems of the system $\mathfrak{V}$.

A set $\{\mathfrak{V}_\tau\}$ $(\tau \in T)$ of the vector subsystems of $\mathfrak{V}$ is *linearly independent* if each $\mathfrak{V}_\tau$ is different from (0) and if it follows that all f_τ are equal to zero when

$$\sum_\tau f_\tau = 0, \tag{6}$$

where $f_\tau \in \mathfrak{V}_\tau$, and the number of terms which differ from zero is finite.

The linear independence of a set $\{\mathfrak{V}_\tau\}$ can be formulated in the following form:

A set $\{\mathfrak{V}_\tau\}$ is linearly independent if for each $\mathfrak{V}_\sigma$ $(\sigma \in T)$ the intersection of the system $\mathfrak{V}_\sigma$ and any finite vector sum of systems $\mathfrak{V}_\tau$, with $\tau \neq \sigma$, is equal to (0).

Indeed, by writing (6) in the form

$$f_\sigma = -\sum_{\tau \neq \sigma} f_\tau$$

we show the equivalance of the two formulations. In particular, *two vector subsystems $\mathfrak{V}'$ and $\mathfrak{V}''$ are linearly independent if their intersection is* (0).

A vector sum $\mathfrak{V}'$ of vector subsystems $\mathfrak{V}_\tau$ is called *direct* if $\{\mathfrak{V}_\tau\}$ is a linearly independent set. Then we write

$$\mathfrak{B}' = \sum_{\tau} \cdot \mathfrak{B}_{\tau}.$$

In general, to denote a direct sum a point will be added to the usual symbol of summation, i.e. we write $\sum \cdot$ or $\dot{+}$.

If $\mathfrak{B}'$ is the direct sum of the subsystems $\mathfrak{B}_{\tau}$, then any vector $f \in \mathfrak{B}'$ has a unique representation of the form

$$f = \sum_{\tau} f_{\tau}, \tag{7}$$

where $f_{\tau} \in \mathfrak{B}_{\tau}$ and the number of non-zero terms is finite. Conversely, a vector sum $\mathfrak{B}'$ of vector subsystems $\mathfrak{B}_{\tau}$ is direct if the representation (7) is unique. Indeed, the uniqueness of (7) for $f = 0$ is the same as the linear independence of $\{\mathfrak{B}_{\tau}\}$.

Thus, the vector system $\mathfrak{B}$ is the direct sum of its vector subsystems $\mathfrak{B}_{\tau}$ if any vector f in $\mathfrak{B}$ has a unique representation of the form (7), where $f_{\tau} \in \mathfrak{B}_{\tau}$. In particular, *each vector system $\mathfrak{B}$ is a direct sum of vector subsystems (f_{τ}) of rank* 1, *where $N = \{f_{\tau}\}$ is a basis of $\mathfrak{B}$.*

THEOREM 1.2.3. *If $\mathfrak{B}^{(1)}$ is a proper vector subsystem of the system $\mathfrak{B}$, then there exists a vector subsystem $\mathfrak{B}^{(2)} \neq (0)$ such that $\mathfrak{B}$ is the direct sum of $\mathfrak{B}^{(1)}$ and $\mathfrak{B}^{(2)}$*:

$$\mathfrak{B} = \mathfrak{B}^{(1)} \dot{+} \mathfrak{B}^{(2)}.$$

Let $N^{(1)}$ be a basis of the vector system $\mathfrak{B}^{(1)}$ and let N be its continuation to a basis of the system $\mathfrak{B}$. The linearly independent set $N^{(2)} = N \backslash N^{(1)}$ is a basis of the linear envelope $\mathscr{L}(N^{(2)}) = \mathfrak{B}^{(2)}$, and $\mathfrak{B}$ is the direct sum of the vector systems $\mathfrak{B}^{(1)}$ and $\mathfrak{B}^{(2)}$. By virtue of the linear independence of N, the intersection $\mathfrak{B}^{(1)} \cap \mathfrak{B}^{(2)}$ is (0) and the representation

$$f = \sum_{\tau} \alpha_{\tau} h_{\tau} \tag{8}$$

of any vector $f \in \mathfrak{B}$ as a linear combination of elements of the basis $N = \{h_{\tau}\}$ gives

$$f = f^{(1)} + f^{(2)}, \qquad f^{(1)} \in \mathfrak{B}^{(1)}, \qquad f^{(2)} \in \mathfrak{B}^{(2)},$$

where $f^{(k)}$ $(k = 1, 2)$ is the sum of those terms in (8) which belong to $\mathfrak{B}^{(k)}$.

§ 3. Isomorphisms and Automorphisms

3.1. Φ-systems. Isomorphisms of vector systems. A Φ-system is a set $\mathfrak{A}$ with a collection Φ of relations defined in $\mathfrak{A}$. Examples of Φ-systems are: (1) an ordered set Z with the relation $\geqslant$, (2) a group with a three-term relation $\theta[A; B; C]$, where $C = AB$, (3) a vector system $\mathfrak{B}$ over a field K. In the last case, Φ contains the three-term relation $\theta[f; g; h]$ where $h = f + g$, and the collection of relations $\theta_\alpha[f, g]$, where $g = \alpha f$, $\alpha \in K$. In each of these three cases the given relations have the following properties: in the first case, the properties of a proper order relation; in the second case, the properties of group operation; and in the third case, linear operations in the vector system. A collection of similar properties of a relation will be called its *description*.

In a system $\mathfrak{A}$ let there be given relations which are in a one-to-one correspondence with the relations of a Φ-system $\mathfrak{A}'$, where corresponding relations have identical descriptions. Then $\mathfrak{A}$ and $\mathfrak{A}'$ will be said to be *systems with identical relations*, and both $\mathfrak{A}$ and $\mathfrak{A}'$ will be called Φ-systems. For example, two ordered sets, two groups, or two vector systems over the same field K are examples of systems with identical relations.

A mapping of the Φ-system $\mathfrak{A}$ onto the Φ-system $\mathfrak{A}'$ will be called an *isomorphic mapping* or an *isomorphism* if the following two conditions are satisfied:

(1) The mapping is one-to-one.

(2) Elements $\xi^{(1)}, \ldots, \xi^{(n)}$ of the system $\mathfrak{A}$ stand in a given relation $\theta \in \Phi$ if and only if their images $\eta^{(1)}, \ldots, \eta^{(n)} \in \mathfrak{A}'$ stand in the relation θ.

The inverse mapping, which assigns to any element in $\mathfrak{A}'$ its unique preimage in $\mathfrak{A}$, maps $\mathfrak{A}'$ isomorphically onto $\mathfrak{A}$. Two Φ-systems $\mathfrak{A}$ and $\mathfrak{A}'$ are by definition *isomorphic* if an isomorphism can be established between them.

From an abstract point of view isomorphic Φ-systems are indistinguishable: corresponding elements of such systems stand in the same relations $\theta \in \Phi$, which are considered to be unique.

In connection with ordered sets Z and Z', in addition to isomorphism we define the dual of an isomorphism: a mapping of Z onto Z' is a *dual isomorphism* if it is an isomorphism of Z onto the set Z'_d, which is the dual of Z, i.e. if for a one-to-one mapping of Z onto Z', $\xi \geqslant \eta$ always implies that $\eta' \geqslant \xi'$, where ξ' and η' are the images of the elements ξ and η of Z. Two ordered sets are called dualy isomorphic if a dual isomorphism can be established between them.

An isomorphic mapping S of the Φ-system $\mathfrak{A}$ onto itself ($\mathfrak{A}' = \mathfrak{A}$) is called an *automorphism* of the Φ-system $\mathfrak{A}$. The inverse mapping, which assigns to each element of the Φ-system $\mathfrak{A}$ its inverse image under the automorphism, is also an automorphism of the Φ-system; this automorphism is called the *inverse* of S and is denoted by S^{-1}. An automorphism is called *involute* or an *involution* if it coincides with its inverse.

In addition to automorphisms, we will also consider dual automorphisms in an ordered set Z, i.e. dual isomorphisms of Z onto Z.

REMARK. If a mapping of a Φ-system $\mathfrak{A}$ onto a Φ-system $\mathfrak{A}'$ is such that whenever elements $\xi^{(1)}, \ldots, \xi^{(n)}$ of the system $\mathfrak{A}$ stand in a relation θ, their images $\eta^{(1)}, \ldots, \eta^{(n)} \in \mathfrak{A}'$ also stand in the relation θ, then this mapping is called a *homomorphic mapping* of $\mathfrak{A}$ onto $\mathfrak{A}'$ or a *homomorphism* of the Φ-systems $\mathfrak{A}$ and $\mathfrak{A}'$.

In the special case when the Φ-systems $\mathfrak{A}$ and $\mathfrak{A}'$ are vector systems $\mathfrak{B}$ and $\mathfrak{B}'$ over the same field K, the condition of isomorphism can be written in the form

$$S(f+g)=Sf+Sg, \quad S(\alpha f)=\alpha Sf \quad (f,\ g\in\mathfrak{B},\ \alpha\in K). \tag{1}$$

Under an isomorphism S the zero of the system $\mathfrak{B}$ is mapped into the zero of the system $\mathfrak{B}'$, and linearly independent (dependent) elements of $\mathfrak{B}$ are transformed into linearly independent (dependent) elements of $\mathfrak{B}'$.

If $\{h_\tau\}$ is a basis of the system $\mathfrak{B}$, then the elements $h'_\tau = Sh_\tau$ form a basis of the system $\mathfrak{B}'$, and thus isomorphic vector systems have the same rank. The converse is also true. Indeed, if two vector systems $\mathfrak{B}$ and $\mathfrak{B}'$ have the same rank $\mathfrak{n}$, then their bases can be written in the form $\{h_\tau\}$, $\{h'_\tau\}$, where τ ranges over the same set T of cardinality $\mathfrak{n}$, and the equations

$$f=\sum_\tau \alpha_\tau h_\tau\in\mathfrak{B}, \qquad f'=\sum_\tau \alpha_\tau h'_\tau\in\mathfrak{B}'$$

(see (1.2.8)) obviously define an isomorphism $f\to f'$ of the system $\mathfrak{B}$ onto $\mathfrak{B}'$.

Thus, *two vector systems* $\mathfrak{B}$ *and* $\mathfrak{B}'$ *are isomorphic if and only if they have the same rank*. In other words, the rank of vector system characterizes it to within an isomorphism, i.e. *the rank forms a complete system of invariants with respect to isomorphism.*

The system $\mathfrak{G}^{(0)}(T;K)$, with the basis $e_\rho(\tau)$ $(\rho\in T)$ has rank $\mathfrak{n}$ equal to the cardinality of the set T. By considering the system $\mathfrak{G}^{(0)}(T;K)$ for sets T of arbitrary cardinality, we thus obtain all vector systems to within an isomorphism, i.e. the system $\mathfrak{G}^{(0)}(T;K)$ is a *universal model of a vector system.* A natural isomorphism between the vector system $\mathfrak{B}$ and the corresponding model $\mathfrak{G}^{(0)}(T;K)$ can be established in the following manner.

Let $\{h_\tau\}$ be a basis of the vector system $\mathfrak{B}$ and let

$$f=\sum_\tau c_\tau h_\tau=\sum_\tau c_\tau(f)\,h_\tau \tag{2}$$

be the representation of an element $f\in\mathfrak{B}$ as a linear combination of elements of the basis. Equation (2) defines a one-to-one correspondence between vectors $f\in\mathfrak{B}$ and functions $c_\tau = c(\tau)$ which differ from zero only at a finite number of points of the set T, i.e. functions in $\mathfrak{G}^{(0)}(T;K)$. This mapping $f\to c(\tau)$ is an isomorphism of the vector systems $\mathfrak{B}$ and $\mathfrak{G}^{(0)}(T;K)$, since

$$c_\tau(f+g)=c_\tau(f)+c_\tau(g); \qquad c_\tau(\alpha f)=\alpha c_\tau(f). \tag{3}$$

Under this isomorphism the basis $\{h_\rho\}$ of the system $\mathfrak{B}$ is mapped into the basis $e_\rho(\tau)$ of the system $\mathfrak{G}^{(0)}(T;K)$.

3.2. The direct sum of vector systems. The concept of isomorphisms will form the basis of the generalization of the definition of the direct sum of vector systems $\mathfrak{B}_\tau$ ($\tau \in T$) to that case when the $\mathfrak{B}_\tau$ are not vector subsystems of a given system $\mathfrak{B}$, i.e. when the $\mathfrak{B}_\tau$ are given independently.

In the vector system $\mathfrak{G}^{(0)}(T; \mathfrak{B}_\tau)$ (see §1.2) we consider vector subsystems $\mathfrak{B}_\tau'$ of those functions $f_\tau(\rho)$ ($\rho \in T$) for which

$$f_\tau(\rho) = 0, \quad \rho \neq \tau.$$

The system $\mathfrak{B}_\tau'$ is isomorphic to the vector system $\mathfrak{B}_\tau$, namely:

$$f_\tau(\rho) \mapsto f_\tau(\tau)$$

is an isomorphism of $\mathfrak{B}_\tau'$ onto $\mathfrak{B}_\tau$.

The vector system $\mathfrak{G}^{(0)}(T; \mathfrak{B}_\tau)$ *is the direct sum of its vector subsystems* $\mathfrak{B}_\tau'$.

Indeed, let $f \in \mathfrak{G}^{(0)}(T; \mathfrak{B}_\tau)$ and let $\tau_1, \tau_2, \ldots, \tau_n$ be those values of the variable $\rho \in T$ for which $f(\rho) \neq 0$. The equations

$$f_{\tau_k}(\rho) = 0, \quad \rho \neq \tau_k; \quad f_{\tau_k}(\tau_k) = f(\tau_k)$$

define functions $f_{\tau_k} \in \mathfrak{G}^{(0)}(T; \mathfrak{B}_\tau)$, $k = 1, \ldots, n$ such that

$$f(\rho) = \sum_{k=1}^{n} f_{\tau_k}(\rho), \tag{4}$$

and thus we find that $\mathfrak{G}^{(0)}(T; \mathfrak{B}_\tau')$ is the sum of the systems $\mathfrak{B}_\tau'$. This sum is direct since the representation (4) is unique: the substitution of $f(\rho) = 0$ and $\rho = \tau_k$ in (4) gives $f_{\tau_k}(\tau_k) = 0$, i.e. $f_{\tau_k}(\rho) = 0$.

By identifying $\mathfrak{B}_\tau'$ with its isomorphic system $\mathfrak{B}_\tau$, we will consider $\mathfrak{G}^{(0)}(T; \mathfrak{B}_\tau)$ to be the direct sum of the vector systems $\mathfrak{B}_\tau$ and we will write

$$\mathfrak{G}^{(0)}(T; \mathfrak{B}_\tau) = \sum_\tau \cdot \mathfrak{B}_\tau.$$

3.3. Vector systems with involutions. A vector system $\mathfrak{B}$ with a given involutive automorphism is called a *vector system with involution.*

In the case of a vector system $\mathfrak{B}$ with involution we will denote the image of a vector $f \in \mathfrak{B}$ under the involution by $f^\times$, and will also call $\mathfrak{B}$ a *vector* $\times$*-system.* We have

$$(f+g)^\times = f^\times + g^\times, \quad (\alpha f)^\times = \alpha f^\times, \quad (f^\times)^\times = f.$$

An element f of a vector $\times$*-system* $\mathfrak{B}$ is said to be *symmetric* if $f^\times = f$, or *skew-symmetric* if $f^\times = -f$.

The symmetric elements of the system $\mathfrak{B}$ obviously form a vector subsystem, and the same is true of skew-symmetric elements. We will denote these vector systems by $\mathfrak{B}^+$ and $\mathfrak{B}^-$, respectively.

A vector system $\mathfrak{B}$ *with involution is the direct sum of its vector subsystems* $\mathfrak{B}^+$ *and* $\mathfrak{B}^-$.

Indeed,

(1) $\mathfrak{B}^+ \cap \mathfrak{B}^- = (0)$, since for an element $f \in \mathfrak{B}^+ \cap \mathfrak{B}^-$,

$$f=f^{\times}=-f,$$

so that $f=0$;

(2) We will represent a vector $f\in\mathfrak{V}$ in the form

$$f=f_1+f_2,$$

where $f_1\in\mathfrak{V}^+$ and $f_2\in\mathfrak{V}^-$. In fact, let

$$f_1=\frac{f+f^{\times}}{2}, \qquad f_2=\frac{f-f^{\times}}{2};$$

then

$$f_1^{\times}=\frac{f^{\times}+f}{2}=f_1, \qquad f_2^{\times}=\frac{f^{\times}-f}{2}=-f_2.$$

§ 4. Linear Transformations

4.1. Linear equivalence relations. Suppose that a mapping $f'=Af$ ($f\in\mathfrak{V}, f'\in\mathfrak{V}'$) of a vector system $\mathfrak{V}$ *into* a vector system $\mathfrak{V}'$ satisfies the linearity conditions (see (1.3.1))

$$A(f+g)=Af+Ag, \qquad A(\alpha f)=\alpha Af. \tag{1}$$

Then A is called a *linear transformation.* The set $\mathfrak{V}$ is called the *domain* and the set $A\mathfrak{V}$ the *range* of the transformation A. The domain and range of a transformation A will usually be denoted by Ω_A and Ω_A^{-1}. The range Ω_A^{-1} is a vector subsystem $\mathfrak{V}''$ of the system $\mathfrak{V}'$, and if whenever $f\neq g$, $Af\neq Ag$, then the mapping of the system $\mathfrak{V}$ onto $\mathfrak{V}''$ is one-to-one. In this case there exists an inverse transformation $A'=A^{-1}$ which associates with any image f' $(=Af)$ in $\mathfrak{V}''$ its (unique) inverse image $f=A'f'=A^{-1}f'$ in $\mathfrak{V}$. It is easy to see that A^{-1} is a linear transformation which maps $\mathfrak{V}''$ onto $\mathfrak{V}$, moreover

$$\Omega_{A^{-1}}=\mathfrak{V}''=\Omega_A^{-1}.$$

Since A and A^{-1} are one-to-one and linear, the mapping A is an isomorphism of the system $\mathfrak{V}$ onto $\mathfrak{V}''$. It will also be said that A is an *isomorphism of the system A onto the system A'*. The set $\mathfrak{N}(A)$ of elements $f\in\mathfrak{V}$ which satisfy the equation

$$Af=0,$$

where A is a linear transformation and $\Omega_A=\mathfrak{V}$, is a vector subsystem of the system $\mathfrak{V}$ and is called the *null system* of the transformation A. It is obvious that $Af\neq Ag$ when $f\neq g$, i.e. the existence of an inverse transformation A^{-1} is equivalent to the condition

$$\mathfrak{N}(A)=(0). \tag{2}$$

REMARK. In the general case when (2) is not necessarily satisfied, $f'=Af$ is a homomorphism (see the remark in Sect. 1.3.1) of the vector system $\mathfrak{V}$

onto $\mathfrak{V}'' = \Omega_A^{-1}$ and, in particular, it is *onto* $\mathfrak{V}'$ if $\mathfrak{V}'' = \mathfrak{V}'$. If $\mathfrak{V}'' \neq \mathfrak{V}'$, then $f' = Af$ is a homomorphism of $\mathfrak{V}$ into $\mathfrak{V}'$.

Every linear transformation A (and, in particular, every homomorphism) generates an equivalence relation θ_A in the vector system $\mathfrak{V}$, namely

$$f \equiv g \; (\theta_A)$$

if $Af = Ag$. The relation $\theta = \theta_A$ has the following properties:

(1) if $f_1 \equiv g_1$ and $f_2 \equiv g_2$ with respect to θ, then

$$f_1 + f_2 \equiv g_1 + g_2 \qquad (\theta), \tag{3}$$

i.e. congruences can be added;

(2) if $f \equiv g(\theta)$, then

$$\alpha f \equiv \alpha g \qquad (\theta), \tag{4}$$

i.e. congruences with respect to θ can be multiplied by elements α of a field K.

Any equivalence relation θ in a vector system $\mathfrak{V}$ with the properties (1) and (2) is said to be *linear*.

The types $\mathring{f} = \theta f$ * with respect to any linear equivalence relation θ form a vector system $\mathring{\mathfrak{V}}$ if we define *linear operations* in $\mathring{\mathfrak{V}}$ by the equations

$$\theta f_1 + \theta f_2 = \theta (f_1 + f_2); \qquad \alpha(\theta f) = \theta(\alpha f). \tag{5}$$

These definitions are valid, since by (3) and (4), the right-hand equation in (5) is independent of the choice of the representatives f, f_1 and f_2 of the types θf, θf_1 and θf_2 respectively. Thus θ is a linear transformation (a homomorphism of $\mathfrak{V}$ onto the vector system $\mathring{\mathfrak{V}}$) with $\Omega_\theta = \mathfrak{V}$ and range $\mathring{\mathfrak{V}}$. The null system $\mathfrak{N}(\theta)$ of this transformation, i.e. the collection of elements f which satisfy the condition

$$f \equiv 0 \; (\theta)$$

is by definition the *kernel* $\mathfrak{V}_0$ *of the linear equivalence relation* (or of the *homomorphism*) θ, moreover,

$$f \equiv g \qquad (\theta) \tag{6}$$

if and only if

$$f - g \in \mathfrak{V}_0. \tag{7}$$

Conversely, *every vector subsystem* $\mathfrak{V}_0$ *of the system* $\mathfrak{V}$ *generates a linear equivalence relation in* $\mathfrak{V}$ *which has* $\mathfrak{V}_0$ *as its kernel.*

In fact, we will define the congruence (6) by the relation (7). This equivalence relation is then linear, i.e. it satisfies the conditions (3) and (4) since if

---

* Here and below we denote the type $\mathring{f}$ of an element f with respect to the equivalence relation θ by θf. By the same token, θ will be considered as the transformation which assigns to each element $f \in \mathfrak{V}$ its type $\mathring{f}$ (see 1.1.1).

$f - g \in \mathfrak{V}_0$, then $\alpha(f-g) \in \mathfrak{V}_0$, and if $f_1 - g_1$ and $f_2 - g_2$ belong to $\mathfrak{V}_0$, then the vector $f_1 + f_2 - (g_1 + g_2)$ belongs to $\mathfrak{V}_0$.

The vector system $\mathring{\mathfrak{V}}$ of types with respect to $\mathfrak{V}_0$ (or equivalently, with respect to θ) is also called the *factor-system* of the system $\mathfrak{V}$ modulo $\mathfrak{V}_0$ and is denoted by $\mathfrak{V}/\mathfrak{V}_0$.

In the case when the equivalence relation $\theta = \theta_A$ is generated by a linear transformation A, the kernel $\mathfrak{V}_0$ coincides with $\mathfrak{N}(A)$; the transformation A induces in the factor-system $\mathring{\mathfrak{V}}$ a linear transformation $\mathring{A}$ which maps $\mathring{\mathfrak{V}}$ into $\mathfrak{V}'$ *isomorphically* (onto $\mathfrak{V}'$ if $\mathfrak{V}'' = \mathfrak{V}'$). Indeed, we let

$$\mathring{A}\mathring{f} = f' = Af, \tag{8}$$

where f is a representative of the type $\mathring{f}$. Then the vector $f' \in \mathfrak{V}'$ is independent of the choice of the representative f of the type $\mathring{f}$, and

$$\mathring{A}(\mathring{f}_1 + \mathring{f}_2) = A(f_1 + f_2) = Af_1 + Af_2 = \mathring{A}\mathring{f}_1 + \mathring{A}\mathring{f}_2,$$
$$\mathring{A}(\alpha\mathring{f}) = A(\alpha f) = \alpha Af = \alpha\mathring{A}\mathring{f}$$

(here f_1, f_2 and f are representatives of the types $\mathring{f}_1$, $\mathring{f}_2$ and $\mathring{f}$). Finally, if $\mathring{A}\mathring{f} = \mathring{A}\mathring{g}$, where $\mathring{f} = \theta f$ and $\mathring{g} = \theta g$, then $Af = Ag$; hence, $\mathring{f} = \mathring{g}$, i.e. $\mathfrak{N}(\mathring{A}) = 0$.

4.2. The systems $\mathfrak{A}(\mathfrak{V}, \mathfrak{V}')$ and $\mathfrak{A}(\mathfrak{V})$. The set of all linear transformations whose domain is $\mathfrak{V}$ and whose range is $\mathfrak{V}'$ can be converted into a vector system $\mathfrak{A}(\mathfrak{V}, \mathfrak{V}')$ over a field K, if we define the linear operations $A + B$, αA ($\alpha \in K$) for such transformations A and B by

$$(A+B)f = Af + Bf, \quad (\alpha A)f = \alpha Af, \tag{9}$$

where the vector f ranges over all of $\mathfrak{V}$ (it is easy to verify that the conditions (2) and (3) in the definition of vector system in §2 are satisfied).

When $\mathfrak{V}' = \mathfrak{V}$, we will write $\mathfrak{A}(\mathfrak{V})$ instead of $\mathfrak{A}(\mathfrak{V}, \mathfrak{V}')$ and we will call the transformations in $\mathfrak{A}(\mathfrak{V})$ *linear operators* in $\mathfrak{V}$. By definining operator multiplication, we convert $\mathfrak{A}(\mathfrak{V})$ into a ring. With this aim we note that in the collection $\Xi(Z)$ of all mappings of any set Z into itself, it is possible to introduce the concept of a product, namely the *product* of two mappings A and B in $\Xi(Z)$ is by definition the mapping $C = AB$ in $\Xi(Z)$ for which

$$Cf = A(Bf) \tag{10}$$

for any $f \in Z$. For every $f \in Z$ and any A, B, C in $\Xi(Z)$

$$(AB)Cf = A(B(Cf)) = A(BC)f,$$

and we have

$$(AB)C = A(BC).$$

i.e. multiplication is associative. This multiplication has an identity, since under the identity mapping E: $Ef = f\,(f \in Z)$ we have

$$EA = AE = A.$$

If $Z = \mathfrak{B}$, then for A and $B \in \mathfrak{A}(\mathfrak{B})$, their product $C = AB$ is an operator in $\mathfrak{A}(\mathfrak{B})$, i.e. $\mathfrak{A}(\mathfrak{B})$ is closed with respect to multiplication. Furthermore, for arbitrary $f \in \mathfrak{B}$

$$A(B_1 + B_2)f = A(B_1 f + B_2 f) = (AB_1)f + (AB_2)f,$$
$$(A_1 + A_2)Bf = (A_1 + A_2)(Bf) = (A_1 B)f + (A_2 B)f$$

and, hence

$$A(B_1 + B_2) = AB_1 + AB_2, \qquad (A_1 + A_2)B = A_1 B + A_2 B,$$

i.e. multiplication is not only associative in $\mathfrak{A}(\mathfrak{B})$, but it is distributive with respect to addition. Thus $\mathfrak{A}(\mathfrak{B})$ is converted into a *ring*. The identity mapping E belongs to $\mathfrak{A}(\mathfrak{B})$, i.e. the ring $\mathfrak{A}(\mathfrak{B})$ has an identity.

If the null system $\mathfrak{N}(A)$ of an operator $A \in \mathfrak{A}(\mathfrak{B})$ is zero, then an inverse transformation A^{-1} exists. However, it belongs to $\mathfrak{A}(\mathfrak{B})$ if and only if the range Ω_A^{-1} coincides with $\mathfrak{B}$. Thus the conditions

$$\mathfrak{N}(A) = (0), \qquad \Omega_A^{-1} = \mathfrak{B} \tag{11}$$

are necessary and sufficient for the operator $A \in \mathfrak{A}(\mathfrak{B})$ to have an inverse which also belongs to $\mathfrak{A}(\mathfrak{B})$, i.e. for A to be an automorphism of the vector system $\mathfrak{B}$.

We also note the following simple property: if $A \in \mathfrak{A}(\mathfrak{B}, \mathfrak{B}')$, if A^{-1} exists and $\alpha \neq 0$, then $(\alpha A)^{-1} = \dfrac{1}{\alpha} A^{-1}$.

REMARK 1. If A is a Φ-system, then the product of two automorphisms is again an automorphism. The identity mapping E is an automorphism, and if $A' = A^{-1}$ is the inverse of an automorphism A, then

$$A'A = AA' = E.$$

Thus, the set of all automorphisms of any Φ-system is a group under multiplication. In particular, the automorphisms of a vector system form a group.

REMARK 2. The concept of multiplication is not restricted to operators. Let $\mathfrak{B}$, $\mathfrak{B}'$ and $\mathfrak{B}''$ be three vector systems, B be a transformation in $\mathfrak{A}(\mathfrak{B}, \mathfrak{B}')$, and A be a transformation in $\mathfrak{A}(\mathfrak{B}', \mathfrak{B}'')$. Then for $f \in \mathfrak{B}$ equation (10) defines a transformation $C = AB$ in $\mathfrak{A}(\mathfrak{B}, \mathfrak{B}'')$ which is the *product of the transformations A and B*. Here, $C(AB) = (CA)B$, if $B \in \mathfrak{A}(\mathfrak{B}, \mathfrak{B}')$, $A \in \mathfrak{A}(\mathfrak{B}', \mathfrak{B}'')$, and $C \in \mathfrak{A}(\mathfrak{B}', \mathfrak{B}'')$, i.e. associativity is preserved.

For any (linear) transformation $A \in \mathfrak{A}(\mathfrak{B}, \mathfrak{B}')$ we can use this definition of product to write a decomposition

$$A = \mathring{A}\theta_A, \tag{12}$$

where θ_A is a transformation in $\mathfrak{A}(\mathfrak{B}, \mathring{\mathfrak{B}})$ and $\mathring{A}$ belongs to $\mathfrak{A}(\mathring{\mathfrak{B}}, \mathfrak{B}')$. Indeed, $\mathring{f} = \theta_A f$ in (8).

The multiplication of transformations is used in the following definition. Let $A \in \mathfrak{A}(\mathfrak{B}, \mathfrak{B}')$, and let U and V be isomorphisms of the systems $\mathfrak{B}$ and $\mathfrak{B}'$ onto $\mathfrak{W}$ and $\mathfrak{W}'$ respectively ($\mathfrak{W} = U\mathfrak{B}$, $\mathfrak{W}' = V\mathfrak{B}'$). Then the mapping $f' = Af$ of the system $\mathfrak{B}$ into $\mathfrak{B}'$ induces a mapping $g' = Bg$ of the system $\mathfrak{W}$ into $\mathfrak{W}'$, where

$$g = Uf, \quad g' = Vf'.$$

Obviously, B is the transformation in $\mathfrak{A}(\mathfrak{W}, \mathfrak{W}')$ such that

$$B = VAU^{-1}. \tag{13}$$

Indeed,

$$Bg = g' = Vf' = VAf,$$

and if we let $f = U^{-1}g$, we obtain

$$Bg = g' = VAU^{-1}g,$$

which proves (13). In (13)

$$U^{-1} \in \mathfrak{A}(\mathfrak{W}, \mathfrak{B}), \quad A \in \mathfrak{A}(\mathfrak{B}, \mathfrak{B}'), \quad V = \mathfrak{A}(\mathfrak{B}', \mathfrak{W}')$$

and, hence $B \in \mathfrak{A}(\mathfrak{W}, \mathfrak{W}')$.

By definition, B is a transformation *isomorphic* to A. It is obvious that $A = V^{-1}BU$, i.e. A is also *isomorphic* to B.

In the special case when $\mathfrak{B}' = \mathfrak{B}$, $U = V$ and $\mathfrak{W}' = \mathfrak{W}$, we obtain

$$B = UAU^{-1}, \quad A = U^{-1}BU. \tag{14}$$

If, in addition, $\mathfrak{W} = \mathfrak{B}$, then U is an automorphism of the system $\mathfrak{B}$, and the operators A and B are said to be *similar*.

4.3. Linear forms. Let a vector system $\mathfrak{B}'$ coincide with a field K. Then elements of the system $\mathfrak{A}(\mathfrak{B}, \mathfrak{B}') = \mathfrak{A}(\mathfrak{B}, K)$ are homomorphisms of $\mathfrak{B}$ into K; in this case they are called *linear forms* in the vector system $\mathfrak{B}$. A linear form $l = A \in \mathfrak{A}(\mathfrak{B}, K)$ satisfies the linearity conditions.

$$l(f+g) = l(f) + l(g), \quad l(\alpha f) = \alpha l(f) \qquad (f, g \in \mathfrak{B}).$$

The coefficients $c_\tau = c_\tau(f)$ in the representation (1.3.2) of a vector $f \in \mathfrak{B}$, in terms of a fixed basis $\{h_\tau\}$ of the vector system $\mathfrak{B}$, are linear forms in $\mathfrak{B}$ (cf. (1.3.3)).

The set of linear forms $\{c_\tau\}$ generated by the basis $\{h_\tau\}$ is linearly independent in $\mathfrak{A}(\mathfrak{B}, K)$. Indeed, for the linear forms c_τ:

$$c_\tau(h_\sigma) = 0, \quad \sigma \neq \tau; \quad c_\tau(h_\tau) = 1, \tag{15}$$

so that by letting $f = h_\sigma$ in the relation

$$\sum_\tau \alpha_\tau c_\tau(f) = 0,$$

by (15) we obtain $a_\sigma = 0$ for any σ.

With the help of the form c_τ it is possible to determine the general form of a linear transformation A in $\mathfrak{A}(\mathfrak{B}, \mathfrak{B}')$:

$$Af = \sum_\tau c_\tau(f) g'_\tau, \tag{16}$$

where $f \in \mathfrak{B}$ and the g'_τ are arbitrary elements of $\mathfrak{B}'$.

In fact, if A is a linear transformation and $Ah_\tau = g'_\tau$, then by virtue of the representation (1.3.2) and the linearity of A,

$$Af = \sum_\tau c_\tau(f) Ah_\tau = \sum_\tau c_\tau(f) g'_\tau.$$

Conversely, for fixed f, the function $c(\tau) = c_\tau(f)$ differs from zero only at a finite number of points $\tau = \tau_1, \ldots, \tau_n$ and (16) uniquely defines the value Af for every $f \in \mathfrak{B}$. A is a linear transformation in $\mathfrak{A}(\mathfrak{B}, \mathfrak{B}')$, and by (15), $Ah_\tau = g'_\tau$.

Thus A is uniquely determined by its values g'_τ in the basis $\{h_\tau\}$, and these g'_τ can be assigned in an arbitrary manner.

In particular, when $\mathfrak{B}' = K$, i.e. for the case of a system $\mathfrak{A}(\mathfrak{B}, K)$ of linear forms

$$g'_\tau = Ah_\tau = l(h_\tau) = \alpha_\tau \in K,$$

and hence the equation

$$l(f) = \sum_\tau \alpha_\tau c_\tau(f), \tag{17}$$

where the α_τ are any elements of the field K, gives a general expression for a linear form in $\mathfrak{B}$.

We note that the representations (16) and (17) can be written in the form

$$A = \sum_\tau c_\tau g'_\tau, \quad l = \sum_\tau \alpha_\tau c_\tau,$$

assuming here that Af and $l(f)$ are calculated according to (16) and (17).

§ 5. Pseudolinear Transformations. Algebras

5.1. Vector systems with pseudoinvolutions. A vector system over the field $K_c(K_r)$ of complex (real) numbers is called a *complex* (*real*) *vector system.* In accordance with this we will call a vector system of functions with values in the field K_c (K_r) a *complex* (*real*) *vector system of functions.*

The automorphism $\alpha \to \bar{\alpha}$ (transition to the conjugate) of the field of complex numbers leads to the consideration not only of linear transformations in the complex vector system $\mathfrak{B}$, but also of transformations A (with domain $\mathfrak{B}$ and range in another complex vector system $\mathfrak{B}'$), which satisfy the conditions

$$A(f+g) = Af + Ag, \qquad A(\alpha f) = \bar{\alpha} Af. \tag{1}$$

The equations in (1) will be called the conditions or properties of *pseudolinearity*. A transformation A which has these properties is said to be *pseudolinear*.

By definition, a pseudolinear transformation A is a *pseudohomomorphism* of $\mathfrak{B}$ into $\mathfrak{B}'$ (*onto* $\mathfrak{B}'$, if the range $\mathfrak{B}''$ coincides with $\mathfrak{B}'$). A mapping of the system $\mathfrak{B}$ onto the range $\mathfrak{B}''$ is a pseudoisomorphism if it is one-to-one. In this case the inverse transformation $A' = A^{-1}$ exists and associates to every vector of the system $\mathfrak{B}''$ its unique inverse image in $\mathfrak{B}$; A^{-1} is a pseudoisomorphism of the system $\mathfrak{B}''$ onto $\mathfrak{B}$. In the case when $\mathfrak{B}'' = \mathfrak{B}'$ the transformations A and A^{-1} establish a pseudoisomorphism between the vector systems $\mathfrak{B}$ and $\mathfrak{B}'$.

The collection $\overline{\mathfrak{A}}(\mathfrak{B}, \mathfrak{B}')$ of all pseudolinear transformations with domain $\mathfrak{B}$ and range in $\mathfrak{B}'$ becomes a complex vector system if, for elements in $\overline{\mathfrak{A}}(\mathfrak{B}, \mathfrak{B}')$, linear operations are again defined by the relations (1.4.9): $(A+B)f = Af + Bf$, $(\alpha A)f = \alpha Af$.

Indeed

$$(\alpha A)(\beta f) = \alpha A(\beta f) = \alpha\bar{\beta} Af = \bar{\beta}(\alpha A) f.$$

If $\mathfrak{B}'$ is the field of complex numbers K_c, then the elements of the system $\overline{\mathfrak{A}}(\mathfrak{B}, K_c)$ are by definition *pseudolinear forms* in the vector system $\mathfrak{B}$.

We now consider separately the case when $\mathfrak{B}' = \mathfrak{B}$. Then the pseudolinear transformations in $\overline{\mathfrak{A}}(\mathfrak{B}, \mathfrak{B}) = \overline{\mathfrak{A}}(\mathfrak{B})$ will be called *pseudolinear operators*.

The application of the general notion of product of transformations given above to pseudolinear operators gives:

The product AB is pseudolinear if one of the operators A, B is pseudolinear and the other is linear. It is linear if A and B are both pseudolinear.

In particular, the product of an automorphism by a pseudoautomorphism is a pseudoautomorphism and the product of two pseudoautomorphisms is an automorphism.

Thus automorphisms and pseudoautomorphisms together form a group which contains the group of automorphisms as a subgroup.

A pseudoautomorphism is called *involutive* or a *pseudoinvolution* if it coincides with its inverse (i.e., if its square equals E).

A vector system $\mathfrak{B}$ with an involutive pseudoautomorphism is called a *vector system with a pseudoinvolution*. In a vector system with a pseudoinvolution we will denote the image of a vector $f \in \mathfrak{B}$ under the given involutive pseudoautomorphism by f^*, and the system $\mathfrak{V}$ itself will also be called a *vector $*$-system*. We have:

$$(f+g)^* = f^* + g^*; \quad (\alpha f)^* = \bar{\alpha} f^*; \quad (f^*)^* = f. \tag{2}$$

An element f of a vector $*$-system is said to be Hermitian (or *Hermitian-symmetric*) if

$$f^* = f.$$

In view of (2) the Hermitian elements of $\mathfrak{V}$ form a real vector system $\mathfrak{V}_r$.

The elements f and if are both Hermitian if and only if $f = 0$.

In fact, if we let $g = 0$ in the first of the equations in (2), we obtain

$$f^* = f^* + 0^*$$

so that $0 = 0^*$. However

$$(if)^* = -if$$

and if f and if are Hermitian, it follows that $if = -if$, i.e. $f = 0$.

THEOREM 1.5.1. *Every element f of a vector $*$-system $\mathfrak{V}$ can be represented in a unique manner in the form*

$$f = f_1 + if_2, \tag{3}$$

where f_1 and f_2 are Hermitian elements.

Indeed, we let

$$f_1 = \frac{f+f^*}{2}, \quad f_2 = \frac{f-f^*}{2i}.$$

Then (3) holds and by virtue of (2)

$$f_1^* = \frac{f^*+f}{2} = f_1, \quad f_2^* = \frac{f^*-f}{-2i} = f_2.$$

If $f = g_1 + ig_2$ is some representation of the element f, where g_1 and g_2 are Hermitian, then $g_1+ig_2 = f_1+if_2$ and $g_1 - f_1 = i(f_2-g_2)$. Thus, the elements $f_2 - g_2$ and $i(f_2-g_2)$ are both Hermitian, and hence $f_2-g_2 = 0$ and $g_1 - f_1 = 0$, i.e. $g_2 = f_2$ and $g_1 = f_1$.

If a vector subsystem $\mathfrak{V}'$ of a $*$-system $\mathfrak{V}$ contains the element f^* whenever it contains f, then $\mathfrak{V}'$ is called a $*$-subsystem of the system $\mathfrak{V}$. The linear equivalence relation which generates such a $*$-subsystem $\mathfrak{V}'$ in $\mathfrak{V}$ satisfies the condition

$$f^* \equiv g^* \, (\theta), \qquad \text{if} \qquad f - g \in \mathfrak{V}, \tag{4}$$

and the set of types $\mathring{f}$ with respect to this equivalence becomes a vector $*$-system $\mathring{\mathfrak{V}} = \theta\mathfrak{V}$, if we set

$$(\mathring{f})^* = (\theta f)^* = \theta f^*. \tag{5}$$

The condition (4) is obviously equivalent to the condition that the vector subsystem $\mathfrak{V}'$ be a $*$-subsystem of the system $\mathfrak{V}$.

5.2. Algebras. Regular elements. The commutator. A ring $\mathfrak{A}$ over a field K is called an *algebra* if it is a vector system over the field K and if multiplication in $\mathfrak{A}$ satisfies the condition

$$(\alpha A)B = A(\alpha B) = \alpha(AB), \tag{6}$$

where A, $B \in \mathfrak{A}$ and $\alpha \in K$.

The algebra $\mathfrak{A}$ has an *identity* if there exists an element $E \in \mathfrak{A}$ for which

$$EA = AE = A$$

for any $A \in \mathfrak{A}$.

The identity E is unique since if E' were another element satisfying the condition $E'A = AE' = A$, then

$$EE' = E'E = E'; \quad E'E = EE' = E$$

and

$$E' = E.$$

The system $\mathfrak{A}(\mathfrak{B})$ of linear operators in a vector system $\mathfrak{B}$ is an algebra with identity, and the identity of the algebra is the identity (unit) operator E. The algebra $\mathfrak{A}$ is called a *real* (*complex*) *algebra of functions*, if $\mathfrak{A}$ is a real (complex) vector system of functions (see 1.4.3) and is also a ring with respect to the usual operations of addition and multiplications of functions.

Two elements A and B of the algebra $\mathfrak{A}$ by definition permute (or commute) if

$$AB = BA. \tag{7}$$

If (7) is true for all A and B in $\mathfrak{A}$, then $\mathfrak{A}$ is called a *commutative algebra.* For example, algebras of functions are commutative.

Let A and B be elements of the algebra $\mathfrak{A}$ (with identity) and let

$$AB = E.$$

Then B is called a right inverse of the element A and A is called a left inverse of the element B. If the element A has both a left inverse B_1 and a right inverse B_2, then they coincide. Indeed,

$$B_1 = B_1(AB_2) = (B_1A)B_2 = B_2.$$

In this case, i.e. when $B_1 = B_2 = B$, the element B is called the inverse of A and is written

$$B = A^{-1}.$$

In contrast to the inverse of an element, a left (right) inverse element is in general not unique. *If there exists a unique left* (*right*) *inverse element* of

A, then it is the inverse A^{-1}. Indeed, let $B_1A = E$, and hence

$$AB_1A = A, \quad AB_1A - A = 0, \quad AB_1A - A + B_1A = E.$$

Then $(AB_1 - E + B_1)A = E$, and since the left inverse is unique,

$$AB_1 - E + B_1 = B_1,$$

i.e. $AB_1 - E = 0$ or $AB_1 = E$. Thus the element B_1 is not only a left but also a right inverse of A, and hence $B_1 = A^{-1}$.

An element $A \in \mathfrak{A}$ for which the inverse A^{-1} exists is called a *regular element of the algebra* $\mathfrak{A}$. Whenever A is regular, the element A^{-1} is regular.

The product AB of two regular elements is a regular element of the algebra $\mathfrak{A}$, moreover

$$(AB)^{-1} = B^{-1}A^{-1}. \tag{8}$$

In fact, since multiplication in the ring is associative

$$(B^{-1}A^{-1})(AB) = E, \quad (AB)(B^{-1}A^{-1}) = E.$$

If A is a regular element of the algebra $\mathfrak{A} = \mathfrak{A}(\mathfrak{B})$ and $B = A^{-1}$, then the equations

$$BA = E, \quad AB = E,$$

which define the inverse element B imply that for any f and g in $\mathfrak{B}$

$$BAf = f, \quad ABg = g.$$

Thus the equations

$$g = Af, \quad f = Bg$$

define mutually inverse mappings A and B of the vector system $\mathfrak{B}$ onto $\mathfrak{B}$. Hence, *regular elements of the algebra* $\mathfrak{A}(\mathfrak{B})$ *coincide with* the *automorphisms of the vector system* $\mathfrak{B}$ *and the concept of inverse element in the algebra* $\mathfrak{A}(\mathfrak{B})$ *agrees with the concept of inverse operator in the system* $\mathfrak{B}$ (cf. (1.4.11)). It follows that regular elements of the algebra $\mathfrak{A}(\mathfrak{B})$ form a multiplicative group. This is true for the case of an algebra $\mathfrak{A}$ with identity.

A subset $\mathfrak{B}$ of the algebra $\mathfrak{A}$ is called a *subalgebra* of $\mathfrak{A}$ if it is closed under the operations αA, $A+B$, and AB.

The set $\mathfrak{C}(\mathfrak{M})$ *of elements of the algebra* $\mathfrak{A}$ *which permute with all elements of the set* $\mathfrak{M} \subset \mathfrak{A}$ *is a subalgebra of the algebra* $\mathfrak{A}$.

Indeed, if $C \in \mathfrak{M}$ and if C permutes with A and B then C will permute with $A+B$, αA and AB $(CAB = ACB = ABC)$.

The subalgebra $\mathfrak{C}(\mathfrak{M})$ is called the commutator of the set $\mathfrak{M}$ in $\mathfrak{A}$. If $\mathfrak{M} = \mathfrak{A}$, then the commutator $\mathfrak{C}(\mathfrak{A})$, i.e. the set of elements of the algebra $\mathfrak{A}$ which permute with all of elements of $\mathfrak{A}$, is by definition the *center* of $\mathfrak{A}$. *The center is a commutative algebra.*

REMARK 1. An algebra over a field K is a Φ-system (Sec. 1.3.1), and hence we may speak of both isomorphisms and homomorphisms of an algebra $\mathfrak{A}$ onto an algebra $\mathfrak{A}'$, when $\mathfrak{A}$ and $\mathfrak{A}'$ are algebras over the same field K. This means that if

$$\left.\begin{array}{l} A\mapsto A', \quad B\mapsto B', \\ \text{then} \\ A+B\mapsto A'+B', \quad \alpha A\mapsto \alpha A', \quad AB\mapsto A'B', \end{array}\right\} \tag{9}$$

moreover, in the case of an isomorphism, the mapping is also assumed to be one-to-one. In non-commutative algebras, in addition to isomorphisms and homomorphisms, we will introduce so-called *inverse isomorphisms and homomorphisms*. In this case the last relation $AB\mapsto A'B'$ in (9) is replaced by

$$AB\mapsto B'A'. \tag{9'}$$

Two algebras $\mathfrak{A}$ and $\mathfrak{A}'$ over the same field K are called *inversely-isomorphic* if an inverse isomorphism can be established between them.

REMARK 2. To each element B of the algebra $\mathfrak{A}$ we can associate two operators $\mathscr{S}_B$ and $\mathscr{T}_B$ in the vector system $\mathfrak{A}$, which are defined by the formulas

$$\mathscr{S}_B A = BA, \quad \mathscr{T}_B A = AB,$$

where A ranges through the system $\mathfrak{A}$. The operators $\mathscr{S}_B$ and $\mathscr{T}_B$, which are called, respectively, the *operator of left multiplication* and *operator of right multiplication*, are linear. This follows from (6) and the distributive laws. They coincide if and only if the element B belongs to the center of the algebra $\mathfrak{A}$.

It follows from the associative law that $\mathscr{S}_B\mathscr{S}_C=\mathscr{S}_{BC}$ and $\mathscr{T}_C\mathscr{T}_B=\mathscr{T}_{BC}$. Thus for the mappings $B\mapsto\mathscr{S}_B$ and $B\mapsto\mathscr{T}_B$,

$$BC\mapsto\mathscr{S}_B\mathscr{S}_C, \quad BC\mapsto\mathscr{T}_C\mathscr{T}_B.$$

If the algebra $\mathfrak{A}$ has an identity, then it follows from $\mathscr{S}_B=\mathscr{S}_C$ or $\mathscr{T}_B=\mathscr{T}_C$ that $B=C$, and the one-to-one mappings $B\mapsto\mathscr{T}_B$ and $B\mapsto\mathscr{T}_B$ turn out to be, respectively, an isomorphism and an inverse-isomorphism. Every homomorphism of the algebra $\mathfrak{A}$ into the algebra of linear operators $\mathfrak{A}(\mathfrak{B})$ of some vector system $\mathfrak{B}$ is said to be a *representation of the algebra* $\mathfrak{A}$ *by linear operators*. The mapping $B\mapsto\mathscr{S}_B$ is called a *regular representation* of the algebra $\mathfrak{A}$.

5.3. Algebras with involution and pseudoinvolution. An algebra $\mathfrak{A}$ is called an algebra with involution or $\times$-algebra if an involution $A\mapsto A^{\times}$ has been given on the vector system $\mathfrak{A}$ such that

$$(AB)^{\times}=B^{\times}A^{\times}. \tag{10}$$

Symmetric (skew-symmetric) elements of the vector $\times$-system $\mathfrak{A}$ are said to be *symmetric (skew-symmetric) elements of the $\times$-algebra $\mathfrak{A}$.*

The identity E of a $\times$-algebra is a symmetric element, i.e. $E^\times = E$.

Indeed, (10) gives

$$A^\times = (AE)^\times = E^\times A^\times;$$
$$A^\times = (EA)^\times = A^\times E^\times.$$

Since both $A^\times$ and A range over the entire $\times$-algebra $\mathfrak{A}$, $E^\times$ is the identity on $\mathfrak{A}$, and hence coincides with E.

If A *is a regular element of the $\times$-algebra $\mathfrak{A}$, then $A^\times$ is a regular element of $\mathfrak{A}$, moreover*

$$(A^\times)^{-1} = (A^{-1})^\times. \tag{11}$$

In fact, by use of the involution we find from the equations $A^{-1}A = E$ and $AA^{-1} = E$ that

$$A^\times (A^{-1})^\times = E;$$
$$(A^{-1})^\times A^\times = E.$$

These are equivalent to (11).

In a $\times$-algebra $\mathfrak{A}$, a regular element A *is symmetric (skew-symmetric) whenever its inverse is symmetric (skew-symmetric) and vice-versa.*

Indeed, for $A^\times = A (A^\times = -A)$, (11) gives

$$(A^{-1})^\times = (A^\times)^{-1} = A^{-1} \quad ((A^{-1})^\times = (-A)^{-1} = -A^{-1}).$$

When $K = K_c$ is the field of complex numbers, we define analogously:

An algebra $\mathfrak{A}$ over the field of complex numbers is called an *algebra with pseudo-involution* or a *$*$-algebra*, if a pseudoinvolution is given on the vector system $\mathfrak{A}$ such that

$$(AB)^* = B^* A^*. \tag{12}$$

Hermitian elements of a vector $*$-system $\mathfrak{A}$ are by definition Hermitian elements of the $*$-algebra $\mathfrak{A}$.

Just as in the case of a $\times$-algebra we can prove that:

1. *The identity E is a Hermitian element of the $*$-algebra $\mathfrak{A}$.*
2. *If A is regular, then A^* is also regular, moreover*

$$(A^*)^{-1} = (A^{-1})^*. \tag{13}$$

3. *If the Hermitian element A is regular, then A^{-1} is Hermitian.*

THEOREM 1.5.2. *The product AB of two Hermitian elements A and B of a $*$-algebra $\mathfrak{A}$ is Hermitian if and only if A and B commute.*

Indeed, if A and B are Hermitian, then

$$(AB)^* = B^* A^* = BA$$

and the condition $BA = AB$, i.e. A commutes with B, is necessary and sufficient for AB to be Hermitian.

REMARK. This theorem obviously remains valid for a $\times$-algebra $\mathfrak{A}$ if "Hermitian" is replaced by "symmetric". The theorem also holds in the case when the elements A and B are assumed to be skew-symmetric.

In particular, it follows from Theorem 1.5.2 that whenever A is Hermitian, the operator $A^2 = AA$ is also Hermitian. The operator $A^n = A^{n-1}A$, where n is a non-negative integer, is also Hermitian. The exponent n can also be negative if A is a regular element of the $*$-algebra $\mathfrak{A}(A^{-n} = (A^{-1})^n)$. Whence it follows that when A is Hermitian, the operator

$$\varphi(A) = \sum_{k=0}^{n} \alpha_k A^k$$

is Hermitian, where

$$\varphi(z) = \sum_{k=0}^{n} \alpha_k z^k$$

is a polynomial with real coefficients α_k.

In the case of an algebra $\mathfrak{A}$ with pseudoinvolution (or involution) it is natural to restrict the notion of subalgebra in a corresponding manner. Namely, a $*$-*subalgebra* $\mathfrak{B}$ of a $*$-algebra $\mathfrak{A}$ is a subalgebra $\mathfrak{B}$ which contains the element A^* whenever it contains A (correspondingly, $A^\times$ for a $\times$-*subalgebra* of a $\times$-algebra A). Clearly a $*$-subalgebra $\mathfrak{B}$ is in turn a $*$-algebra.

In a $*$-algebra $\mathfrak{A}$, we introduce a new notion of commutativity, namely we will say that an element B in $\mathfrak{A}$ $*$-commutes with $A \in \mathfrak{A}$ if $BA = AB$ and

$$B^* A = AB^*, \tag{14}$$

i.e. if B and B^* commute with A.

Of course, it is possible to give a similar definition of $\times$-commutativity.

By the application of the $*$-operation to (14) we obtain $A^*B = BA^*$ so that the element A $*$-commutes with B, i.e. *$*$-commutativity is a symmetric relation.* We note that an element A need not $*$-commute with itself, and the equation

$$A^* A = AA^* \tag{15}$$

imposes a condition on the element A which is satisfied *if and only if the Hermitian elements A_1 and A_2 in the decomposition $A = A_1 + iA_2$ commute.*

In fact, since $A^* = A_1 - iA_2$, (15) follows from $A_1A_2 = A_2A_1$, and in a similar manner the equation $A_1A_2 = A_2A_1$ follows from (15) if we express A_1 and A_2 in (15) in terms of A and A^*.

The set $\mathfrak{C}^*(\mathfrak{M})$ of elements of the $*$-algebra $\mathfrak{A}$, which $*$-commute with all elements of the set $\mathfrak{M} \subset \mathfrak{A}$, is by definition the *$*$-commutator of the set $\mathfrak{M}$ in $\mathfrak{A}$.* The commutator $\mathfrak{C}^*(\mathfrak{M})$ is a $*$-subalgebra of the $*$-algebra $\mathfrak{A}$.

5.4. Ideals. A vector subsystem $\mathfrak{a}$ of the algebra $\mathfrak{A}$ is called a *left* (*right*) *ideal* in $\mathfrak{A}$, if

$$BA \in \mathfrak{a}\ (AB \in \mathfrak{a}) \qquad \textit{for any} \qquad A \in \textit{and } B \in \mathfrak{A}.$$

If $\mathfrak{A}$ is an algebra with pseudoinvolution (involution), then by definition $\mathfrak{a}$ is a $*$-ideal ($\times$-ideal) when the following condition is also satisfied: whenever an element A belongs to $\mathfrak{a}$, the element A^* ($A^\times$) also belongs to $\mathfrak{a}$.

If $\mathfrak{a}$ is both a left and right ideal ($*$-ideal, $\times$-ideal), then $\mathfrak{a}$ is called a *two-sided ideal* ($*$-ideal, $\times$-ideal) in $\mathfrak{A}$. The entire algebra ($*$-algebra, $\times$-algebra) $\mathfrak{A}$ and the zero element in $\mathfrak{A}$ are trivial examples of such ideals.

Since every left (right) ideal $\mathfrak{a}$ is a vector system, it generates a linear equivalence relation θ in the algebra $\mathfrak{A}$:

$$A \equiv B\,(\theta), \tag{16}$$

if $A - B \in \mathfrak{a}$, moreover the congruence (16) admits multiplication from the left (right) by elements of $\mathfrak{A}$, i.e. it follows from (16) that

$$CA \equiv CB\,(\theta) \qquad (AC \equiv BC\,(\theta)),$$

where C is any element of $\mathfrak{A}$.

In the case of a two-sided ideal $\mathfrak{a}$, multiplication is possible both from the left and right, and therefore *congruences can be multiplied*, i.e. in this case it follows from

$$\left.\begin{aligned} &A \equiv B\,(\theta), \quad C \equiv D\,(\theta) \\ &\text{that} \\ &AC \equiv BD\,(\theta). \end{aligned}\right\} \tag{17}$$

Indeed, it is sufficient to multiply the first congruence on the right by C, the second on the left by B, and use the transitivity of the equivalence relation.

The types $\mathring{A}$ with respect to the equivalence θ form a vector system (see Sec. 1.4.1), and (17), which holds in the case of a two-sided ideal, permits multiplication to be introduced in the set of $\mathring{\mathfrak{A}}$-types, converting $\mathring{\mathfrak{A}}$ into an algebra. The product of the types $\mathring{A} = \theta A$ and $\mathring{B} = \theta B$ of the elements A and B is defined to be the type of the element AB, i.e.

$$(\theta A)(\theta B) = \theta\,(AB). \tag{18}$$

By virtue of (17) this definition is independent of the choice of the representatives A and B of the types $\mathring{A}$ and $\mathring{B}$. Since by (4.10)

$$\theta\,(A + B) = \theta A + \theta B, \qquad \theta\,(\alpha A) = \alpha\,(\theta A), \tag{19}$$

θ homomorphically maps $\mathfrak{A}$ onto the algebra $\mathring{\mathfrak{A}} = \theta\mathfrak{A}$.

If an ideal $\mathfrak{a}$ is a $*$-ideal, then $\mathfrak{a}$ is a $*$-system, and hence it follows from (16) that $A^* \equiv B^*(\theta)$.

Therefore the formula $(\theta A)^* = \theta A^*$ defines a $*$-operation in the algebra $\mathfrak{A}$ since

$$(\theta A \cdot \theta B)^* = (\theta(AB))^* = \theta((AB)^*) = \theta(B^* A^*) = \theta B^* \cdot \theta A^*.$$

Thus θ now becomes a homomorphism of the $*$-algebra $\mathfrak{A}$ *onto* the $*$-algebra $\hat{\mathfrak{A}}$ (cf. (19) and (18)).

§ 6. Lattices and Directed Sets

6.1. Complete lattices. Let Z be an ordered set. If the collection of majorants (minorants) of a set $N \subset Z$ has a least (largest) element η, then η is called the *upper* (*lower*) *bound* of the set N. The upper bound and the lower bound will be denoted, respectively, by sup N and inf N.

The existence of an upper (lower) bound of a set N assumes a non-empty set of majorants (minorants), i.e. that N is bounded above (below).

An ordered set Z is called a lattice if any pair $\{\xi, \eta\}$ in Z has both an upper bound sup $\{\xi, \eta\}$ and a lower bound inf $\{\xi, \eta\}$. In a lattice Z, $\xi \geqslant \eta$ if and only if $\xi = \sup\{\xi, \eta\}$, or equivalently $\eta = \inf\{\xi, \eta\}$.

If an upper bound exists for any set in Z which is bounded above, and a lower bound exists for any set in Z which is bounded below, then Z is called a *complete lattice*. The upper and lower bounds of any subset of a complete lattice Z exist if and only if the lattice Z is bounded, i.e. it has greatest and least elements.

If every set in Z which is bounded above has an upper bound, and if the set Z itself has a least element, then Z is a complete lattice.

In fact, in this case the set of minorants of any subset N of the set Z has an upper bound which is also a lower bound of the set N.

Examples of complete lattices are:

1) The ordered set $\mathfrak{P}(T)$ is a complete bounded lattice. Indeed, if $\mathfrak{M} = \{M_\rho\}$ is any subset of $\mathfrak{P}(T)$, then obviously

$$\sup \mathfrak{M} = \bigcup_\rho M_\rho; \quad \inf \mathfrak{M} = \bigcap_\rho M_\rho,$$

where $\bigcup_\rho M_\rho$ and $\bigcap_\rho M_\rho$ denote the union and intersection of the sets $M_\rho \in \mathfrak{M}$.

In accordance with these formulas, in place of sup and inf we will make use of the notation:

$$\sup\{\xi, \eta\} = \xi \cup \eta; \quad \inf\{\xi, \eta\} = \xi \cap \eta; \qquad \sup N = \bigcup_{\xi_\rho \in N} \xi_\rho, \quad \inf N = \bigcap_{\xi_\rho \in N} \xi_\rho. \tag{1}$$

for the lattice $\mathfrak{P}(T)$ (and sometimes for arbitrary lattices).

2) The set K_r of real numbers, ordered by the relation $\xi \geqslant \eta$, is a lattice, moreover sup $\{\xi, \eta\}$ and inf $\{\xi, \eta\}$ are equal, respectively, to the greatest and least of the numbers ξ, η. We note that K_r is a complete, but unbounded

lattice, where sup N and inf N are the usual upper and lower bound of the set N of real numbers.

3) The ordered sets $\mathfrak{G}_r(T)$ and $\tilde{\mathfrak{G}}_r(T)$ (see Sec. 1.2.2) are lattices. The elements

$$\Phi(\tau) = \sup\{F, G\}, \qquad \Psi(\tau) = \inf\{F, G\},$$

where the functions $F = F(\tau)$ and $G = G(\tau)$ belong to $\mathfrak{G}_r(T)$ or $\tilde{\mathfrak{G}}_r(T)$, are defined for each $\tau \in T$ by the formulas

$$\Phi(\tau) = \sup\{F(\tau), G(\tau)\}, \qquad \Psi(\tau) = \inf\{F(\tau), G(\tau)\}.$$

The lattices $\mathfrak{G}_r(T)$ and $\tilde{\mathfrak{G}}_r(T)$ are complete.

Indeed, let $\{F_\rho(\tau)\}$ be an arbitrary set in $\mathfrak{G}_r(T)$ which is bounded above in $\mathfrak{G}_r(T)$ (by the element $G(\tau) \in \mathfrak{G}_r(T)$ to which all elements of the set $\{F_\rho(\tau)\}$ are subordinate). Then for all $\tau \in T$,

$$F_\rho(\tau) \leqslant \sup_\rho\{F_\rho(\tau)\} \leqslant G(\tau)$$

and the function

$$\Phi(\tau) = \sup_\rho\{F_\rho(\tau)\},$$

which belongs to $\mathfrak{G}_r(T)$, is the least majorant of the set $\{F_\rho(\tau)\}$, i.e.

$$\Phi(\tau) = \sup_\rho\{F_\rho\}.$$

In the same way, $\Psi = \inf_\rho\{F_\rho\}$ is defined for all $\tau \in T$ by the formula

$$\Psi(\tau) = \inf_\rho\{F_\rho(\tau)\}.$$

The proof is similar for the set $\tilde{\mathfrak{G}}_r(T)$.

We note that our arguments apply to any family $\mathfrak{G}(T, Z)$ of functions $F(\tau)$ on T with values in a complete lattice Z, if a proper order relation $F \geqslant G$ in $\mathfrak{G}(T, Z)$ is defined by the formula $F(\tau) \geqslant \mathfrak{G}(\tau)$ (τ ranges over T). A set $\mathfrak{G}(T, Z)$ which has been ordered in this manner becomes a complete lattice if Z is a complete lattice.

4) The set $\mathfrak{S}(\mathfrak{B})$ of vector subsystems of a fixed vector system is a complete bounded lattice if we let $\mathfrak{B}' \geqslant \mathfrak{B}''$ ($\mathfrak{B}', \mathfrak{B}'' \in \mathfrak{S}(\mathfrak{B})$) when $\mathfrak{B}''$ is contained in $\mathfrak{B}'$. The upper and lower bounds $\sup_\rho\{\mathfrak{B}_\rho\}$ and $\inf_\rho\{\mathfrak{B}_\rho\}$ ($\mathfrak{B}_\rho \in \mathfrak{S}(\mathfrak{B})$) now coincide, respectively, with vector sum and intersection of the vector system $\mathfrak{B}_\rho$. The entire system $\mathfrak{B}$ and (0) are, respectively, the greatest and least elements of the lattice $\mathfrak{S}(\mathfrak{B})$.

A subset Z' of a lattice Z is called a sublattice if whenever ξ and $\eta \in Z'$, the elements $\sup\{\xi, \eta\}$ and $\inf\{\xi, \eta\} \in Z'$.

The lattice $\mathfrak{G}_r(T)$ is a sublattice of the lattice $\tilde{\mathfrak{G}}_r(T)$. Any sublattice of the lattice $\tilde{\mathfrak{G}}_r(T)$ will be called a lattice of functions.

We note that a subset Z' of a lattice Z can be a lattice with respect to the order relation induced in Z' by the order relation in Z, but not be a sublattice in the sense of our definition. Examples are $\mathfrak{P}(\mathfrak{V})$ and $\mathfrak{S}(\mathfrak{V})$, where $\mathfrak{V}$ is a vector system.

A sublattice Z' of a lattice Z is said to be *σ-closed* if each of its countable subsets has an upper bound in Z which belongs to Z'.

The set Z_d which is dual to the lattice Z is again a lattice, moreover it is complete if the lattice Z is complete. Indeed, if one passes over to the dual order relation in the lattice Z, then majorants and minorants change roles, and hence

$$\sup_d \{\xi, \eta\} = \inf \{\xi, \eta\}; \qquad \inf_d \{\xi, \eta\} = \sup \{\xi, \eta\}, \tag{2}$$

where $\sup_d$ and $\inf_d$ denote, respectively, the upper and lower bounds in Z_d. Similarly, for any set $N \in Z$

$$\sup_d N = \inf N, \qquad \inf_d N = \sup N. \tag{2'}$$

It follows from (2) and (2′) that:

Under a dual isomorphism of the lattice Z onto Z', the upper (lower) bound of a set N in Z is mapped into the lower (upper) bound of the image N' of the set N in Z'.

6.2. Distributive lattices and Boolean algebras. A lattice Z is said to be *distributive* if

$$\xi \cap (\eta \cup \zeta) = (\xi \cap \eta) \cup (\xi \cap \zeta) \tag{3}$$

for all elements in Z. If a lattice Z is distributive, then by (2)

$$\xi \cup (\eta \cap \zeta) = (\xi \cup \eta) \cap (\xi \cup \zeta). \tag{4}$$

is identically satisfied for elements in the dual lattice Z_d. If, in addition, (4) is satisfied (identically) for Z, then (3) is satisfied identically for Z_d, i.e. the lattice Z_d is distributive.

A lattice Z_d which is the dual of a distributive lattice Z is distributive.

Indeed, according to the last remark it is sufficient to show that (4) is satisfied in the distributive lattice Z. The application of (3) gives

$$(\xi \cup \eta) \cap (\xi \cup \zeta) = ((\xi \cup \eta) \cap \xi) \cup ((\xi \cup \eta) \cap \zeta) = \xi \cup (\eta \cap \zeta),$$

for $(\xi \cup \eta) \cap \xi = \xi$ since $\xi \cup \eta \geqslant \xi$, and $\xi \cup (\xi \cap \zeta) = \xi$ since $\xi \geqslant \xi \cap \xi$.

Whence it follows that (3) and (4) are equivalent. In fact, we have proved that, in Z, (3) implies (4). Conversely, if (4) holds in Z, then the lattice Z_d is distributive, and hence its dual lattice Z is also distributive, i.e. (4) implies (3).

Lattices of sets and lattices of functions are distributive lattices.

In a lattice Z with least element (zero) 0 and greatest element (identity) I, an element ξ' is said to be the *complement* of ξ if

$$\xi \cup \xi' = I; \qquad \xi \cap \xi' = 0. \tag{5}$$

The complement is unique in a distributive lattice Z.

Indeed, if ξ_1' and ξ_2' are both complements of ξ, then $\xi \cap \xi_1' = 0$ and $\xi \cup \xi_2' = I$. Then by (4),

$$\xi_2' = \xi_2' \cup 0 = \xi_2' \cup (\xi \cap \xi_1') = (\xi_2' \cup \xi) \cap (\xi_2' \cup \xi_1') = \xi_2' \cup \xi_1'.$$

Therefore $\xi_2' \geqslant \xi_1'$; similarly, by interchanging ξ_1' and ξ_2' we find $\xi_1' \geqslant \xi_2'$, and thus $\xi_1' = \xi_2'$.

A distributive lattice Z with least and greatest elements, where each element ξ has a complement ξ', is called a *Boolean algebra*.

In a Boolean algebra Z, for each pair of elements ξ, η such that $\eta \leqslant \xi$, there exists a unique *difference*, i.e. an element ζ such that $\eta \cup \zeta = \xi$ and $\eta \cap \zeta = 0$. Indeed, the sublattice of the lattice Z which consists of elements subordinate to ξ is obviously again a Boolean algebra, and the difference of interest to us is the complement of the element η in this Boolean algebra.

The mapping $\xi \mapsto \xi'$ is an involutive dual automorphism of the Boolean algebra Z.

In fact, the element ξ' is uniquely defined, and since the relations in (5) are symmetric, $(\xi')' = \xi$. Furthermore, if $\xi \geqslant \eta$, then

$$\xi' = \xi' \cap I = \xi' \cap (\eta \cup \eta') = (\xi' \cap \eta) \cup (\xi' \cap \eta') = \xi' \cap \eta',$$

since $\xi' \cap \eta \leqslant \xi' \cap \xi = 0 \leqslant \xi' \cap \eta'$, and hence $\xi' \leqslant \eta'$. The last proposition can be written in the form:

$$(\xi \cup \eta)' = \xi' \cap \eta', \qquad (\xi \cap \eta)' = \xi' \cup \eta', \; (\xi')' = \xi.$$

The lattice $\mathfrak{P}(T)$ is an example of a Boolean algebra, and in this case $M' = T \setminus M$ is the complement of the set $M \in \mathfrak{P}(T)$.

6.3. Complete ordered sets. A subset N of an ordered set Z is *directed above* (*below*) if any two elements of the set N have a majorant (minorant) which belongs to N. In particular, the set N can coincide with Z.

Obviously, every chain $\mathfrak{Z}$ in an ordered set Z, and also any lattice Z, is directed both above and below.

For a function $\xi(\tau)$ whose domain is a set $T (\tau \in T)$ which is directed above, and whose range is an ordered set Z, the concept of monotonicity is naturally introduced, namely: such a function $\xi(\tau)$ is *non-decreasing* or *non-increasing*, if $\tau' \geqslant \tau$ implies that $\xi(\tau') \geqslant \xi(\tau)$ or, respectively, that $\xi(\tau') \leqslant \xi(\tau)$. If $T = \{n\}$ is the (directed above) set of all natural numbers, then the functions $\xi(\tau) = \xi(n)$ are sequences. By considering functions on directed sets as a generalization of sequences, we will introduce another concept which for this generalization corresponds to the notion of a subsequence.

A subset T' of a set T which is directed above is said to be *co-final* to T if for each element $\tau \in T$ there can be found a majorant τ' in T'*. Whenever T is directed above, any co-final subset T' is directed above.

A function $\xi'(\tau)$ is said to be *co-final* to a function $\xi(\tau)$ on a set T which is directed above, if $\xi'(\tau)$ has been defined on a subset T' co-final to T and $\xi'(\tau) = \xi(\tau)$ for $\tau \in T$. Co-final functions correspond to subsequences.

An ordered set Z is said to be *complete* if: (1) Z is directed both above and below; (2) every subset $Z' \subset Z$ which is directed above (below) and bounded above (below) has an upper (lower) bound.

THEOREM 1.6.1. *Let the values of a non-decreasing (non-increasing) function $\xi(\tau)$ on a set T which is directed above belong to a complete ordered set Z, and let the function $\xi(\tau)$ be bounded above (below)*** *Then the upper (lower) bound* $\sup\limits_{\tau} \{\xi(\tau)\}$ $\left(\inf\limits_{\tau} \{\xi(\tau)\}\right)$ *exists.*

Indeed, the set of values $\{\xi(\tau)\}$ is directed above when the function $\xi(\tau)$ does not decrease and it is directed below if $\xi(\tau)$ does not increase. To complete the proof we need only use the completeness of Z.

We note that in order for a lattice Z to be a complete ordered set, it is necessary and sufficient that Z be complete.

§ 7. Ordered Vector Systems

7.1. Positive elements. A real vector system Z is said to be *ordered* if Z is an ordered set and if the order relation given in Z satisfies the conditions: for any $\xi, \eta, \zeta \in Z$ such that $\xi > \eta$, and arbitrary, real $\alpha > 0$

$$1)\ \xi + \zeta > \eta + \zeta; \qquad 2)\ \alpha\xi > \alpha\eta. \tag{1}$$

Elements ω of an ordered vector system which satisfy the condition $\omega > 0$ ($\omega \geqslant 0$) are said to be *positive* (*non-negative*).

In an ordered vector system, $\xi > \eta$ if and only if

$$\xi - \eta > 0,$$

i.e. if $\xi - \eta$ is positive.

Positive elements of an ordered vector system have the following properties:

(a) a sum of positive elements is positive; in fact, if $\omega_1 > 0$ and $\omega_2 > 0$, then by (1)

$$\omega_1 + \omega_2 > 0 + \omega_2 > 0 + 0 = 0;$$

---

* The concept of co-finality is essential only in that case when T does not have a greatest element: otherwise it simply means that T' contains the greatest element of the set T.

** That is, the set of values of the function is bounded above (below).

(b) if $\omega > 0$, then $\alpha\omega > 0$, when $\alpha > 0$ (this follows from the inequality (2) of (1)).

Conversely, let the concept of positive element be introduced in a real vector system Z, i.e. some non-zero elements in Z are declared to be positive; moreover, they are assumed to satisfy the properties (a) and (b) (writing $\omega > 0$ means that ω is a positive element). We will write $\xi > \eta$ if $\xi - \eta > 0$, and $\xi \geqslant \eta$, if $\xi > \eta$ or $\xi = \eta$. In this way Z is converted into an ordered vector system, since $\xi \geqslant \eta$ turns out to be a proper order relation in Z which satisfies the conditions in (1). Indeed, the relation $\xi > \eta$ is irreflexive ($\xi > \xi$ is equivalent to $0 > 0$, and 0 is not a positive element) and transitive (it follows from $\xi > \eta$ and $\eta > \zeta$ that $\xi > \zeta$, since $\xi - \zeta = (\xi - \eta) + (\eta - \zeta) > 0$ by (a)). For $\xi > \eta$ and $\alpha > 0$, (1) holds: $\xi + \zeta - (\eta + \zeta) = \xi - \eta > 0$, and in view of (b), $\alpha\xi - \alpha\eta = \alpha(\xi - \eta) > 0$.

An ordered vector system Z can be considered to be a Φ-system (Sec. 1.3.1), so that the definitions of isomorphism, automorphism and of isomorphic systems given in §3 can be applied to it. A mapping of the system Z onto an ordered vector system Z' is called an *isomorphism* if it is an isomorphism both for the vector systems Z and Z' and for the ordered sets Z and Z'. If the mapping of Z onto Z' is both an isomorphism of the vector systems Z and Z', and a dual isomorphism of the ordered sets Z and Z', then it will be called a *dual isomorphism* of the ordered vector systems Z and Z'. Two ordered vector systems are called *dually-isomorphic* if a dual isomorphism can be established between them.

In the case $Z' = Z$ a (dual) isomorphism is said to be a (*dual*) *automorphism* of the ordered vector system Z.

Let N be a subset of an ordered vector system Z. Let $N + \xi_0$ and αN denote the corresponding sets of elements of the form $\xi + \xi_0$ and $\alpha\xi$, where ξ ranges over the set N. Then

$$\left.\begin{aligned} \sup(N + \xi_0) = \xi_0 + \sup N, \quad \sup(\alpha N) = \alpha \sup N, \\ \inf(-N) = -\sup N, \end{aligned}\right\} \tag{2}$$

if $\alpha \geqslant 0$ and if $\sup N$ exists, and

$$\left.\begin{aligned} \inf(N + \xi_0) = \xi_0 + \inf N, \quad \inf \alpha N = \alpha \inf N, \\ \sup(-N) = -\inf N, \end{aligned}\right\} \tag{2'}$$

if $\alpha \geqslant 0$ and if $\inf N$ exists. The last of the equations in (2) and (2′) follow from the fact that multiplication by -1 is a dual isomorphism of the system Z, and the remaining equations are obtained immediately from the inequalities in (1).

7.2. Vector orders. An ordered vector system Z is called a *vector order* if the ordered set Z is directed above (and thus also below in view of the existence of the dual automorphism $\xi' = -\xi$).

An ordered vector system Z is a vector order if and only if one of the two following conditions is satisfied:

1) each element $\xi \in Z$ has a non-negative majorant;

2) each element in Z can be represented as a difference of non-negative elements.

In fact, if Z is a vector order and if $\xi \in Z$, then the elements ξ and 0 have a majorant ω which is non-negative. Conversely, let 1) be satisfied for an ordered vector system Z and let the elements ξ and η belong to Z. Then, by assumption, the difference $\xi - \eta$ has a non-negative majorant ω:

$$\omega \geqslant \xi - \eta, \quad \omega \geqslant 0,$$

and by letting $\zeta = \eta + \omega$, we obtain

$$\zeta \geqslant \xi, \quad \zeta \geqslant \eta,$$

i.e. ζ is a majorant of ξ and η, and Z is a vector order.

Thus, 1) has been proven equivalent to the definition, and it is sufficient to show that 2) is equivalent to 1). We assume that 1) is satisfied in Z. Then an arbitrary element $\xi \in Z$ has a non-negative majorant ω, i.e. $\omega \geqslant \xi$ and $\omega \geqslant 0$. We let $\xi_1 = \omega$ and $\xi_2 = \omega - \xi$. Then $\xi_1 \geqslant 0$, $\xi_2 \geqslant 0$, and

$$\xi = \xi_1 - \xi_2. \tag{3}$$

Conversely, if we represent each element $\xi \in Z$ in the form (3) with ξ_1 and ξ_2 non-negative, then since $\xi_1 = \xi + \xi_2$, the element $\omega = \xi_1$ is a non-negative majorant of ξ.

A non-negative element ω is called an *absolute majorant* of the element ξ if ω is a majorant of the elements ξ and $-\xi$. In this case, $\omega \geqslant \xi$ and $\omega \geqslant -\xi$, or equivalently

$$-\omega \leqslant \xi \leqslant \omega.$$

In a vector order Z, any element ξ has an absolute majorant.

Indeed, let η be a non-negative majorant of the element ξ and let ω be a majorant of the elements η and $-\xi$. Then

$$\omega \geqslant \eta \geqslant \xi, \quad \omega \geqslant -\xi,$$

i.e. ω is an absolute majorant of ξ.

Obviously, elements of a set $N \subset Z$ have a common absolute majorant if and only if N is bounded in the ordered set Z.

LEMMA 1.7.1. *If ω is an absolute majorant of ξ, then there exists a representation of the element ξ in the form*

$$\xi = \xi^+ - \xi^-, \tag{4}$$

where

$$0 \leqslant \xi^+ \leqslant \omega; \quad 0 \leqslant \xi^- \leqslant \omega. \tag{5}$$

Indeed, it is sufficient to let

$$\xi^+ = \frac{\omega+\xi}{2}, \qquad \xi^- = \frac{\omega-\xi}{2}.$$

For this choice of the elements ξ^+ and ξ^-, we have: $\omega = \xi^+ + \xi^-$.

A vector order Z is called a lattice of vectors if the ordered set Z is a lattice*.

7.3. Linear forms in Z. In particular, a vector order Z is a vector system, and therefore it is possible to consider linear forms $l(\xi)$ $(\xi \in Z)$ on Z. A linear form $l(\xi)$ is by definition *non-negative* if $l(\xi) \geqslant 0$ for all $\xi \geqslant 0$, and it is *positive* if it is not identically equal to zero.

The concept of a positive element in the vector system $\mathfrak{B}(Z, K_r)$ of linear forms in Z defines a proper order relation (positive forms have the properties (a) and (b) Sec. 1.7.1), and thereby $\mathfrak{B}(Z, K_r)$ is converted into an ordered vector system.

We say that the linear form $l = l(\xi)$ is *bounded* if it has an absolute majorant, i.e. if there exists a non-negative linear form $p = p(\xi)$ such that

$$-p(\xi) \leqslant l(\xi) \leqslant p(\xi)$$

for all $\xi \in Z$, $\xi \geqslant 0$.

The set Λ *of bounded linear forms in* Z *forms a vector order.*

In fact, if the linear forms l_1, l_2 and l belong to Λ, and if p_1, p_2, and p are their respective absolute majorants, then p_1+p_2 and αp will be absolute majorants for the forms l_1+l_2 and αl. Thus, the linear forms l_1+l_2 and αl are bounded, i.e. Λ is a vector subsystem of the system of all linear forms in Z.

The ordered set Λ contains all non-negative linear forms and thus, by its very definition, it contains an absolute majorant of any one of its elements, and hence is a vector order. Thus every linear form l in Λ can be represented in the form $l = l_1 - l_2$, where l_1 and l_2 are non-negative linear forms in Z, and conversely, each difference of two such forms belongs to Λ.

7.4. Vector ∗-orders. A set Z is called an *ordered vector ∗-system, vector ∗-order*, or *vector ∗-lattice* if Z is a vector ∗-system and the collection Z_r of Hermitian elements of Z (considered as a linear set over the field of real numbers) is, respectively, an ordered set, vector order or vector lattice.

Any element ξ of an ordered vector ∗-system Z (as in general vector ∗-systems) can be represented uniquely in the form

$$\xi = \xi' + i\xi'', \tag{6}$$

where ξ' and ξ'' are Hermitian elements, i.e. $\xi' \in Z_r$ and $\xi'' \in Z_r$. A non-negative element ω is called an *absolute majorant* of $\xi \in Z$, if ω is an absolute majorant (in Z_r) of the elements ξ' and ξ'' in the representation (6).

* As is easily shown, the lattice Z is distributive.

LEMMA 1.7.2. *If ω is an absolute majorant of an element ξ of a vector $*$-order Z, then there exists a representation of the form*

$$\xi = \xi_1 - \xi_2 + i(\xi_3 - \xi_4), \tag{7}$$

where

$$0 \leqslant \xi_1 \leqslant \omega; \quad 0 \leqslant \xi_2 \leqslant \omega; \quad 0 \leqslant \xi_3 \leqslant \omega; \quad 0 \leqslant \xi_4 \leqslant \omega. \tag{8}$$

This follows from (6) and Lemma 1.7.1.

Let $l(\xi)$ be a linear form in the vector $*$-system Z and let

$$l^*(\xi) = \overline{l(\xi^*)} \tag{9}$$

for any $\xi \in Z$. Then

$$l^*(\xi + \eta) = \overline{l((\xi + \eta)^*)} = \bar{l}(\xi^*) + \bar{l}(\eta^*) = l^*(\xi) + l^*(\eta),$$

$$l^*(\alpha\xi) = \bar{l}(\bar{\alpha}\xi^*) = \bar{\bar{\alpha}}\bar{l}(\xi^*) = \alpha l^*(\xi),$$

i.e. $l^*(\xi)$ is also a linear form in Z.

The mapping $l \to l^*$ represents a pseudoinvolution in the vector system $\mathfrak{V}(Z, K_c)$ of linear forms in Z; it converts $\mathfrak{V}(Z, K_c)$ into a vector $*$-system.

Hermitian elements of $\mathfrak{V}(Z, K_c)$ (defined by the equation $l^* = l$) *coincide with those linear forms which have real values on Z_r.*

Indeed, for $l^* = l$ and $\xi^* = \xi$, (9) gives $l(\xi) = \bar{l}(\xi)$, i.e. the value of $l(\xi)$ is real. Conversely, if $l(\xi)$ is real for $\xi \in Z_r$, then for any $\xi = \xi' + i\xi'' \in Z$ ($\xi' \in Z_r$, $\xi'' \in Z_r$).

$$\overline{l(\xi^*)} = \bar{l}(\xi' - i\xi'') = \overline{l(\xi') - il(\xi'')} = l(\xi') + il(\xi'') = l(\xi),$$

i.e. $l^* = l$.

In particular, non-negative linear forms $p(\xi) = l(\xi)$ are Hermitian (i.e. forms for which $p(\xi) \geqslant 0$ for $\xi \geqslant 0$).

The linear form $l = l' + il''$, where l' and l'' are Hermitian forms, is said to be *bounded* if it has an absolute majorant, i.e. if there exists a non-negative linear form p such that

$$-p(\xi) \leqslant l'(\xi) \leqslant p(\xi), \qquad -p(\xi) \leqslant l''(\xi) \leqslant p(\xi)$$

for all ξ in Z_r.

Whenever the forms

$$l(\xi) = l'(\xi) + il''(\xi)$$

are bounded (the forms $l'(\xi)$ and $l''(\xi)$ are Hermitian), the form $l^*(\xi) = l'(\xi) - il''(\xi)$, $\xi \in Z_r$, is also bounded. Therefore the set Λ of bounded forms in $\mathfrak{V}(Z, K_c)$ now forms a vector $*$-order, and hence any form $l(\xi)$ in Λ can be represented in the form

$$l(\xi) = p_1(\xi) - p_2(\xi) + i(p_3(\xi) - p_4(\xi)).$$

The non-negative linear forms $p_k(\xi)$ ($k = 1, 2, 3, 4$) can be chosen so that a given absolute majorant $p(\xi)$ of the form $l(\xi)$ is also an (absolute) majorant for the forms $p_k(\xi)$.

§ 8. Convergence in Complete Vector Orders (∗-orders)

8.1. Infinitely small. A vector order Z (and, in particular, a vector lattice Z) is said to be *complete*, if the ordered set Z is complete*). The families of functions $\tilde{\mathfrak{G}}_r(T)$ and $\mathfrak{G}_r(T)$ are complete vector lattices.

In a complete vector order Z it is possible to introduce the concept of a *limit*, and in particular, of the *limit* $\lim \xi_n = \xi$ *of a sequence* $\{\xi_n\}$ *in* Z. Naturally, it is necessary to require that the following conditions be fulfilled:

(a) linearity

$$\lim(\xi_n + \eta_n) = \lim \xi_n + \lim \eta_n, \quad \lim(\alpha\xi_n) = \alpha \lim \xi_n,$$

if the limits on the right side exist;

(b) each subsequence of a sequence which converges to the limit ξ, also converges to ξ;

(c) $\lim \xi_n = \xi$, if $\xi_n = \xi$ for $n = 1, 2, 3, \ldots$.

If the limit has properties (a), (b), and (c), and if $\xi_n = \xi + \rho_n$, then the equation $\lim \rho_n = 0$ is equivalent to the limit relation $\lim \xi_n = \xi$. Therefore it is sufficient to establish in what case the sequence $\{\rho_n\}$ should converge to 0, i.e. be *infinitely small*. We first consider non-increasing sequences.

A non-increasing sequence $\{\omega_n\}$ in Z is said to be *infinitely small*, or more precisely, *monotone infinitely small*, if

$$\inf_n \{\omega_n\} = 0.$$

We will also use the notation $\omega_n \to 0$ for monotone infinitely small.

We note that the terms ω_n of a monotone infinitely small sequence are non-negative, and that for every non-increasing sequence $\{\omega_n\}$ of non-negative terms there exists, by virtue of the completeness of the vector order Z, a lower bound $\inf_n \{\omega_n\} \geqslant 0$.

Monotone infinitely small sequences have the following properties:

(1) *a subsequence of a monotone infinitely small sequence is monotone infinitely small*;

(2) *monotone infinitely small sequences form a semivector system,* i.e. *whenever* $\{\omega_n\}$ *and* $\{\omega_n'\}$ *are monotone infinitely small, the sequences* $\{\omega_n + \omega_n'\}$ and $\{\alpha\omega_n\}$, *for* $\alpha \geqslant 0$, *are also monotone infinitely small.*

Indeed, $\{\alpha\omega_n\}$ does not increase and (cf. (1.7.2′))

---

* This definition agrees with the remark at the end of Sec. 1.6.3.

$$\inf_n \{\alpha\omega_n\} = \alpha \inf_n \{\omega_n\} = 0.$$

Furthermore, if $\omega''_n = \omega_n + \omega'_n$, then the sequence $\{\omega''_n\}$ is non-negative, does not increase, and

$$\omega''_n = \omega_n + \omega'_n \leqslant \omega_m + \omega'_n$$

for $n \geqslant m$. Therefore (see (1.7.2′))

$$0 \leqslant \inf_n \{\omega''_n\} \leqslant \inf_n \{\omega_m + \omega'_n\} = \omega_m,$$

and since $\omega_m \to 0$, we have $\inf_n \{\omega''_n\} = 0$.

We turn now to the construction of the general concept of an infinitely small sequence (not necessarily monotone).

By letting (for sequences in Z) $\{\xi_n\} \geqslant \{\eta_n\}$, if $\xi_n \geqslant \eta_n$ for each n, we define a proper order relation (see Sec. 1.2.2, example 3°) in the vector system $\mathfrak{G}(N, Z)$* of all sequences in Z. This converts $\mathfrak{G}(N, Z)$ into a vector order, in which the sequence $\{\omega_n\}$ is an absolute majorant for $\{\xi_n\}$ in $\mathfrak{G}(N, Z)$ if and only if ω_n, for each n, is an absolute majorant for ξ_n in Z.

Lemma 1.8.1. *If $\{\omega_n\}$ is an absolute majorant of $\{\xi_n\}$, and if $|\alpha_n| \leqslant \gamma_n$, then $\{\gamma_n \omega_n\}$ is an absolute majorant of $\{\alpha_n \xi_n\}$.*

Indeed, for $\alpha_n \geqslant 0$

$$\alpha_n \xi_n \leqslant \alpha_n \omega_n \leqslant \gamma_n \omega_n;$$

if $\alpha_n < 0$, then

$$\alpha_n \xi_n = (-\alpha_n)(-\xi_n) \leqslant \gamma_n \omega_n,$$

and moreover, we always have

$$-\alpha_n \xi_n = (-\alpha_n) \xi_n \leqslant \gamma_n \omega_n.$$

A sequence $\{\rho_n\}$ is said to be *infinitely small*, if a monotone infinitely small $\{\omega_n\}$ can be found among its absolute majorants.

Some properties of infinitely small sequences are:

1) *A subsequence of an infinitely small sequence $\{\rho_n\}$ is again infinitely small.*

Indeed, let a monotone infinitely small $\{\omega_n\}$ be an absolute majorant for $\{\rho_n\}$ and let $\{\rho_{n_k}\}$ be a subsequence of $\{\rho_n\}$. Then the monotone infinitely small $\{\omega_{n_k}\}$ will be an absolute majorant of $\{\rho_{n_k}\}$.

2) *The infinitely small $\{\rho_n\}$ form a vector order.*

In fact, let $\{\omega_n\}$ and $\{\omega'_n\}$ be monotone infinitely small and be respectively absolute majorants of infinitely small $\{\rho_n\}$ and $\{\rho'_n\}$. Then $\{\omega_n + \omega'_n\}$ and $\{|\alpha| \omega_n\}$ will be monotone infinitely small and will be absolute majorants of $\{\rho_n + \rho'_n\}$ and $\{\alpha\rho_n\}$, and hence $\{\rho_n + \rho'_n\}$ and $\{\alpha\rho_n\}$ are infinitely small. Thus,

* $\mathfrak{G}(N, Z)$ is the vector system of all functions, defined on the set of natural numbers N, with values in Z; see Sec. 1.2.2.

the collection of infinitely small sequences forms a vector subsystem of the vector order $\mathfrak{G}(N, Z)$, and is itself a vector order, since every infinitely small $\{\rho_n\}$ has an absolute majorant in it.

3) *Every infinitely small sequence can be represented as the difference of two non-negative infinitely small sequences.*

This follows from (2) and Lemma 1.7.1.

Having defined

$$\lim \xi_n = \xi$$

by the condition that $\xi_n - \xi$ be infinitely small, we arrive at a concept of *limit of a sequence*, which obviously satisfies the conditions (a), (b), and (c), the fulfillment of which we require above.

The limit $\lim \xi_n = \xi$, and the corresponding convergence of the sequence $\{\xi_n\}$ to ξ, is said to be the *order limit and order convergence*, respectively.

We turn to the consideration of vector $*$-orders. A vector $*$-order Z is said to be *complete* if the vector order Z_r of Hermitian elements of Z is complete.

The definition of infinitely small $\{\rho_n\}$ given previously is preserved in a complete vector $*$-order Z. It is equivalent to the fact that in the decomposition

$$\rho_n = \rho_n' + i\rho_n''; \quad \rho_n' \in Z_r, \quad \rho_n'' \in Z_n,$$

$\{\rho_n'\}$ and $\{\rho_n''\}$ are infinitely small in the vector order Z_r. Therefore $\lim \xi_n = \xi$ if and only if $\lim \xi_n' = \xi'$ and $\lim \xi_n'' = \xi''$, where $\xi_n = \xi_n' + i\xi_n''$ and $\xi = \xi' + i\xi''$ ($\xi_n', \xi_n'', \xi', \xi'' \in Z_r$).

Infinitely small sequences now form a vector $*$-order, since whenever $\rho_n = \{\rho_n' + i\rho_n''\}$ is infinitely small, the sequence $\rho_n^* = \{\rho_n' - i\rho_n''\}$ is also infinitely small. Infinitely small $\{\rho_n\}$ can therefore be represented in the form

$$\{\rho_n\} = \{\omega_n^{(1)} - \omega_n^{(2)} + i\omega_n^{(3)} - i\omega_n^{(4)}\}, \tag{1}$$

where the $\{\omega^{(k)}\}$ ($k = 1, 2, 3, 4$) are non-negative and infinitely small.

Remark. The notions of infinitely small and limit can be extended, by considering a sequence $\{\rho_n\}$ to be infinitely small in a new (extended) sense if, in each of its subsequences, there can be found in turn a subsequence which is infinitely small in the old sense. Such an extension has a number of advantages, particularly in relation to the topology induced by order convergence.

8.2. Uniformly infinitely small. In this section we define an important kind of convergence in complete vector $*$-orders which is a natural generalization of ordinary uniform convergence.

Lemma 1.8.2. *A set $\{\beta\omega\}$ is not bounded above if ω is a fixed positive element of a complete vector $*$-order Z and β ranges over a set of real numbers which is not bounded above.*

Indeed, the set $\{\beta\omega\}$ is directed above, and if it were bounded above, then by completeness there would exist an upper bound $\eta = \sup_{\beta} \{\beta\omega\}$ which obviously would coincide with

$$\sup_{\beta} \{(\beta+1)\omega\} = \omega + \sup_{\beta} (\beta\omega) = \omega + \eta.$$

Thus $\omega+\eta = \eta$, and hence $\omega = 0$, which contradicts the condition $\omega > 0$.

A sequence

$$\{\omega_n\} = \{\alpha_n \omega\} \tag{2}$$

is monotone infinitely small if

$$0 < \omega \in Z, \quad \alpha_n \geqslant \alpha_{n+1}, \quad \lim \alpha_n = 0. \tag{3}$$

In fact, the sequence $\{\alpha_n\omega\}$ does not increase and is bounded below, and therefore

$$\inf_n (\alpha_n \omega) = \eta \geqslant 0$$

exists. It remains to show that $\eta = 0$. For this, it is possible to assume that all α_n are positive, since otherwise our assertion is trivial. From $\alpha_n\omega \geqslant \eta$ it then follows that $\frac{1}{\alpha_n} \eta \leqslant \omega$ and since the set $\{\beta_n\} = \left\{\frac{1}{\alpha_n}\right\}$ is not bounded, the possibility $\eta > 0$ is excluded by Lemma 1.8.2. Hence $\eta = 0$.

A sequence $\{\rho_n\}$ in a complete vector $*$-order is said to be *uniformly infinitely small* if it has an absolute majorant of the form (2)—(3)*.

As is easily seen, the uniformly infinitely small $\{\rho_n\}$ form a vector $*$-order, and hence the representation (1) holds for them, where the $\omega_n^{(k)}$ are non-negative uniformly infinitely small. Moreover, we have:

LEMMA 1.8.3. *If a sequence $\{\xi_n\}$ in Z is bounded, and if* $\lim \alpha_n = 0$, *then $\{\alpha_n \xi_n\}$ is uniformly infinitely small.*

Indeed, in view of the boundedness of $\{\xi_n\}$ there exists a common absolute majorant ω for all of the ξ_n. We now let $\gamma_n = \sup_{k \geq n} |\alpha_k|$. Then

$$\gamma_n \geqslant \gamma_{n+1}, \quad \lim \gamma_n = 0, \quad |\alpha_n| \leqslant \gamma_n,$$

and hence by Lemma 1.8.1, which holds in the case of a $*$-order Z, the sequence $\{\gamma_n\omega\}$ will be an absolute majorant of the form (2)–(3) for $\{\xi_n\}$.

If $\xi_n - \xi$ is uniformly infinitely small, then we say that the sequence $\{\xi_n\}$ in Z *converges uniformly* to ξ and the element ξ is its *uniform limit*.

---

* We will keep this definition for "double" sequences $\rho_{m,n}$ when the set of indices is ordered thus: $[m', n'] \geqslant [m, n]$, if $m' \geqslant m$ and $n' \geqslant n$.

THEOREM 1.8.1. *If* $\lim \xi_n = \xi$ $(\xi_n \in Z)$, *and if* $\lim \alpha_n = \alpha$ $(\alpha_n \in K_c)$, *then*

$$\lim \alpha_n \xi_n = \alpha \xi, \tag{4}$$

moreover if $\{\xi_n\}$ *converges uniformly, then* $\{\alpha_n \xi_n\}$ *converges uniformly.*

In fact, in the decomposition

$$\alpha_n \xi_n - \alpha \xi = \alpha (\xi_n - \xi) + (\alpha_n - \alpha) \xi_n$$

by Lemma 1.8.3, the second term on the right is uniformly infinitely small and the first term is infinitely small, or even uniformly infinitely small, when $\{\xi_n\}$ converges uniformly.

If a concept of limit $\lim \xi_n$ for sequences $\{\xi_n\}$ is introduced in a (real or complex) vector system $\mathfrak{B}$, and if the limit satisfies the conditions (a), (b), and (c) of Sec. 1.8.1, and the additional requirement

(d) *for* $\lim \xi_n = \xi$ *and* $\lim \alpha_n = \alpha$, *Equation* (4) *is true, then* $\mathfrak{B}$ *is called a vector space* (complex and real). Thus, complete vector orders or $*$-orders Z, with an order limit generated by an order relation in Z, are examples of vector spaces.

8.3. The convergence of series. For order convergence (in a complete vector $*$-order) of the series

$$\sum_{n=1}^{\infty} \xi_n \tag{5}$$

with non-negative terms $\xi_n \in Z$, the same criteria are true that hold in the case of real numbers:

The series (5) *converges if and only if its partial sums* $\eta_n = \sum_{k=1}^{n} \xi_k$ *are bounded above.*

In fact, the sequence $\{\eta_n\}$ does not decrease, and it converges if and only if the set $\{\eta_n\}$ is bounded above, moreover $\lim \eta_n = \sup_n \{\eta_n\}$.

From the criterion indicated above we have:

The series (5), *with non-negative terms, converges if one of its majorant series* $\sum_{n=1}^{\infty} \eta_n$ *converges* $(\eta_n \geqslant \xi_n)$.

The series (5) *converges* absolutely if the sequence of its terms has an absolute majorant $\{\omega_n\}$ such that the majorant series $\sum_{n=1}^{\infty} \omega_n$ converges.

It follows from the representation of elements of the sequence $\{\xi_n\} \in \mathfrak{G}(N,Z)$ in the form $\{\xi_n\} = \{\omega_n^{(1)} - \omega_n^{(2)} + i\omega_n^{(3)} - i\omega_n^{(4)}\}$, where $\omega_n^{(k)} \geqslant 0$ and $\omega_n^{(k)} \leqslant \omega_n$, $k = 1, 2, 3, 4$ and $n = 1, 2, 3, \ldots$, that the series (5) has the representation

$$\sum_{n=1}^{\infty} \xi_n = \sum_{n=1}^{\infty} \omega_n^{(1)} - \sum_{n=1}^{\infty} \omega_n^{(2)} + i \sum_{n=1}^{\infty} \omega_n^{(3)} - i \sum_{n=1}^{\infty} \omega_n^{(4)} \tag{6}$$

as a combination of series $\sum_{n=1}^{\infty} \omega_n^{(k)}$ $(k = 1, 2, 3, 4)$ with non-negative terms and common absolute majorant $\{\omega_n\}$. Therefore, in the case of absolute convergence of the series (5), i.e. in the case when the series $\sum_{n=1}^{\infty} \omega_n$ converges, all four of the series $\sum_{n=1}^{\infty} \omega_n^{(k)}$ converge, and this means, in view of (6), that the series (5) converges. Thus, an absolutely convergent series converges (in the sense of order convergence).

In an absolutely convergent series, the order of succession of the terms can be altered in any manner without altering convergence and without changing the sum of the series.

In series with non-negative terms, the usual proof of this proposition carries over from the theory of number series, and in view of the representation (6), the general case can be reduced to this special case. Similarly, we can carry over the proof of the well known theorem on double series:

THEOREM 1.8.2. *If the partial sums*

$$\sigma_{mn} = \sum_{r=1}^{m} \sum_{s=1}^{n} \xi_{rs}$$

of a double series $\sum_{r,s}^{\cdot} \xi_{rs}$, *with non-negative terms* ξ_{rs} *in a complete vector* $*$*-order Z, are bounded, then the series* $\sum_{s=1}^{\infty} \xi_{rs}$ *and* $\sum_{r=1}^{\infty} \xi_{rs}$, *and also the series* $\sum_{r=1}^{\infty} \sigma_r$ *and* $\sum_{s=1}^{\infty} \tau_s$ *formed from the sums* $\sigma_r = \sum_{s=1}^{\infty} \xi_{rs}$ *and* $\tau_s = \sum_{r=1}^{\infty} \xi_{rs}$, *converge and*

$$\sum_{r=1}^{\infty} \sigma_r = \sum_{s=1}^{\infty} \tau_s = \lim \sigma_{mn},$$

where the limit $\lim \sigma_{mn} = \sup_{m,n} \sigma_{mn}$ *is taken for* $m, n \to \infty$, *i.e. according to the directed set of pairs of indices*: $[m', n'] \geqslant [m, n]$, *if* $m' \geqslant m$ *and* $n' \geqslant n$.

This theorem is extended in an obvious manner to absolutely convergent series.

8.4. The Cauchy criterion for uniform convergence.

THEOREM 1.8.3. *If the sequence* $\{\xi_n - \xi_m\}$ *converges uniformly to zero for* $m,n \to \infty$, *then the sequence* $\{\xi_n\}$ *from a complete vector* $*$*-order Z converges uniformly to some element* $\xi \in Z$.

Indeed, the uniform convergence of $\{\xi_n - \xi_m\}$ to 0 means that for some fixed $\omega > 0$ in Z, and any (real) $\varepsilon > 0$, the element $\varepsilon\omega$ is an absolute majorant of $\xi_n - \xi_m$ for $m \geqslant n_0$ (ε), $n \geqslant n_0$ (ε), i.e.

$$-\varepsilon\omega \leqslant \xi_n' - \xi_m' \leqslant \varepsilon\omega; \qquad -\varepsilon\omega \leqslant \xi_n'' - \xi_m'' \leqslant \varepsilon\omega, \tag{7}$$

where $\xi_m = \xi_m' + i\xi_m''$ $(\xi_m', \xi_m'' \in Z_r)$. Now let $\varepsilon_k > \varepsilon_{k+1} > 0$ and suppose the series $\sum_{k=1}^{\infty} \varepsilon_k$ converges. Then $\varepsilon_k\omega$ is an absolute majorant of $\xi_{n_{k+1}} - \xi_{n_k}$, where

$n_k = n_0\ (\varepsilon_k)$, and the series

$$\xi_{n_1} + \sum_{k=1}^{\infty} (\xi_{n_{k+1}} - \xi_{n_k})$$

converges absolutely to

$$\xi = \lim \xi_{n_k} = \xi' + i\xi''.$$

By substituting $m = n_k$ in the inequality (7) and passing to the limit for $k \to \infty$, we find that for $n \geqslant n_0\ (\varepsilon)$

$$-\varepsilon\omega \leqslant \xi'_n - \xi' \leqslant \varepsilon\omega, \quad -\varepsilon\omega \leqslant \xi''_n - \xi'' \leqslant \varepsilon\omega,$$

i.e. the sequence $\{\xi_n\} = \{\xi'_n + i\xi''_n\}$ converges uniformly to $\xi = \xi' + i\xi''$.

§ 9. Positive, Completely Linear Functionals

9.1. Positive linear functionals. For a positive linear form $p(\xi)$ in a complete vector $*$-order Z, the sequence $\{p(\omega_n)\}$ does not increase when $\{\omega_n\}$ does not increase. A positive linear form is said to be a *positive linear functional* in Z if whenever $\{\omega_n\}$ is monotone infinitely small, $\{p(\omega_n)\}$ is also monotone infinitely small, i.e. if

$$\lim_{n \to \infty} p\,(\omega_n) = \inf p\,(\omega_n) = 0, \tag{1}$$

when (ω_n) is monotone infinitely small. Equation (1) is equivalent to the condition

$$\lim_{n \to \infty} p\,(\xi_n) = p\,(\xi), \quad \text{when} \quad \lim_{n \to \infty} \xi_n = \xi. \tag{2}$$

Indeed, let

$$\xi_n = \xi'_n + i\xi''_n, \quad \xi = \xi' + i\xi'' \quad (\xi'_n, \xi''_n, \xi', \xi'' \in Z_r)$$

and

$$\xi' - \omega_n \leqslant \xi'_n \leqslant \xi' + \omega_n, \quad \xi'' - \omega_n \leqslant \xi''_n \leqslant \xi'' + \omega_n,$$

where ω_n is monotone infinitely small. Then in view of the positivity of the form $p(\xi)$,

$$p\,(\xi') - p\,(\omega_n) \leqslant p\,(\xi'_n) \leqslant p\,(\xi') + p\,(\omega_n),$$
$$p\,(\xi'') - p\,(\omega_n) \leqslant p\,(\xi''_n) \leqslant p\,(\xi'') + p\,(\omega_n),$$

so that (2) follows from (1).

Thus, *positive linear functionals are positive linear forms which are continuous with respect to the order convergence of sequences in* Z.

We strengthen this condition of continuity considerably by generalizing the notion of monotone infinitely small in a suitable manner. Namely, in place of sequences $\{\xi_n\}$, functions of an integral variable n, we will consider functions $\xi(\tau) \in \mathfrak{G}(N, Z)$ defined on all possible directed above sets $T(\tau \in T)$ *which have no greatest element*. When this convergence is considered the class of linear functionals is, in general, smaller.

A non-increasing function $\omega(\tau) \in \mathfrak{G}(T; Z)$ (see Sec. 1.6.3) for which $\inf_{\tau} \{\omega(\tau)\} = 0$ is said to be *monotone infinitely small along the directed set* T). We then write

$$\lim \omega(\tau) = 0.$$

If $\omega(\tau)$ is monotone infinitely small and if $p(\xi)$ is a positive linear form in Z, then the function $p(\omega(\tau))$ is non-negative and does not increase in T. Hence, by the completeness of Z, the lower bound $\inf \{p(\omega(\tau))\} \geqslant 0$ exists.

We now replace (1) by a condition which is more restrictive:

$$\inf_{\tau} p(\omega(\tau)) = 0, \tag{3}$$

where $\omega(\tau)$ is arbitrary and monotone infinitely small. Positive linear forms in Z which satisfy (3) will be called *positive, completely linear functionals.*

9.2. Additional conditions on a vector *-order Z. A set Π of positive linear forms $p(\xi) \geqslant 0$ in Z is said to be *sufficient* if $p(\xi) \geqslant 0$, $\xi \in Z$, for all $p \in \Pi$ implies that $\xi \geqslant 0$.

If a sufficient set Π of positive linear forms has been given, then $\xi \geqslant \eta$ is equivalent to the inequality $p(\xi) \geqslant p(\eta)$ for all $p \in \Pi$, and the equation $\xi = \eta$ is thus equivalent to the equation $p(\xi) = p(\eta)$ for any $p \in \Pi$.

For some applications we impose still another condition on a complete vector *-order Z:

(I) *A set of positive, completely linear functionals is sufficient.*

In the theory of extensions of linear operations with values in Z (see, for example, Chapter II), in place of (I) another condition turns out to be useful.

(II) *If* $\{N_k\}$ *is a sequence of directed below subsets of a *-order* Z *and if* $\inf N_k = 0$ *for every* k, *then it is possible to choose elements* $\xi_k \in N_k$ *such that the series*

$$\sum_{k=1}^{\infty} \xi_k \tag{4}$$

(*with non-negative terms*) *will converge.*

The vector order (*-order) of real (complex) numbers satisfies (II). Neither of the conditions (I) or (II) is a consequence of the other*.

LEMMA 1.9.1. *Let a complete vector *-order satisfy* (II), *and let* $\{N_k\}$ *have the properties indicated in the formulation of this condition. Then the set* N *of*

* For example, in $\mathfrak{G}([0, 1], K_r)$, (I) is satisfied, but (II) is not. For proof of the latter assertion it is sufficient to take $N_k = N$, where N is the set of all functions which have only the values 0 and 1 and which equal zero only at a finite number of points. The space of all measurable functions, finite almost everywhere on [0, 1], is an example of a complete vector order in which (II) is satisfied, but (I) is not.

elements in Z which are represented by sums of convergent series (4), *where* $\xi_k \in N_k$, *is directed below, and*

$$\inf N = 0. \tag{5}$$

Indeed, if ξ' and ξ'' belong to N, if

$$\xi' = \sum_{k=1}^{\infty} \xi'_k,$$
$$\xi'_k \in N_k;$$
$$\xi'' = \sum_{k=1}^{\infty} \xi''_k,$$
$$\xi''_k \in N_k,$$

and if ξ_k is a minorant of the pair ξ'_k, ξ''_k in N_k, then the series $\sum_{k=1}^{\infty} \xi_k$ converges to some element $\xi \in N$, with $\xi \leqslant \xi'$, and $\xi \leqslant \xi''$. Thus the set N is directed below. In view of the convergence of the series (4) we have

$$\xi = \sum_{k=1}^{n} \xi_k + \sum_{k=n+1}^{\infty} \xi_k = \sum_{k=1}^{n} \xi_k + \eta_n \tag{6}$$

and $\lim \eta_n = \inf_n \eta_n = 0$. For fixed n, (6) gives

$$\inf N = \inf_{\xi \in N} \{\xi\} \leqslant \eta_n,$$

and, finally, $\inf N = 0$.

We note that in (II) the requirement that $\inf N_k = 0$ can be replaced by the following condition: $\inf N_k = \omega_k$ and the series $\sum_{k=1}^{\infty} \omega_k$ converges. Then, in place of (5), we have

$$\inf N = \sum_{k=1}^{\infty} \omega_k. \tag{5'}$$

9.3. A generalization of the concept of convergence. The theory of convergence in Z presented in the preceding section is entirely preserved in such vector orders or $*$-orders in which the condition of completeness of Z has been replaced by the considerably weaker requirement of the existence of upper bounds for increasing and bounded above sequences of elements in Z. Namely, in such *countable complete* vector $*$-orders, the consideration of limits of sequences is in some sense exhausting. In the case of a complete vector $*$-order, it is natural to consider limits of functions with values in Z which have been defined on a directed above set (without a greatest element). For such functions the notion of monotone infinitely small has already been defined. From this is defined a general concept of infinitely small $\rho(\tau)$ ($\rho(\tau)$ has some monotone infinitely small $\omega(\tau)$ as an absolute majorant), and the limit $\lim \xi(\tau) = \xi$ of a function $\xi(\tau)$ on a directed set

$T(\xi(\tau)-\xi$ is infinitely small). With the following changes, the arguments of §8 carry over to this more general case. The formation of the sum of two infinitely small functions is possible only if they are defined on the same set T. Thus infinitely small functions form a vector $*$-order only for fixed T. Here, co-final functions (see Sec. 1.6.3) correspond to subsequences. Obviously, a function $\omega'(\tau)$ which is co-final to a monotone infinitely small function (on a set T' which is co-final T), is also monotone infinitely small, and therefore a function $\rho'(\tau)$ which is co-final to an infinitely small function $\rho(\tau)$, is infinitely small.

It is not difficult to convince oneself that positive completely linear functionals are positive linear forms which are continuous with respect to convergence of functions on directed (above) sets in Z.

CHAPTER 2

Integral and Measure with Values in a Complete Vector $*$-order

§ 1. Semifields of Functions

1.1. Lebesgue sets. A family $\mathfrak{F}$ of real functions $F(\tau)$ defined on a fixed set $T(\tau \in T)$ will be called a *semivector system of functions* if whenever $F(\tau)$ and $G(\tau)$ belong to $\mathfrak{F}$, the functions

$$F(\tau)+G(\tau); \quad \alpha F(\tau), \quad \alpha \geqslant 0,$$

also belong to $\mathfrak{F}$. If a real vector system or semivector system of functions $\mathfrak{F}$ is, in addition, a lattice of functions, i.e. if whenever $F(\tau)$ and $G(\tau)$ belong to $\mathfrak{F}$, the functions

$$\Phi(\tau)=\sup\{F, G\}, \qquad \Psi(\tau)=\inf\{F, G\}, \tag{1}$$

also belong to $\mathfrak{F}$, then we will call $\mathfrak{F}$, respectively, a *field* or *semifield* of functions. A field or semifield of functions is *regular* if it contains all (real) constants.

A real vector system of functions $\mathfrak{F}$ is a field of functions if and only if whenever $F(\tau)$ belongs to $\mathfrak{F}$, the function $|F(\tau)|$ also belongs to $\mathfrak{F}$.

Indeed, if $\mathfrak{F}$ is a field of functions, if $F(\tau) \in \mathfrak{F}$, and if

$$F^{+}(\tau)=\sup\{F, 0\}, \quad F^{-}(\tau)=\sup\{-F, 0\}=-\inf\{F, 0\},$$

then $F^{+}(\tau)$, $F^{-}(\tau)$, and also the function $|F(\tau)| = F^{+}(\tau)+F^{-}(\tau)$ belong to $\mathfrak{F}$. Conversely, if whenever $F(\tau)$ belongs to $\mathfrak{F}$, the function $|F(\tau)|$ always belongs to $\mathfrak{F}$, then the identity

$$\sup\{F, G\}=\sup\{F-G, 0\}+G= =\frac{F-G+|F-G|}{2}+G=\frac{F+G+|F-G|}{2},$$

$$\inf\{F, G\}=\frac{F+G-|F-G|}{2}$$

shows that $\mathfrak{F}$ is a lattice of functions.

To each real function $F(\tau)$ there corresponds its system of *Lebesgue sets*

$$M = T(F > \alpha) = \{\tau \mid \tau \in T,\ F(\tau) > \alpha\} \text{ (*)}$$

where α is an arbitrary real number. *To each regular semifield of functions* $\mathfrak{F}$ we associate a system $\mathfrak{M} = \mathfrak{M}(\mathfrak{F})$ which consists of the Lebesque sets M of all functions in $\mathfrak{F}$. We note that $T \in \mathfrak{M}(\mathfrak{F})$. Since whenever $F(\tau)$ belongs to $\mathfrak{F}$, the function $F(\tau)-\alpha$ also belongs to $\mathfrak{F}$, the system $\mathfrak{M}(\mathfrak{F})$ is already exhausted by sets of the form

$$M = T(F > 0). \tag{2}$$

The system $\mathfrak{M}(\mathfrak{F})$ is closed under the formation of sums and intersections of a finite number of its sets, i.e. $\mathfrak{M}(\mathfrak{F})$ is a sub-lattice of the lattice $\mathfrak{P}(T)$. Indeed, if the functions $\Phi(\tau)$ and $\Psi(\tau)$ are defined by (1), then

$$T(\Phi > 0) = T(F > 0) \cup T(G > 0),$$
$$T(\Psi > 0) = T(F > 0) \cap T(G > 0).$$

1.2. σ-closed families; σ-extensions. A family of functions $\mathfrak{F}$ is *σ-closed* if the limit $\lim F_n(\tau)$ of every non-decreasing sequence in $\mathfrak{F}$ which satisfies the condition

$$F_n(\tau) \leqslant F'(\tau) \in \mathfrak{F}, \tag{3}$$

belongs to $\mathfrak{F}$.

We note that *if a lattice of functions* $\mathfrak{F}$ *is σ-closed, then the function*

$$\Phi(\tau) = \sup_n \{F_n(\tau)\},$$

where $\{F_n(\tau)\}$ *is an arbitrary sequence in* $\mathfrak{F}$ *satisfying the condition* (3), *belongs to* $\mathfrak{F}$. In fact, the functions

$$\Phi_n(\tau) = \sup\{F_1, F_2, \ldots, F_n\}$$

belong to $\mathfrak{F}$, $\Phi_n(\tau) \leqslant \Phi_{n+1}(\tau) \leqslant F'(\tau) \in \mathfrak{F}$, and $\lim \Phi_n(\tau) = \Phi(\tau)$.

A system $\mathfrak{M}$ of subsets of the set T is said to be *σ-closed* if a countable sum (union) $\bigcup_n M_n$ of sets M_n in $\mathfrak{M}$ satisfying the condition

$$M_n \subset M' \in \mathfrak{M}, \tag{4}$$

always belongs to $\mathfrak{M}$.

In the case when the sum of all sets of the system $\mathfrak{M}$ is contained in $\mathfrak{M}$, (4) is always satisfied. For example, this is true for the system $\mathfrak{M}(\mathfrak{F})$, where the sum coincides with T. By replacing T, if necessary, by the sum of the sets of the system $\mathfrak{M}$, it is always possible to assume that this sum equals T.

* We recall that the expression $\{x \mid P(x)\}$ denotes the set of all elements x for which the proposition $P(x)$ holds.

THEOREM 2.1.1. *If a regular semifield $\mathfrak{F}$ is σ-closed, then the system of Lebesgue sets $\mathfrak{M}(\mathfrak{F})$ is σ-closed and consists precisely of those sets whose characteristic functions belong to $\mathfrak{F}$.*

In fact, let $M_n \in \mathfrak{M}(\mathfrak{F})$, $M_n = T(G_n > 0)$, where $G_n(\tau) \in \mathfrak{F}$, and let $F_n(\tau) = \inf \{G_n, 1\}$. Then

$$F_n(\tau) \in \mathfrak{F}, \quad M_n = T(F_n > 0), \quad F_n(\tau) \leqslant 1 \in \mathfrak{F},$$

and hence (see above) $\Phi(\tau) = \sup_n \{F_n\} \in \mathfrak{F}$. Since

$$T(\Phi > 0) = \bigcup_n T(F_n > 0),$$

we have

$$\bigcup_n M_n = T(\Phi > 0) \in \mathfrak{M}(\mathfrak{F}),$$

for any sequence $\{M_n\}$ in $\mathfrak{M}(\mathfrak{F})$, i.e. the system $\mathfrak{M}(\mathfrak{F})$ is σ-closed.

Now let M be an arbitrary set in $\mathfrak{M}(\mathfrak{F})$, and let $F(\tau)$ be a function in $\mathfrak{F}$ which satisfies (2). We let

$$F_n(\tau) = \inf \{nF^+, 1\}, \text{ where } F^+(\tau) = \sup \{F, 0\}.$$

Then $F_n(\tau) \in \mathfrak{F}$,

$$F_n(\tau) \leqslant F_{n+1}(\tau) \leqslant 1 \in \mathfrak{F}$$

and $\lim F_n(\tau) = \chi(\tau)$, where $\chi(\tau)$ is the characteristic function of the set M. Since the family $\mathfrak{F}$ is σ-closed, the function $\chi(\tau)$ belongs to $\mathfrak{F}$.

Conversely, if $\chi(\tau)$ is a characteristic function in $\mathfrak{F}$, then by the very definition of the system $\mathfrak{M}(\mathfrak{F})$, the set $M = T(\chi > 0)$ belongs to $\mathfrak{M}(\mathfrak{F})$, and the theorem is proved.

If the family $\mathfrak{F}$ is not σ-closed, then it admits *σ-closed extensions*. In order to construct such extensions we will use the notion of a majorant system of functions.

Every semivector system $\mathfrak{I}$ of non-negative functions on T will be called a *majorant system* and, in particular, a *majorant system of the family $\mathfrak{F}$* if

$$F(\tau) \leqslant F'(\tau) \in \mathfrak{I} \tag{5}$$

for each function $F(\tau)$ in $\mathfrak{F}$.

The collection $\mathfrak{I}_1$ of all non-negative constants is a majorant system; moreover, $\mathfrak{I}_1$ is a majorant for the family $\mathfrak{F}$ if and only if all functions of this family are bounded above. For any semifield $\mathfrak{F}$, *the family $\mathfrak{F}^+$ of all non-negative functions* in $\mathfrak{F}$ is the natural majorant system for $\mathfrak{F}$, and the system of non-negative real functions $\tilde{\mathfrak{G}}_r^+(T)$ is a majorant system for any family $\mathfrak{F}$.*

If we replace (3) by

---

* In what follows, the superscript $^+$ by a symbol which denotes a family of functions designates the subfamily of all non-negative functions of that family.

$$F_n(\tau) \leqslant F'(\tau) \in \mathfrak{J}, \tag{3'}$$

where $\mathfrak{J}$ is a majorant system for $\mathfrak{F}$, then we obtain a generalization of the notion that a family be σ-closed; namely, we say that *the family $\mathfrak{F}$ is σ-closed with respect to the majorant system $\mathfrak{J}$*. If the family $\mathfrak{F}$ is σ-closed with respect to $\mathfrak{J}$, then it is obviously σ-closed in the original sense. In the event that $\mathfrak{J} = \mathfrak{F}^+$, for a semifield of functions $\mathfrak{F}$, the two notions are equivalent.

Now let $\mathfrak{F}$ be any family of functions on T, let $\mathfrak{J}$ be a majorant system of $\mathfrak{F}$, and let $\mathfrak{F}_\sigma^{(\mathfrak{J})}$ be the collection of limits $\lim F_n(\tau) = F(\tau)$ of all possible *non-decreasing* sequences $\{F_n(\tau)\}$ in $\mathfrak{F}$ which satisfy a condition of the form (3′). $\mathfrak{F}_\sigma^{(\mathfrak{J})}$ is called the *σ-extension of the family $\mathfrak{F}$ with respect to $\mathfrak{J}$.*

Since, for the limit $F(\tau)$ of a non-decreasing sequence $\{F_n(\tau)\}$, (5) follows from (3′), *the system $\mathfrak{J}$ is also a majorant for the σ-extension $\mathfrak{F}_\sigma^{(\mathfrak{J})}$.*

THEOREM 2.1.2. *The σ-extension $\mathfrak{F}_\sigma^{(\mathfrak{J})}$ of a semifield $\mathfrak{F}$ is a σ-closed semifield of functions.*

PROOF. (1) The σ-extension $\mathfrak{F}_\sigma^{(\mathfrak{J})}$ of a semivector system $\mathfrak{F}$ is a semivector system of functions. Indeed, for non-decreasing sequences of functions $\{F_n(\tau)\}$ and $\{G_n(\tau)\}$ which satisfy (3), and also the condition $G_n(\tau) \leqslant G'(\tau) \in \mathfrak{J}$, the sequences $\{F_n(\tau)+G_n(\tau)\}$ and $\{\alpha F_n(\tau)\}$ ($\alpha \geqslant 0$) in $\mathfrak{F}$ are non-decreasing, and

$$F_n(\tau) + G_n(\tau) \leqslant F'(\tau) + G'(\tau) \in \mathfrak{J}, \quad \alpha F_n(\tau) \leqslant \alpha F'(\tau) \in \mathfrak{J}.$$

Moreover, passage to the limit is linear, and hence whenever the family $\mathfrak{F}_\sigma^{(\mathfrak{J})}$ contains $F(\tau) = \lim F_n(\tau)$ and $G(\tau) = \lim G_n(\tau)$, it also contains the functions $F(\tau) + G(\tau)$ and $\alpha F(\tau)$ ($\alpha \geqslant 0$).

(2) *The σ-extension $\mathfrak{F}_\sigma^{(\mathfrak{J})}$ of a lattice $\mathfrak{F}$ is a σ-closed lattice of functions.* In fact, given sequences $\{F_n(\tau)\}$ and $\{G_n(\tau)\}$, we form the sequences

$$\Phi_n(\tau) = \sup\{F_n, G_n\}, \quad \Psi_n(\tau) = \inf\{F_n, G_n\}.$$

Then, for the non-decreasing sequences $\{\Phi_n(\tau)\}$ and $\{\Psi_n(\tau)\}$ in $\mathfrak{F}$,

$$\Phi_n(\tau) \leqslant \sup\{F', G'\} \leqslant F'(\tau) + G'(\tau) \in \mathfrak{J},$$
$$\Psi_n(\tau) \leqslant F'(\tau) + G'(\tau) \in \mathfrak{J},$$

and since

$$\sup\{F, G\} = \lim \Phi_n(\tau), \quad \inf\{F, G\} = \lim \Psi_n(\tau)$$

whenever $F(\tau)$ and $G(\tau)$ belong to $\mathfrak{F}_\sigma^{(\mathfrak{J})}$, the functions $\sup\{F, G\}$ and $\inf\{F, G\}$ also belong to $\mathfrak{F}_\sigma^{(\mathfrak{J})}$. Thus $\mathfrak{F}_\sigma^{(\mathfrak{J})}$ is a lattice of functions.

Now let $\{F_n(\tau)\}$ be a non-decreasing sequence in $\mathfrak{F}_\sigma^{(\mathfrak{J})}$ which satisfies (3′), and let $F(\tau) = \lim F_n(\tau)$ be its limit.
Then

$$F_n(\tau) = \lim_{k \to \infty} F_{n,k}(\tau) = \sup_k \{F_{n,k}\},$$

where

$$
\left.
\begin{aligned}
F_{n,\,k}(\tau) &\leqslant F_{n,\,k+1}(\tau) \in \mathfrak{F}, \\
F_{n,\,k}(\tau) &\leqslant F_n(\tau) \leqslant F'(\tau) \in \mathfrak{I}.
\end{aligned}
\right\}
\tag{6}
$$

Hence,

$$
F(\tau) = \sup \{F_n\} = \sup_n \sup_k F_{n,\,k} = \sup_{n,\,k} F_{n,\,k}.
\tag{7}
$$

But the upper bound of a countable set $\{F_{n,k}\}$ in $\mathfrak{F}$ can also be represented as the limit of a non-decreasing sequence of functions

$$
F'_m(\tau) = \sup_{n+k \leqslant m} \{F_{n,\,k}\} \in \mathfrak{F},
\tag{8}
$$

Moreover, by (6)

$$
F'_m(\tau) \leqslant F'(\tau) \in \mathfrak{I}.
\tag{9}
$$

By (7), (8), and (9), $F(\tau) = \sup_n \{F_n(\tau)\} = \lim_{m\to\infty} F'_m(\tau)$ belongs to $\mathfrak{F}_\sigma^{(\mathfrak{I})}$, i.e. the lattice $\mathfrak{F}_\sigma^{(\mathfrak{I})}$ is σ-closed.

COROLLARY 1. *A semifield (or lattice) of functions $\mathfrak{F}$ is σ-closed with respect to a majorant system $\mathfrak{I}$ of $\mathfrak{F}$ if and only if $\mathfrak{F}_\sigma^{(\mathfrak{I})} = \mathfrak{F}$.*

COROLLARY 2. *The characteristic function $\chi(\tau)$ of a set $M \in \mathfrak{M}(\mathfrak{F})$, where $\mathfrak{F}$ is a regular semifield of functions, belongs to each of the σ-extensions $\mathfrak{F}_\sigma^{(\mathfrak{I})}$.*
Since $\mathfrak{F}_\sigma^{(\mathfrak{I})} \supset \mathfrak{F}$ is σ-closed, this follows from Theorem 2.1.1.

The dependence of the σ-extension $\mathfrak{F}_\sigma^{(\mathfrak{I})}$ on the majorant system $\mathfrak{I}$ can be removed in the following manner. We will say that a majorant system $\mathfrak{I}$ is *subordinate* to a majorant system $\mathfrak{I}'$ if the system $\mathfrak{I}'$ is a majorant for $\mathfrak{I}$, and we say that two majorant systems are *equivalent* if each is subordinate to the other. For example, the system $\mathfrak{I}_1$ of all non-negative constants, and the system of all bounded non-negative real functions $\mathfrak{G}_r^+(T)$ are equivalent, and both are subordinate to the system $\tilde{\mathfrak{G}}_r^+(T)$.

If $\mathfrak{I}$ and $\mathfrak{I}'$ are majorant systems for a family of functions $\mathfrak{F}$, and if the system $\mathfrak{I}$ is subordinate to $\mathfrak{I}'$, then for corresponding σ-extensions $\mathfrak{F}_\sigma^{(\mathfrak{I})}$ and $\mathfrak{F}_\sigma^{(\mathfrak{I}')}$ with respect to $\mathfrak{I}$ and $\mathfrak{I}'$, we have $\mathfrak{F}_\sigma^{(\mathfrak{I})} \subset \mathfrak{F}_\sigma^{(\mathfrak{I}')}$, and hence

$$
\mathfrak{F}_\sigma^{(\mathfrak{I}')} = \mathfrak{F}_\sigma^{(\mathfrak{I})},
$$

when the systems $\mathfrak{I}$ and $\mathfrak{I}'$ are equivalent.

In particular, *for the case of a semifield $\mathfrak{F}$, the system $\mathfrak{F}^+$ is subordinate to any majorant system of $\mathfrak{F}$, and the σ-extension of the semifield $\mathfrak{F}$ with respect to $\mathfrak{I} = \mathfrak{F}^+$ is the smallest σ-extension of the family $\mathfrak{F}$.*

For a majorant system $\mathfrak{I}$, we define $[\mathfrak{I}]$ to be the set of functions $F(\tau)$ in $\mathfrak{G}_r^+(T)$ which satisfy an inequality of the form (5). The family of functions $[\mathfrak{I}]$ contains $\mathfrak{I}$ and is obviously a majorant system which is equivalent to $\mathfrak{I}$. For example, $[\mathfrak{I}_1] = \mathfrak{G}_r^+(T)$.

Two majorant systems $\mathfrak{J}$ and $\mathfrak{J}'$ are equivalent if and only if $[\mathfrak{J}'] = [\mathfrak{J}]$. Thus, in the set of mutually equivalent majorant systems, we have selected one representative which has the following characteristic property: if $F(\tau) \in [\mathfrak{J}]$, then every non-negative function $G(\tau) \leqslant F(\tau)$ also belongs to $[\mathfrak{J}]$.

A majorant system $\mathfrak{J}$ is said to be *regular* if $1 \in [\mathfrak{J}]$.

§ 2. Baire Fields of Functions

2.1. Fundamental properties. *If whenever the family $\mathfrak{F}$ contains $F(\tau)$, it also contains $-F(\tau)$ (in particular, if $\mathfrak{F}$ is a field of functions), and if $\mathfrak{J}$ is a majorant system for $\mathfrak{F}$, then**

$$|F(\tau)| \leqslant F'(\tau) \in \mathfrak{J}. \tag{1}$$

A field of functions $\mathfrak{F}$ is said to be *Baire with respect to a majorant system* $\mathfrak{J}$ if it is σ-closed with respect to $\mathfrak{J}$. Moreover, if $\mathfrak{J} = \mathfrak{F}^+$, then $\mathfrak{F}$ will be called simply a *Baire field of functions.*

A field $\mathfrak{F}$ is a Baire field of functions with respect to $\mathfrak{J}$ if and only if the limit $F(\tau)$ of a sequence $\{F_n(\tau)\}$ which satisfies the conditions

$$F_n(\tau) \in \mathfrak{F}, \quad |F_n(\tau)| \leqslant F'(\tau) \in \mathfrak{J}, \quad \lim F_n(\tau) = F(\tau), \tag{2}$$

belongs to $\mathfrak{F}$. In fact, if this requirement is satisfied, then the field $\mathfrak{F}$ is certainly σ-closed with respect to $\mathfrak{J}$. Conversely, let $\mathfrak{F}$ be σ-closed with respect to $\mathfrak{J}$. We first let $\{F_n(\tau)\}$ be a non-increasing sequence in $\mathfrak{F}$ which satisfies (2). Then $-F_n(\tau) \leqslant F'(\tau) \in \mathfrak{J}$, and the limit

$$\lim(-F_n(\tau)) = -\lim F_n(\tau) = -F(\tau)$$

of the sequence $\{-F_n(\tau)\}$ belongs to $\mathfrak{F}$. Whenever $-F(\tau) \in \mathfrak{F}$, $F(\tau) \in \mathfrak{F}$. The general case can be reduced to the case of non-decreasing and non-increasing sequences. Let $\{F_n(\tau)\}$ be any sequence in $\mathfrak{F}$, and let

$$\Phi_{mn}(\tau) = \sup\{F_m, F_{m+1}, \ldots, F_n\},$$
$$\Psi_{mn}(\tau) = \inf\{F_m, F_{m+1}, \ldots, F_n\}.$$

Then $\Phi_{mn}(\tau) \in \mathfrak{F}$, $\Psi_{mn}(\tau) \in \mathfrak{F}$, and if (2) holds, then

$$|\Phi_{mn}(\tau)| \leqslant F'(\tau) \in \mathfrak{J}, \quad |\Psi_{mn}(\tau)| \leqslant F'(\tau) \in \mathfrak{J}.$$

Since, for fixed m the sequence $\{\Phi_{mn}(\tau)\}$ does not decrease, and since $\{\Psi_{mn}(\tau)\}$ does not increase, the limits

$$\lim_{n\to\infty} \Phi_{mn}(\tau) = \Phi_m(\tau), \quad \lim_{n\to\infty} \Psi_{mn}(\tau) = \Psi_m(\tau)$$

belong to $\mathfrak{F}$ and

* It follows from $F(\tau) \leqslant G'(\tau) \in \mathfrak{J}$ and $-F(\tau) \leqslant G''(\tau) \in \mathfrak{J}$ that $|F(\tau)| \leqslant G'(\tau) + G''(\tau) = F'(\tau) \in \mathfrak{J}$.

$$|\Phi_m(\tau)| \leqslant F'(\tau) \in \mathfrak{J}, \qquad |\Psi_m(\tau)| \leqslant F'(\tau) \in \mathfrak{J}.$$

But $\{\Phi_m(\tau)\}$ is non-increasing and $\{\Psi_m(\tau)\}$ is non-decreasing, so that

$$\overline{\lim_{n\to\infty}} F_n(\tau) = \lim \Phi_m(\tau) \in \mathfrak{F}; \qquad \underline{\lim_{n\to\infty}} F_n(\tau) = \lim_{m\to\infty} \Psi_m(\tau) \in \mathfrak{F}. \tag{3}$$

Therefore, in the case of a convergent sequence, the limit $F(\tau) = \lim F_n(\tau)$ belongs to $\mathfrak{F}$.

The collection of functions $F(\tau)$ of the family $\mathfrak{F}$ which satisfy a condition of the form (1) with respect to some majorant system $\mathfrak{J}$ will be denoted by $(\mathfrak{F}/\mathfrak{J})$.

THEOREM 2.2.1. *If $\mathfrak{F}$ is a field or a Baire field of functions, then the family $(\mathfrak{F}/\mathfrak{J})$ is, respectively, a field or a Baire field of functions.*

Indeed, if $F(\tau)$ and $G(\tau)$ belong to $(\mathfrak{F}/\mathfrak{J})$ and satisfy inequalities of the type (1), then

$$|\alpha F(\tau) + \beta G(\tau)| \leqslant |\alpha| F'(\tau) + |\beta| G'(\tau) \in \mathfrak{J}$$

and (the functions $F(\tau)$ and $G(\tau)$ are real)

$$|\sup\{F, G\}| \leqslant \sup\{F', G'\} \leqslant F' + G' \in \mathfrak{J};$$
$$|\inf\{F, G\}| \leqslant F' + G' \in \mathfrak{J}.$$

Thus, whenever the family $(\mathfrak{F}/\mathfrak{J})$ contains $F(\tau)$ and $G(\tau)$, it also contains $\alpha F(\tau) + \beta G(\tau)$, $\sup\{F, G\}$, and $\inf\{F, G\}$, i.e. $(\mathfrak{F}/\mathfrak{J})$ is a field of functions. If, in addition, $\mathfrak{F}$ is a Baire field, and if $\{F_n(\tau)\}$ is a convergent sequence in $(\mathfrak{F}/\mathfrak{J})$ which satisfies

$$|F_n(\tau)| \leqslant G'(\tau) \in \mathfrak{J}', \qquad \mathfrak{J}' = (\mathfrak{F}/\mathfrak{J})^+,$$

then $G'(\tau) \leqslant F'(\tau) \in \mathfrak{J}$, and hence (2) holds. We have

$$F(\tau) \in \mathfrak{F}, \qquad |F(\tau)| \leqslant F'(\tau) \in \mathfrak{J},$$

for the limit $F(\tau) = \lim F_n(\tau)$, since $\mathfrak{F}$ is a Baire field. Therefore $F(\tau) \in (\mathfrak{F}/\mathfrak{J})$, i.e. $(\mathfrak{F}/\mathfrak{J})$ is a Baire field.

REMARK. $(\mathfrak{F}/\mathfrak{J})$ is a regular field if and only if the field $\mathfrak{F}$ and the majorant system $\mathfrak{J}$ are regular.

For characteristics of systems $\mathfrak{M}(\mathfrak{J})$ of Lebesgue sets which correspond to regular Baire fields of functions, we need the notion of a *field of sets*, i.e. a system of sets $\mathfrak{R} \subset \mathfrak{P}(T)$ such that the union, intersection, and difference of two sets in $\mathfrak{R}$ again belong to $\mathfrak{R}$. Moreover, *we will always require that the union of all sets of the field $\mathfrak{R}$ coincide with T.* In particular, this requirement is satisfied if the field $\mathfrak{R}$ itself contains the set T. In this case we will say that the field of sets $\mathfrak{R}$ is *regular*. For example, $\mathfrak{P}(T)$ is a regular field of sets.*

* In the literature, in addition to the terms "field" and "regular field", the terms "ring" and "algebra" of sets are also used.

Whenever a set M belongs to a regular field $\mathfrak{R}$, its complement $M^c = T\setminus M$ also belongs to $\mathfrak{R}$. Since

$$(M_1 \cap M_2)^C = M_1^C \cup M_2^C, \qquad M_2 \setminus M_1 = M_2 \cap M_1^C$$

a regular field of sets can also be defined as follows:

A system $\mathfrak{R}$ of subsets of a set T is a regular field of sets if it contains T and if $\mathfrak{R}$ is closed with respect to the operations of finite union and complement.

REMARK. A regular field of sets is obviously a Boolean algebra.

A field of sets $\mathfrak{B}$ is called a *Borel field* if it is σ-closed.

Whenever the sets M_n belong to a regular Borel field, their union $\bigcup_{n=1}^{\infty} M_n$ and intersection $\bigcap_{n=1}^{\infty} M_n$ always belong to the regular Borel field. A Borel field $\mathfrak{B}$ is closed with respect to the formation of intersections $\bigcap_{n=1}^{\infty} M_n$ even if it is not regular. Indeed, we let $M_n' = \bigcap_{k=1}^{n} M_k$. Then

$$\bigcap_{n=1}^{\infty} M_n = \bigcap_{n=1}^{\infty} M_n' = M_1 \setminus \bigcup_{n=1}^{\infty} (M_1 \setminus M_n')$$

and whenever M_n belongs to $\mathfrak{B}$, the sets M_n' and $M_1 \setminus M_n'$ also belong to $\mathfrak{B}$. Thus the sum of the latter also belongs to $\mathfrak{B}$, since $M_1\setminus M_n' \subset M_1 \in \mathfrak{B}$.

THEOREM 2.2.2. *For a regular Baire field of functions $\mathfrak{F}$, the system of Lebesgue sets $\mathfrak{M}(\mathfrak{F})$ is a regular Borel field of sets.*

In fact, since $\mathfrak{F}$ is σ-closed, it contains the characteristic function $\chi(\tau)$ of every set $M \in \mathfrak{M}(\mathfrak{F})$ (Theorem 2.1.1) and, in addition to $\chi(\tau)$, $\mathfrak{F}$ contains the characteristic function $1-\chi(\tau)$ of the set $T\setminus M$. Thus $T\setminus M$ belongs to $\mathfrak{M}(\mathfrak{F})$, and the lattice $\mathfrak{M}(\mathfrak{F})$ is a field of sets which is σ-closed by Theorem 2.1.1.

2.2. Measurable functions. Theorem 2.2.2 associates a regular Borel field $\mathfrak{B} = \mathfrak{M}(\mathfrak{F})$ with every regular Baire field of functions $\mathfrak{F}$. We shall see that all regular Borel fields of sets can be obtained in this way.

For an arbitrary regular Borel field $\mathfrak{B} \in \mathfrak{P}(T)$, we form the set of functions $\mathfrak{F}$ in $\mathfrak{G}_r(T)$ such that for any real α their Lebesgue sets,

$$M_\alpha = T(F > \alpha)$$

belong to the Borel field $\mathfrak{B}$. These functions will be said to be ($\mathfrak{B}$)-*measurable.*

We note that whenever M_α belongs to $\mathfrak{B}$, the sets

$$N_\alpha = T(F < \alpha), \quad M_\alpha' = T(F \geqslant \alpha), \quad N_\alpha' = T(F \leqslant \alpha)$$

also belong to $\mathfrak{B}$. Indeed, for $\alpha < \beta$,

$$M_{\alpha\beta} = M_\alpha \setminus M_\beta = T(\alpha < F \leqslant \beta) \in \mathfrak{B},$$

and hence if $\alpha_n < \alpha_{n+1}$ and $\lim \alpha_n = \beta$,

$$\bigcap_{n=1}^{\infty} M_{\alpha_n\beta} = T(F = \beta) \in \mathfrak{B}.$$

Therefore the set $M'_\alpha = M_\alpha \cup T(F = \alpha)$ belongs to $\mathfrak{B}$. It follows from the decomposition

$$N_\alpha = M_{\alpha\beta_1} \bigcup \left(\bigcup_{n=1}^{\infty} M_{\beta_n\beta_{n+1}} \right),$$

where $\alpha < \beta_1$, $\beta_n < \beta_{n+1}$, and $\lim \beta_n = \infty$, that $N_\alpha \in \mathfrak{B}$, and finally that $N'_\alpha = N_\alpha \cup T(F = \alpha) \in \mathfrak{B}$.

As is easily seen, each of the families $\{N_\alpha\}$, $\{M'_\alpha\}$, $\{N'_\alpha\}$ (or $M_{\alpha\beta}$) may replace $\{M_\alpha\}$ in the definition of ($\mathfrak{B}$)-measurability of a function $F(\tau)$.

We note the following properties of ($\mathfrak{B}$)-measurable functions:

1) *Whenever $F(\tau)$ is ($\mathfrak{B}$)-measurable, the function—$F(\tau)$ is also ($\mathfrak{B}$)-measurable.*

Indeed, $T(-F > \alpha) = T(F < -\beta) \in \mathfrak{B}$.

2) *A family $\mathfrak{F}$ of ($\mathfrak{B}$)-measurable functions is a lattice of functions which is σ-closed with respect to $\mathfrak{I} = \mathfrak{G}_r^+(T)$.*

In fact, let $F(\tau)$ and $G(\tau)$ be functions in $\mathfrak{F}$ and let $\Phi(\tau) = \sup\{F, G\}$ and $\Psi(\tau) = \inf\{F, G\}$. Then

$$T(\Phi > \alpha) = T(F > \alpha) \cup T(G > \alpha) \in \mathfrak{B};$$

$$T(\Psi > \alpha) = T(F > \alpha) \cap T(G > \alpha) \in \mathfrak{B}.$$

If $\Phi(\tau) = \sup_n \{F_n(\tau)\}$, where $\{F_n(\tau)\}$ is a sequence in $\mathfrak{F}$, and if $\Phi(\tau)$ is finite for all τ, then

$$T(\Phi > \alpha) = \bigcup_n T(F_n > \alpha) \in \mathfrak{B}.$$

In particular, if the sequence $\{F_n(\tau)\}$ is non-decreasing, then the limit $\lim F_n(\tau) = \sup\{F_n(\tau)\}$ is ($\mathfrak{B}$)-measurable, i.e. the lattice $\mathfrak{F}$ is σ-closed with respect to $\mathfrak{G}_r^+(T)$.

3) *For a sequence $\{F_n(\tau)\}$ of ($\mathfrak{B}$)-measurable functions, the upper (lower) limit $\overline{\lim}\, F_n(\tau)$ ($\underline{\lim}\, F_n(\tau)$) is also ($\mathfrak{B}$)-measurable if it is finite.* In particular, *any (finite) limit of a convergent sequence of ($\mathfrak{B}$)-measurable functions is ($\mathfrak{B}$)-measurable.*

Indeed, in addition to the limits of non-decreasing sequences, the (finite) limits of non-increasing sequences of ($\mathfrak{B}$)-measurable functions are also ($\mathfrak{B}$)-measurable, since whenever $F(\tau)$ is ($\mathfrak{B}$)-measurable, $-F(\tau)$ is ($\mathfrak{B}$)-measurable. Therefore, it follows from the representation (3) (Sec. 2.1) of the upper limit $\overline{\lim}\, F_n(\tau)$ (lower limit $\underline{\lim}\, F_n(\tau)$) as the limit of a monotone sequence that the upper (lower) limit is ($\mathfrak{B}$)-measurable.

4) *The family F of ($\mathfrak{B}$)-measurable functions is a regular Baire field of functions (with respect to $\mathfrak{I} = \mathfrak{G}_r^+(T)$).*

Indeed, as will be shown in the following section, $\mathfrak{F}$ is a (real) vector system and this, together with 2), will prove the above assertation.

We note that $\mathfrak{M}(\mathfrak{F}) = \mathfrak{B}$, and that $\mathfrak{F}$ is the largest Baire field having these properties.

§ 3. Baire ∗-fields of Functions

3.1. A ∗-field of functions. A family of complex functions $\mathfrak{F} \subset \tilde{\mathfrak{G}}(T)$ is by definition a *∗-field of functions* if:

1) $\mathfrak{F}$ is a vector ∗-system of functions, i.e. $\mathfrak{F}$ is a complex vector system with the pseudoinvolution

$$(F(\tau))^* = \overline{F(\tau)};$$

2) the real functions in $\mathfrak{F}$ (the Hermitian elements of $\mathfrak{F}$) form a lattice of functions $\mathfrak{F}_r$.

A ∗-field of functions is said to be *regular* if it contains the function $F_0(\tau) \equiv 1$.

The collection F_r of real functions of a ∗-field $\mathfrak{F}$ (respectively, of a regular ∗-field $\mathfrak{F}$) is a real field (respectively, a regular field) of functions (see Sec. 2.1 and 2.2).

We again associate a system of sets $\mathfrak{M}(\mathfrak{F})$ with a regular ∗-field of functions $\mathfrak{F}$ by letting $\mathfrak{M}(\mathfrak{F}) = \mathfrak{M}(\mathfrak{F}_r)$. Similarly, we will say that $\mathfrak{J}$ is a *majorant system* for a ∗-field of functions $\mathfrak{F}$ if $\mathfrak{J}$ is a majorant system for the field $\mathfrak{F}_r$.

A ∗-field $\mathfrak{F}$ will be said to be σ-closed with respect to $\mathfrak{J}$ if the field $\mathfrak{F}_r$ is σ-closed. *Baire ∗-fields of functions with respect to* $\mathfrak{J}$, i.e. ∗-fields $\mathfrak{F}$ which are σ-closed with respect to a majorant system $\mathfrak{J}$, are again ∗-fields of functions $\mathfrak{F}$ which are closed with respect to passage to the limit (2) of Sec. 2.2.1.

For ∗-fields $\mathfrak{F}$, we retain the definition of the family $(\mathfrak{F}/\mathfrak{J})$ (where $\mathfrak{J}$ is any majorant system) which was given in Sec. 2.2.1 for fields of functions. For ∗-fields, Theorem 2.2.1 takes the form:

THEOREM 2.3.1. *If* $\mathfrak{F}$ *is a ∗-field (Baire ∗-field), then the family* $(\mathfrak{F}/\mathfrak{J})$ *is a ∗-field (Baire ∗-field) of functions.*

Indeed, according to Theorem 2.2.1, the set of real functions in $(\mathfrak{F}/\mathfrak{J})$, i.e. $(\mathfrak{F}_r/\mathfrak{J})$, is a field of functions, and it is a Baire field if $\mathfrak{F}_r$ is a Baire field. Whenever $F(\tau)$ belongs to the vector system $(\mathfrak{F}/\mathfrak{J})$, $(F(\tau))* = \overline{F(\tau)}$ also belongs to $(\mathfrak{F}/\mathfrak{J})$.

3.2. Elementary $\mathfrak{B}$-functions. For a given regular Borel field of sets $\mathfrak{B}$, a complex function $F(\tau)$ is said to be *($\mathfrak{B}$)-measurable* if both of the real components $F_1(\tau)$ and $F_2(\tau)$ in the decomposition $F(\tau) = F_1(\tau)+iF_2(\tau)$ are ($\mathfrak{B}$)-measurable. In particular, the functions $F(\tau)$ of a regular Borel ∗-field $\mathfrak{F}$ are ($\mathfrak{B}$)-measurable for $\mathfrak{B} = \mathfrak{M}(\mathfrak{F}) = \mathfrak{M}(\mathfrak{F}_r)$. A function $\Phi(\tau) \in \tilde{\mathfrak{G}}(T)$ will be called an *elementary ($\mathfrak{B}$)-function* if it has a finite or countable number of values and if its sets of constancy are sets of the Borel field $\mathfrak{B}$. An elementary ($\mathfrak{B}$)-function is ($\mathfrak{B}$)-measurable.

The collection of elementary ($\mathfrak{B}$)-functions which have only a finite number of values will be denoted by $\mathfrak{E}(\mathfrak{B})$, and the collection of all elementary ($\mathfrak{B}$)-functions will be denoted by $\breve{\mathfrak{E}}(\mathfrak{B})$.

By forming the closures of $\mathfrak{E}(\mathfrak{B})$ and $\breve{\mathfrak{E}}(\mathfrak{B})$ with respect to uniform convergence, we obtain the families $\mathfrak{G}(\mathfrak{B})$ and $\breve{\mathfrak{G}}(\mathfrak{B})$, respectively. Some remarks on partitions of a set T will be necessary for the study of these families of functions. A partition*

$$T = \bigcup_n M_n, \quad M_k \bigcap M_n = \varnothing \qquad (k \neq n) \tag{1}$$

of the set T into a finite or countable number of disjoint sets M_n will be called a ($\mathfrak{B}$)-*partition* if all of the M_n belong to the field $\mathfrak{B}$. A ($\mathfrak{B}$)-partition

$$T = \bigcup_k N_k, \quad N_k \bigcap N_m = \varnothing \qquad (k \neq m) \tag{2}$$

is by definition *subordinate* to the ($\mathfrak{B}$)-partition (1) if each N_k is a subset of some M_n.

For two arbitrary ($\mathfrak{B}$)-partitions, (1) and (2), the intersection

$$M_n \cap N_k = M_{nk} \in \mathfrak{B}$$

is again a ($\mathfrak{B}$)-partition, which is called the *product of the* ($\mathfrak{B}$)-*partitions* (1) and (2). The product of two ($\mathfrak{B}$)-partitions is subordinate to each of them. The notion of a product of ($\mathfrak{B}$)-partitions can be immediately generalized to a finite number of factors.

A ($\mathfrak{B}$)-partition (1) will be said to be *admissible* for an elementary ($\mathfrak{B}$)-function $\varphi(\tau)$ if the value of $\varphi(\tau)$ on each M_n is constant. Whenever (1) is admissible for an elementary ($\mathfrak{B}$)-function $\varphi(\tau)$, any partition subordinate to (1) is also admissible. To each ($\mathfrak{B}$)-partition (1) which is admissible for an elementary ($\mathfrak{B}$)-function $\varphi(\tau)$, there corresponds a representation

$$\varphi(\tau) = \sum_n \alpha_n \chi_n(\tau), \tag{3}$$

where $\chi_n(\tau)$ is the characteristic function of the set $M_n \in \mathfrak{B}$, and $\varphi(\tau) = \alpha_n$ on M_n. Since the product of the admissible ($\mathfrak{B}$)-partitions of a finite number of elementary ($\mathfrak{B}$)-functions $\varphi_k(\tau)$ is admissible for each function $\varphi_k(\tau)$, this product can serve as the common admissible ($\mathfrak{B}$)-partition for all the functions $\varphi_k(\tau)$.

Therefore we have:

LEMMA 2.3.1. *For a finite number of elementary* ($\mathfrak{B}$)-*functions* $\varphi_k(\tau)$, *representations of the form* (3) *by means of a common admissible* ($\mathfrak{B}$)-*partition exist.*

---

* The symbol $\varnothing$ denotes the empty set.

For two functions $\varphi(\tau)$ and $\psi(\tau) \in \tilde{\mathfrak{E}}(\mathfrak{B})$, let

$$\varphi(\tau) = \sum_n \alpha_n \chi_n(\tau), \qquad \psi(\tau) = \sum_n \beta_n \chi_n(\tau)$$

be representations by means of a general $(\mathfrak{B})$-partition (1). Then

$$\varphi(\tau) + \psi(\tau) = \sum_n (\alpha_n + \beta_n) \chi_n(\tau), \quad \alpha\varphi(\tau) = \sum_n \alpha\alpha_n \chi_n(\tau), \tag{4}$$

$$(\varphi(\tau))^* = \overline{\varphi(\tau)} = \sum_n \bar{\alpha}_n \chi_n(\tau)$$

give representations of the functions $\varphi(\tau)+\psi(\tau)$, $\alpha\varphi(\tau)$, and $\overline{\varphi(\tau)}$ in terms the same $(\mathfrak{B})$-partition. Thus $\tilde{\mathfrak{E}}(\mathfrak{B})$ is a vector $*$-system of functions, as is $\mathfrak{E}(\mathfrak{B})$. If the functions $\varphi(\tau)$ and $\psi(\tau)$ are real, and if $\gamma_n = \sup\{\alpha_n, \beta_n\}$ and $\delta_n = \inf\{\alpha_n, \beta_n\}$, then

$$\sup\{\varphi, \psi\} = \sum_n \gamma_n \chi_n(\tau), \qquad \inf\{\varphi, \psi\} = \sum_n \delta_n \chi_n(\tau) \tag{5}$$

are representations of the functions $\sup\{\varphi, \psi\}$ and $\inf\{\varphi, \psi\}$, respectively. Hence $\tilde{\mathfrak{E}}(\mathfrak{B})$ and $\mathfrak{E}(\mathfrak{B})$ are also lattices of functions. *Thus* $\tilde{\mathfrak{E}}(\mathfrak{B})$, $\mathfrak{E}(\mathfrak{B})$, *and their closures* $\tilde{\mathfrak{G}}(\mathfrak{B})$ *and* $\mathfrak{G}(\mathfrak{B})$, *are $*$-fields of functions.*

REMARK. We extend the definition of elementary function to the case of an arbitrary regular field $\mathfrak{K}$, namely, we will say that $\varphi(\tau)$ is an elementary $(\mathfrak{K})$-function if it takes on a finite number of values and if its sets of constancy are sets of the field $\mathfrak{K}$. In connection with this, the partition (1) will be called a $(\mathfrak{K})$-partition if $M_n \in \mathfrak{K}$ and if the number of sets of M is finite. With these changes, the notions of representation of a function, subordination, product, and admissibility of partitions carry over to the case of $(\mathfrak{K})$-partitions. Therefore Lemma 2.3.1 holds, and the family $\mathfrak{E}(\mathfrak{K})$ of elementary $(\mathfrak{K})$-functions, and its closure $\mathfrak{G}(\mathfrak{K})$ (with respect to uniform convergence), are $*$-fields of functions. Like $\mathfrak{E}(\mathfrak{K})$, $\mathfrak{G}(\mathfrak{K})$ contains only bounded functions.

THEOREM 2.3.2. *The family* $\tilde{\mathfrak{G}}(\mathfrak{B})$ *coincides with the system of all* $(\mathfrak{B})$-*measurable functions, and* $\mathfrak{G}(\mathfrak{B})$ *coincides with the system of bounded* $(\mathfrak{B})$-*measurable functions.*

Elementary $(\mathfrak{B})$-functions are $(\mathfrak{B})$-measurable, and therefore all functions in $\tilde{\mathfrak{G}}(\mathfrak{B})$ are $(\mathfrak{B})$-measurable since the system of $(\mathfrak{B})$-measurable functions is closed with respect to uniform convergence (see Sec. 2.2.2, property 3). Moreover, functions of the family $\mathfrak{G}(\mathfrak{B})$ are obviously bounded.

Now let $F(\tau)$ be an arbitrary real $(\mathfrak{B})$-measurable function, and let ε be a positive number. We choose real numbers α_k such that

$$\alpha_k < \alpha_{k+1}, \quad \alpha_{k+1} - \alpha_k < \varepsilon, \quad \lim_{k\to\infty} \alpha_k = \infty, \quad \lim_{k\to-\infty} \alpha_k = -\infty. \tag{6}$$

The sets $N_k = T(F > \alpha_k)$, and the sets

$$M_k = N_k \setminus N_{k+1} = T(\alpha_k < F(\tau) \leqslant \alpha_{k+1})$$

belong to the field $\mathfrak{B}$. The sets M_k are disjoint, and since $F(\tau)$ is finite, they form a ($\mathfrak{B}$)-partition

$$T = \bigcup{}' M_k \tag{1'}$$

of the set T. The prime indicates that the empty sets of M_k are omitted in the union. With the ($\mathfrak{B}$)-partition (1'), we associate the functions

$$\varphi(\tau) = \sum_k{}' \alpha_k \chi_k(\tau), \qquad \psi(\tau) = \sum_k{}' \alpha_{k+1} \chi_k(\tau)$$

in $\mathfrak{E}(\mathfrak{B})$, where $\chi_k(\tau)$ is the characteristic function of the set M_k. By the definition of the functions $\varphi(\tau)$ and $\psi(\tau)$, we have

$$\varphi(\tau) < F(\tau) \leqslant \psi(\tau), \qquad 0 < \psi(\tau) - \varphi(\tau) < \varepsilon$$

and hence

$$0 < F(\tau) - \varphi(\tau) < \varepsilon, \qquad 0 \leqslant \psi(\tau) - F(\tau) < \varepsilon. \tag{7}$$

By virtue of this construction, sequences $\{\varepsilon_n\}$ with $\lim \varepsilon_n = 0$ correspond to sequences $\{\varphi_n(\tau)\}$ and $\{\psi_n(\tau)\}$ in $\mathfrak{E}(\mathfrak{B})$ which converge uniformly to the function $F(\tau)$. Therefore $F(\tau)$ belongs to the family $\tilde{\mathfrak{G}}(\mathfrak{B})$, and the same is true for any complex ($\mathfrak{B}$)-measurable function $F(\tau)$. Thus, $\tilde{\mathfrak{G}}(\mathfrak{B})$ consists precisely of the ($\mathfrak{B}$)-measurable functions.

If the real ($\mathfrak{B}$)-measurable function $F(\tau)$ is bounded, then the ($\mathfrak{B}$)-partition (1') is finite, and the functions $\varphi(\tau)$ and $\psi(\tau)$ belong to $\mathfrak{E}(\mathfrak{B})$. Thus, in the case under consideration, the function $F(\tau)$ belongs to $\mathfrak{G}(\mathfrak{B})$ since it is the uniform limit of the sequences $\varphi_n(\tau)$ and $\psi_n(\tau)$ in $\mathfrak{E}(\mathfrak{B})$. Whence it follows that the family $\mathfrak{G}(\mathfrak{B})$ coincides with the system of bounded ($\mathfrak{B}$)-measurable functions.

THEOREM 2.3.3. *The families $\tilde{\mathfrak{G}}(\mathfrak{B})$ and $\mathfrak{G}(\mathfrak{B})$ are regular Baire *-fields of functions. The product and quotient (with denominator not assuming the value (0)) of two functions in $\tilde{\mathfrak{G}}(\mathfrak{B})$ again belong to $\tilde{\mathfrak{G}}(\mathfrak{B})$.*

In fact, the family $\tilde{\mathfrak{G}}(\mathfrak{B})$ is a *-field of functions and, since it coincides with the system of ($\mathfrak{B}$)-measurable functions, it is σ-closed (see Sec. 2.2.2, property 2). For the family $\mathfrak{G}(\mathfrak{B})$, we have the representation

$$\mathfrak{G}(\mathfrak{B}) = (\tilde{\mathfrak{G}}(\mathfrak{B})/\mathfrak{J}_1),$$

where $\mathfrak{J}_1$ is the majorant system of non-negative constants. Hence, like $\tilde{\mathfrak{G}}(\mathfrak{B})$, $\mathfrak{G}(\mathfrak{B})$ is a Baire *-field of functions (Theorem 2.3.1).

Let the functions $F(\tau)$ and $G(\tau)$ belong to $\tilde{\mathfrak{G}}(\mathfrak{B})$, and let $\{\varphi_n(\tau)\}$ and $\{\psi_n(\tau)\}$ be sequences in $\tilde{\mathfrak{E}}(\mathfrak{B})$ which converge uniformly to $F(\tau)$ and $G(\tau)$, respectively. If $G(\tau)$ does not vanish, then the sequence $\{\psi_n(\tau)\}$ can be chosen so that $\psi_n(\tau) \neq 0$ for all τ, for example, such that $|\psi_n(\tau)| \geqslant |G(\tau)|$. Then the sequences $\{\varphi_n(\tau)\psi_n(\tau)\}$ and $\{\varphi_n(\tau)/\psi_n(\tau)\}$ belong to $\tilde{\mathfrak{E}}(\mathfrak{B})$ and converge, respectively, to the functions $F(\tau)\,G(\tau)$ and $F(\tau)/G(\tau)$, which are ($\mathfrak{B}$)-measurable (as the limits of ($\mathfrak{B}$)-measurable functions) and hence belong to $\tilde{\mathfrak{G}}(\mathfrak{B})$.

Theorems 2.3.2 and 2.3.3 give:

COROLLARY. *The family* $\mathfrak{F}$ *of all* ($\mathfrak{B}$)*-measurable functions is a regular Baire *-field; moreover,* $\mathfrak{F}$ *is closed with respect to multiplication and division (when the latter is possible).*

3.3. The system $\mathfrak{J}(\mathfrak{B})$. We consider the families $(\tilde{\mathfrak{E}}(\mathfrak{B})/\mathfrak{J})$ and $(\tilde{\mathfrak{G}}(\mathfrak{B})/\mathfrak{J})$, where $\mathfrak{J}$ is a regular majorant system, and we will denote the closure of the first of these with respect to uniform convergence by $\mathfrak{J}(\mathfrak{B})$. It follows from $\mathfrak{E}(\mathfrak{B}) \subset (\tilde{\mathfrak{E}}(\mathfrak{B})/\mathfrak{J}) \subset \tilde{\mathfrak{E}}(\mathfrak{B})$ that

$$\mathfrak{G}(\mathfrak{B}) \subset \mathfrak{J}(\mathfrak{B}) \subset \tilde{\mathfrak{G}}(\mathfrak{B}). \tag{8}$$

More precisely,

$$\mathfrak{J}(\mathfrak{B}) = (\tilde{\mathfrak{G}}(\mathfrak{B})/\mathfrak{J}). \tag{9}$$

In fact, by taking the closures with respect to uniform convergence of the families of functions $(\tilde{\mathfrak{E}}(\mathfrak{B})/\mathfrak{J})$ and $(\tilde{\mathfrak{G}}(\mathfrak{B})/\mathfrak{J}) \supset (\tilde{\mathfrak{E}}(\mathfrak{B})/\mathfrak{J})$, we obtain

$$\mathfrak{J}(\mathfrak{B}) \subset (\tilde{\mathfrak{G}}(\mathfrak{B})/\mathfrak{J}). \tag{9'}$$

On the other hand, for a real function $F(\tau)$ in $(\tilde{\mathfrak{G}}(\mathfrak{B})/\mathfrak{J})$, the sequences $\{\varphi_n(\tau)\}$ and $\{\psi_n(\tau)\}$ which *approximate* $F(\tau)$, and which were constructed for the proof of Theorem 2.3.2, belong to $(\tilde{\mathfrak{E}}(\mathfrak{B})/\mathfrak{J})$. Indeed, the inequality (7), together with $|F(\tau)| \leqslant F'(\tau) \in \mathfrak{J}$, gives

$$|\varphi(\tau)| \leqslant |F(\tau)| + \varepsilon \leqslant F'(\tau) + \varepsilon \in \mathfrak{J};$$
$$\psi(\tau) \leqslant F'(\tau) + \varepsilon \in \mathfrak{J}.$$

Therefore $F(\tau) \in \mathfrak{J}(\mathfrak{B})$. This is true for any (complex) function in $(\tilde{\mathfrak{G}}(\mathfrak{B})/\mathfrak{J})$, i.e. $(\tilde{\mathfrak{G}}(\mathfrak{B})/\mathfrak{J}) \subset \mathfrak{J}(\mathfrak{B})$, and together with (9′) this proves (9).

THEOREM 2.3.4. $\mathfrak{J}(\mathfrak{B})$ *is a regular Baire *-field of functions. Whenever* $\mathfrak{J}(\mathfrak{B})$ *contains* $F(\tau)$ *and* $G(\tau)$, *it also contains* $|F(\tau)|$, $F(\tau)G(\tau)$, *and* $F(\tau)/G(\tau)$ *if*

$$|F(\tau)\,G(\tau)| \leqslant F'(\tau) \in \mathfrak{J}; \quad \left|\frac{F(\tau)}{G(\tau)}\right| \leqslant G'(\tau) \in \mathfrak{J}. \tag{10}$$

Indeed, it follows from (9) that $\mathfrak{J}(\mathfrak{B})$ is a regular Baire *-field of functions since the family $\tilde{\mathfrak{G}}(\mathfrak{B})$ is a regular Baire *-field of functions, and since $\mathfrak{J}$ is a regular majorant system. Let $F(\tau) \in \mathfrak{J}(\mathfrak{B})$, and let $\{\varphi_n(\tau)\}$ be a sequence in $(\tilde{\mathfrak{E}}(\mathfrak{B})/\mathfrak{J})$ which converges uniformly to $F(\tau)$. Then $\{|\varphi_n(\tau)|\}$ converges uniformly to $|F(\tau)|$, and like $\varphi_n(\tau)$, $|\varphi_n(\tau)|$ belongs to $(\tilde{\mathfrak{E}}(\mathfrak{B})/\mathfrak{J})$. Therefore $|F(\tau)| \in \mathfrak{J}(\mathfrak{B})$. Finally, by (9) and Theorem 2.3.3, the product $F(\tau)G(\tau)$ and the quotient $F(\tau)/G(\tau)$ belong to $\tilde{\mathfrak{G}}(\mathfrak{B})$, and if (10) is satisfied, then they belong to $\mathfrak{J}(\mathfrak{B}) = (\tilde{\mathfrak{G}}(\mathfrak{B})/\mathfrak{J})$.

REMARK. A majorant system $\mathfrak{J}$ will be said to be *multiplicative* if the product $F'(\tau)G'(\tau)$ of the functions $F'(\tau)$ and $G'(\tau)$ in $\mathfrak{J}$ belongs to the family $[\mathfrak{J}]$. If a regular majorant system $\mathfrak{J}$ is multiplicative, then $\mathfrak{J}(\mathfrak{B})$ is a $*$-algebra of functions.

In fact, in that case the first condition in (10) always holds, since it follows from

$$|F(\tau)| \leqslant F' \in \mathfrak{J}, \quad |G(\tau)| \leqslant G'(\tau) \in \mathfrak{J}$$

that

$$|F(\tau)G(\tau)| \leqslant F'(\tau)G'(\tau) \in [\mathfrak{J}],$$

and the majorant system $[\mathfrak{J}]$ is equivalent to $\mathfrak{J}$.

All regular Baire $*$-fields of functions are exhausted by the families $\mathfrak{J}(\mathfrak{B})$, namely

THEOREM 2.3.5. *If $\mathfrak{F}$ is a regular Baire $*$-field of functions, and if $\mathfrak{B} = \mathfrak{M}(\mathfrak{F})$ and $\mathfrak{J} = \mathfrak{F}^+$, then*

$$\mathfrak{F} = \mathfrak{J}(\mathfrak{B}) = (\tilde{\mathfrak{G}}(\mathfrak{B})/\mathfrak{J}). \tag{11}$$

PROOF. Any elementary function $\varphi(\tau) = \sum_{k=1}^{\infty} \alpha_k \chi_k(\tau)$ in $(\mathfrak{E}(\mathfrak{B})/\mathfrak{J})$ belongs to the $*$-field $\mathfrak{F}$. In fact, let $\varphi_n(\tau) = \sum_{k=1}^{n} \alpha_k \chi_k(\tau)$. Since the $*$-field $\mathfrak{F}$ is σ-closed,

$$|\varphi_n(\tau)| = \left| \sum_{k=1}^{n} \alpha_k \chi_k(\tau) \right| \leqslant \sum_{k=1}^{\infty} |\alpha_k| \chi_k(\tau) = F'(\tau) \in \mathfrak{F}^+$$

and

$$\varphi(\tau) = \lim_{n\to\infty} \varphi_n(\tau) \in \mathfrak{F}.$$

Since the family F is closed with respect to uniform convergence, in addition to $\{\mathfrak{E}(\mathfrak{B})/\mathfrak{J}\} \subset \mathfrak{F}$, we have

$$\mathfrak{J}(\mathfrak{B}) \subset \mathfrak{F}. \tag{11'}$$

On the other hand, functions of the family $\mathfrak{F}$ are $(\mathfrak{B})$-measurable ($\mathfrak{B} = \mathfrak{M}(\mathfrak{F})$), and hence for $\mathfrak{J} = \mathfrak{F}^+$,

$$\mathfrak{F} \subset (\tilde{\mathfrak{G}}(\mathfrak{B})/\mathfrak{J}) = \mathfrak{J}(\mathfrak{B}),$$

which together with (11′) proves (11).

From Theorems 2.3.4 and 2.3.5 we have

THEOREM 2.3.6. *In any regular Baire $*$-field of functions $\mathfrak{F}$, the product and quotient of two functions $F(\tau)$ and $G(\tau)$ in $\mathfrak{F}$ again belong to $\mathfrak{F}$ if, correspondingly,*

$$|F(\tau)G(\tau)| \leqslant F'(\tau) \in \mathfrak{F}; \quad \left| \frac{F(\tau)}{G(\tau)} \right| \leqslant G'(\tau) \in \mathfrak{F}.$$

3.4. Similarity of Baire ∗-fields of functions. We consider the collection $\mathfrak{S} = \mathfrak{S}(T)$ of all regular Baire ∗-fields of functions $\mathfrak{F} \subset \tilde{\mathfrak{G}}(T)$, where T is a fixed set. Two families of functions $\mathfrak{F}$ and $\mathfrak{F}'$ in $\mathfrak{S}(T)$ are said to be *similar* if

$$\mathfrak{M}(\mathfrak{F}) = \mathfrak{M}(\mathfrak{F}'). \tag{12}$$

Since the regular ∗-fields $\mathfrak{F}$ and $\mathfrak{F}'$ are σ-closed, (12) is equivalent to the following condition: *the families $\mathfrak{F}$ and $\mathfrak{F}'$ contain the same characteristic functions.*

Similarity is an equivalence relation in $\mathfrak{S}$. The set of representatives of a type $\mathring{\mathfrak{F}}$ with respect to this equivalence, i.e. the collection of all ∗-fields in $\mathfrak{S}$ which are similar to some $\mathfrak{F} \in \mathfrak{S}$, will be called a *series of similar ∗-fields in* $\mathfrak{S}$. If $\mathfrak{F}$ is any element of such a series, and if $\mathfrak{B} = \mathfrak{M}(\mathfrak{F})$ ($\mathfrak{B}$ is a regular Borel field of sets), then $\mathfrak{B}$ is independent of the choice of the representative $\mathfrak{F}$ in the series, and by Theorem 2.3.5 and (8),

$$\mathfrak{G}(\mathfrak{B}) \subset \mathfrak{F} \subset \tilde{\mathfrak{G}}(\mathfrak{B}), \tag{8'}$$

i.e. a series of similar ∗-fields in $\mathfrak{S}$ has a greatest and least element, namely, $\tilde{\mathfrak{G}}(\mathfrak{B})$ and $\mathfrak{G}(\mathfrak{B})$, and it consists of precisely all ∗-fields $\mathfrak{F}$ in $\mathfrak{S}$ which satisfy (8′). Indeed, the system $\mathfrak{M}(\mathfrak{F})$ is the same for all such ∗-fields $\mathfrak{F}$.

Thus, we have established a one-to-one correspondence between the series of similar ∗-fields in $\mathfrak{S}$ (or types $\mathring{\mathfrak{F}}$ with respect to similarity in $\mathfrak{S}$) and the regular Borel fields of sets $\mathfrak{B} \subset \mathfrak{P}(T)$: The field $\mathfrak{B} = \mathfrak{M}(\mathfrak{F})$, where $\mathfrak{F}$ is a representative of the type $\mathring{\mathfrak{F}}$, corresponds to the type $\mathring{\mathfrak{F}}$, and the series of all ∗-fields $\mathfrak{F}$ in $\mathfrak{S}$ which satisfy (8′) corresponds to a regular Borel field of sets $\mathfrak{B}$.

REMARK. The results of Sec. 3.2, 3.3, and 3.4 could also be formulated for real fields of functions $\mathfrak{F} \subset \tilde{\mathfrak{G}}_r(T)$.

§ 4. Baire Closures of Families of Functions

The concept of majorant system can also be applied for any family of complex functions $\mathfrak{F}$, namely: *the system $\mathfrak{J}$ is a majorant* for $\mathfrak{F}$ if

$$|F(\tau)| \leqslant F'(\tau) \in \mathfrak{J}$$

for each function $F(\tau)$ in $\mathfrak{F}$. In $\mathfrak{F}$, we consider the passage to the limit

$$F_n(\tau) \in \mathfrak{F}, \quad |F_n(\tau)| \leqslant F'(\tau) \in \mathfrak{J}, \quad \lim_{n\to\infty} F_n(\tau) = F(\tau) \tag{1}$$

with respect to a majorant system $\mathfrak{J}$ for a family $\mathfrak{F}$ of complex (or real) functions.

A family $\mathfrak{F}$ is by definition a *Baire system of functions with respect to a majorant system* $\mathfrak{J}$ if $\mathfrak{F}$ is closed with respect to passage to the limit (1)*.

* The choice $\mathfrak{F} = \tilde{\mathfrak{G}}_r^+(T)$ yields the usual concept of Baire systems of (finite) functions.

The intersection $\mathfrak{F}'$ of all families of functions which contain a given family $\mathfrak{F}$, and are Baire systems with respect to $\mathfrak{J}$, is again a Baire system of functions with respect to $\mathfrak{J}$, and it is called the *Baire closure of the family* $\mathfrak{F}$ *(with respect to* $\mathfrak{J}$*)*.

An *n-term relation* $\theta(F^{(1)}, F^{(2)}, \ldots, F^{(n)})$ $(F^{(k)} = F^{(k)}(\tau) \in \mathfrak{F})$ in a Baire system of functions $\mathfrak{F}$ is said to be *continuous in the argument* $F^{(k)}$ if, for fixed $F^{(m)}(\tau)$ for $m \neq k$, the set of functions $F^{(k)}(\tau) \in \mathfrak{F}$ for which $\theta(F^{(1)}, F^{(2)}, \ldots, F^{(n)})$ holds is closed with respect to passage to the limit (1).

LEMMA 2.4.1. *Let* $\mathfrak{F}'$ *be the Baire closure of the family* $\mathfrak{F}$. *If an n-term relation* θ *in* $\mathfrak{F}'$ *is the identity* in* $\mathfrak{F}$ *and is continuous in each of its n arguments, then* θ *is the identity in* $\mathfrak{F}'$.

In fact, let the hypothesis of the lemma be satisfied, and let m be the maximal integer which satisfies the condition: the relation θ holds for each system $[F^{(1)}, F^{(2)}, \ldots, F^{(n)}]$ in which the first m functions belong to $\mathfrak{F}'$, and the remaining $n-m$ functions belong to the family $\mathfrak{F}$. It is obvious that such an $m \geqslant 0$ exists, and the lemma will be proved if we show that $m = n$. We assume that $m < n$, and we let $\mathfrak{F}''$ denote the collection of functions $G(\tau)$ in $\mathfrak{F}'$ which satisfy the condition: the relation θ holds for each system $[F^{(1)}, F^{(2)}, \ldots, F^{(n)}]$ in which $F^{(m+1)}(\tau) = G(\tau)$, the first m functions belong to $\mathfrak{F}'$, and the remaining $n-m-1$ functions belong to the family $\mathfrak{F}$. By the continuity of θ in $F^{(m+1)}$, the family $\mathfrak{F}''$ is closed with respect to passage to the limit (1) (with $\mathfrak{F}$ replaced by $\mathfrak{F}''$), i.e. it is a Baire system of functions with respect to $\mathfrak{J}$. Since $\mathfrak{F}'' \supset \mathfrak{F}$, the family $\mathfrak{F}''$ contains the closure $\mathfrak{F}'$, and hence $\mathfrak{F}'' = \mathfrak{F}'$, which contradicts the assumption that m is a maximal integer.

A family of functions $\mathfrak{F}$ is said to be a *$*$-family* if whenever $F(\tau)$ is contained in $\mathfrak{F}$, the function $(F(\tau))^* = \overline{F(\tau)}$ is also contained in $\mathfrak{F}$.

The Baire closure $\mathfrak{F}'$ *of a $*$-family of functions* $\mathfrak{F}$ *is again a $*$-family.*

Indeed, let $\theta = \theta(F)$ be the following property (one-term relation) in $\mathfrak{F}'$: whenever $F(\tau)$ belongs to $\mathfrak{F}'$, $\overline{F(\tau)}$ also belongs to $\mathfrak{F}'$. θ is the identity in $\mathfrak{F}$ ($\mathfrak{F}$ is a $*$-family), and it is continuous. Therefore, if $\theta(F_n)$ and (1) (with $\mathfrak{F}$ replaced by $\mathfrak{F}'$) hold, then

$$\overline{F_n(\tau)} \in \mathfrak{F}', \quad |\overline{F}_n(\tau)| \leqslant F'(\tau) \in \mathfrak{J}, \quad \lim \overline{F_n(\tau)} = \overline{F(\tau)}$$

and hence $\overline{F(\tau)} \in \mathfrak{F}'$, i.e. $F(\tau)$ has property θ. According to Lemma 2.4.1, all elements of $\mathfrak{F}'$ have the property θ, i.e. $\mathfrak{F}'$ is a $*$-family**

With the help of Lemma 2.4.1 we also obtain:

The Baire closure $\mathfrak{F}'$ *(with respect to* $\mathfrak{J}$*) of a lattice* $\mathfrak{F}$ *of (real) functions is a Baire lattice of functions.*

In fact, let the (two-term) relation $\theta(F, G) = F\,\theta\,G$, where $F, G \in \mathfrak{F}'$, denote

* See Sec. 1.1.1.

** In the case of a one-term relation (property), Lemma 2.4.1 is essentially useless, since the verification of continuity differs little from the proof given in this special case.

$$\sup\{F, G\} \in \mathfrak{F}' \tag{2}$$

The relation $\theta(F, G)$ is the identity in $\mathfrak{F}$ (since if $F(\tau)$ and $G(\tau)$ belong to F, then $\sup\{F, G\} \in \mathfrak{F} \subset \mathfrak{F}'$), and it is continuous in $\mathfrak{F}'$. Indeed, under passage to the limit (1) (with $\mathfrak{F}$ replaced by $\mathfrak{F}'$),

$$\sup\{F, G\} = \lim \sup\{F_n, G\},$$

and if $|G(\tau)| \leqslant G'(\tau) \in \mathfrak{J}$, then

$$|\sup\{F_n, G\}| \leqslant \sup\{F', G'\} \leqslant F'(\tau) + G'(\tau) \in \mathfrak{J}.$$

Since $\mathfrak{F}'$ is a Baire system of functions with respect to $\mathfrak{J}$, whenever $\sup\{F_n, G\}$ belongs to $\mathfrak{F}'$, $\sup\{F, G\}$ also belongs to $\mathfrak{F}'$, and this means that $\theta(F, G)$ is continuous in the argument F. According to Lemma 2.4.1, $\theta(F, G)$ is an identity in $\mathfrak{F}'$, i.e. (2) holds for any $F(\tau)$ and $G(\tau)$ in $\mathfrak{F}'$. An analogous argument leads to the same result with sup replaced by inf.

The Baire closure of a vector system of functions is again a vector system of functions.

This assertion can be proved with the help of Lemma 2.4.1 in a manner which is similar to the preceding remarks; instead of the relation $\sup(F, G) \in \mathfrak{F}'$, we must now consider the two-term relation $\theta(F, G)$ denoting that $F(\tau)+G(\tau) \in \mathfrak{F}'$, and the one-term relation $\theta_\alpha(F)$ denoting that $\alpha F(\tau) \in \mathfrak{F}'$.

In a similar manner we can prove:

THEOREM 2.4.1. *The Baire closure of a *-field (field) of functions $\mathfrak{F}$ with respect to $\mathfrak{J}$ is a Baire *-field (field) of functions (with respect to $\mathfrak{J}$).*

§ 5. Positive Linear Operations

5.1. The extension of a positive linear operation. Let $\mathfrak{F}$ be a vector *-order of functions defined, on a set T which contains the function $F(\tau) \equiv 1$. A linear transformation $L(F)$, defined in the vector system F, which has values in a complete vector *-order Z, will be called a *linear operation.* A linear operation $L(F)$ satisfies the conditions:

1) $L(F+G) = L(F)+L(G)$;

2) $L(\alpha F) = \alpha L(F) \qquad (\alpha \in K_c)$.

If, moreover, the following condition is satisfied:

3) $L(F) \geqslant 0$ for $F(\tau) \geqslant 0$ and $L(F) \not\equiv 0$,

then $L(F)$ is called a *positive linear operation.*

LEMMA 2.5.1. *If a linear operation $L(F)$ is positive, then*

$$(L(F))^* = L(\bar{F}). \tag{1}$$

In fact, let

$$F(\tau) = F_1(\tau) - F_2(\tau) + iF_3(\tau) - iF_4(\tau)$$

be a representation of any function $F(\tau) \in \mathfrak{F}$ in terms of non-negative functions $F_k(\tau)$ in $\mathfrak{F}$. Then

$$L(F) = L(F_1) - L(F_2) + i(L(F_3) - L(F_4)),$$

and the values of $L(F_k)$ are non-negative, and hence are Hermitian. Therefore

$$(L(F))^* = L(F_1) - L(F_2) - i(L(F_3) - L(F_4)) = L(\bar{F}).$$

In particular, it follows from (1) that *for a real function $F(\tau)$ in $\mathfrak{F}$, the value of $L(F)$ is a Hermitian element of the vector $*$-order Z.*

It follows from 3) that the *linear operation $L(F)$ is monotone* in the set $\mathfrak{F}_r$ of real functions in $\mathfrak{F}$:

$$L(G) \geqslant L(F), \quad \text{if} \quad G(\tau) \geqslant F(\tau). \tag{2}$$

It follows from 3) and $G(\tau) - F(\tau) \geqslant 0$ that

$$L(G) - L(F) = L(G - F) \geqslant 0.$$

In particular, for a bounded real function $F(\tau)$ in $\mathfrak{F}$,

$$\alpha L(1) \leqslant L(F) \leqslant \beta L(1), \tag{2'}$$

where $\alpha = \inf_\tau F(\tau)$, $\beta = \sup_\tau F(\tau)$; and $\gamma L(1)$, $\gamma = \sup_\tau |F(\tau)|$, is an absolute majorant for $L(F) \in Z$.

The following assertion is more general:

$L(G)$ is an absolute majorant for $L(F)$ if $|F(\tau)| \leqslant G(\tau)$ and, in particular, if $G(\tau) = |F(\tau)|$.

Indeed, for real $F(\tau)$ we have:

$$G(\tau) - F(\tau) \geqslant 0, \quad G(\tau) + F(\tau) \geqslant 0,$$

hence

$$L(G) - L(F) = L(G - F) \geqslant 0, \quad L(G) + L(F) = L(G + F) \geqslant 0.$$

If $F(\tau) \in \mathfrak{F}$ is a complex function, and if $F(\tau) = F_1(\tau) + iF_2(\tau)$ is a decomposition of F into real components $F_1(\tau)$ and $F_2(\tau)$, then

$$|F_1(\tau)| \leqslant G(\tau), \quad |F_2(\tau)| \leqslant G(\tau)$$

and hence $L(G)$ is an absolute majorant both for $L(F_1)$ and for $L(F_2)$. Thus it is also an absolute majorant for the element $L(F) = L(F_1) + iL(F_2)$ with Hermitian components $L(F_1)$ and $L(F_2)$.

LEMMA 2.5.2. *A positive linear operation $L(F)$ is continuous with respect to uniform convergence in $\mathfrak{F}$.*

Indeed, let a sequence $\{F_n(\tau)\}$ in $\mathfrak{F}$ converge uniformly to a function $F(\tau)\in\mathfrak{F}$, and let $\gamma_n = \sup_\tau | F_n(\tau)-F(\tau) |$. Then $\lim \gamma_n = 0$, and $\gamma_n L(1)$ is uniformly infinitely small in Z (see Sec. 1.8.2), and it is an absolute majorant for $L(F_n)-L(F) = L(F_n-F)$, so that

$$\lim L(F_n) = L(F),$$

moreover, the convergence is uniform in Z.

A linear operation $L'(F)$ defined on $\mathfrak{F}'$ is called an *extension*, or *continuation, of an operation* $L(F)$ on $\mathfrak{F}$ if $\mathfrak{F}'\supset\mathfrak{F}$ and if $L'(F) = L(F)$ for $F(\tau)\in\mathfrak{F}$.

THEOREM 2.5.1. *A positive linear operation $L(F)$, defined on a vector *-order of functions $\mathfrak{F}$, has a positive linear extension $L'(F)$ on the closure $\mathfrak{F}'$ of the family $\mathfrak{F}$ with respect to uniform convergence. The extension $L'(F)$ with the indicated properties is unique.*

In fact, let a sequence $\{F_n(\tau)\}$ in $\mathfrak{F}$ converge uniformly to the function $F(\tau)\in\mathfrak{F}'$, and let $\gamma_n = \sup_\tau | F_n(\tau)-F(\tau) |$. Then $\lim \gamma_n = 0$,

$$|F_n(\tau)-F_m(\tau)| \leqslant |F_n(\tau)-F(\tau)| + |F(\tau)-F_m(\tau)| \leqslant \gamma_n+\gamma_m$$

and $\{(\gamma_n+\gamma_m)L(1)\}$ is an absolute majorant for $\{L(F_n)-L(F_m)\} = \{L(F_n-F_m)\}$. Therefore $\{L(F_n)\}$ is a Cauchy sequence with respect to uniform convergence in Z, and by virtue of Theorem 1.8.3 there exists a uniform limit $\lim L(F_n)$ in Z. The formula

$$L'(F) = \lim L(F_n) \tag{3}$$

determines the value of $L'(F)$ in a unique manner for every function $F(\tau)$ in the closure $\mathfrak{F}'$, since $L'(F)$ is independent of the choice of the sequence $\{F_n(\tau)\}$ which converges uniformly to $F(\tau)$. Indeed, if $\{G_n(\tau)\}$ also converges uniformly to $F(\tau)$, then $L(F_n)-L(G_n) = L(F_n-G_n)$ is uniformly infinitely small in Z.

The operation $L'(F)$ is obviously linear. It is positive, since if $F(\tau)\geqslant 0$, and if $\{F_n(\tau)\}$ is a real sequence in $\mathfrak{F}$ which approximates $F(\tau)$, then it follows from

$$F_n(\tau) \geqslant F_n(\tau) - F(\tau) \geqslant -\gamma_n$$

that $L(F_n) \geqslant -\gamma_n L(1)$, and since $\lim \gamma_n = 0$,

$$L'(F) = \lim_{n\to\infty} L(F_n) \geqslant 0.$$

The operation $L'(F)$ is an extension of the operation $L(F)$ (for $F(\tau)\in\mathfrak{F}$, and $F_n(\tau) = F(\tau)$, we have $L'(F) = L(F)$). This extension is unique, since by virtue of the continuity of $L'(F)$ (Lemma 2.5.2), the value of $L'(F)$ is uniquely determined by (3).

REMARK. It is easy to see from the proof of Theorem 2.5.1 that it can be generalized if uniform convergence is replaced by the following form of convergence: a sequence $\{F_n(\tau)\}$ in $\mathfrak{F}$ *is relatively uniformly convergent to* $F(\tau)$ if

$$|F_n(\tau) - F(\tau)| \leqslant \gamma_n F'(\tau),$$

where $\lim \gamma_n = 0$ and $F'(\tau) \in \mathfrak{F}$ (see Sec. 1.8.2). Here $L(F')$ plays the role of the element $L(1)$.

5.2. Lebesque operations. A function $\xi(F)$, with values in a complete vector $*$-order Z, and which is defined on a semifield, field, or $*$-field of functions $\mathfrak{F}$, is said to be *continuous* if

$$\lim \xi(F_n) = \xi(F) \tag{4}$$

for every convergent sequence in $\mathfrak{F}$ which satisfies:

$$|F_n(\tau)| \leqslant F'(\tau) \in \mathfrak{F}, \quad \lim F_n(\tau) = F(\tau) \in \mathfrak{F}. \tag{5}$$

If $\mathfrak{F}$ is a Baire system of functions, then the second equation in (5) is automatically satisfied.

In the case when $\xi(F) = L(F)$ is a positive linear operation on a $*$-field (or field) $\mathfrak{F}$ we have:

LEMMA 2.5.3. *A positive linear operation $L(F)$ on a Baire $*$-field (field) of functions $\mathfrak{F}$ is continuous if and only if*

$$\lim L(G_n) = 0 \tag{6}$$

for

$$G_n(\tau) \in \mathfrak{F}, \quad G_n(\tau) \geqslant G_{n+1}(\tau), \quad \lim_{n\to\infty} G_n(\tau) = 0. \tag{7}$$

The necessity of the conditions (6–7) is obvious, and it is only necessary to show that these conditions, and the equations in (5), imply the limit equation (4) with $\xi(F) = L(F)$. We let

$$\Phi_n(\tau) = |F_n(\tau) - F(\tau)|, \quad G_n(\tau) = \sup_{k \geqslant n} \{\Phi_k\}.$$

Then $\Phi_n(\tau) \in \mathfrak{F}$, $|\Phi_n(\tau)| \leqslant 2F'(\tau) \in \mathfrak{F}$, $G_n(\tau) \in \mathfrak{F}$, and (7) is satisfied. Therefore (6) is also satisfied. But since $\Phi_n(\tau) \leqslant G_n(\tau)$, we have $L(|F_n - F|) \leqslant L(G_n)$, and hence $\lim L(|F_n - F|) = 0$; thus $\lim L(F_n) = L(F)$, since $L(|F_n - F|)$ is an absolute majorant of the difference $L(F_n) - L(F) = L(F_n - F)$.

REMARK. For a positive linear operation $L(F)$ on a $*$-field (field) of functions $\mathfrak{F}$ the conditions (6) and (7) are equivalent to the following:

$$\lim L(F_n) = L(F), \tag{8}$$

if the sequence $\{F_n(\tau)\}$ in $\mathfrak{F}$ is non-decreasing and converges to the function $F(\tau) \in \mathfrak{F}$. It is sufficient to let $G_n(\tau) = F(\tau) - F_n(\tau)$. Then $G_n(\tau)$ satisfies the conditions in (7) and, since $L(G_n) = L(F) - L(F_n)$, it follows that (6) and (8) are satisfied.

A continuous linear operation $L(F)$ will be called a *Lebesgue operation* if it is defined on a Baire $*$-field (field) of functions $\mathfrak{F}$.

A positive linear operation $L(F)$ on a Baire $*$-field (or field) of functions $\mathfrak{F}$ is a Lebesgue operation if and only if it satisfies the conditions (6) *and* (7).

Positive Lebesgue operations are of particular importance to us. Some facts from the theory of set functions are necessary for their analytic representation (in the case when $\mathfrak{F}$ is a *regular* field).

A *set function* $\mu(M)$, defined on a field $\mathfrak{K} \subset \mathfrak{P}(T)$ $(M \in \mathfrak{K})$, with values in a complete vector $*$-order (order) Z is said to be *additive* if for *disjoint* sets $M_k \in \mathfrak{K}$,

$$\mu\left(\bigcup_{k=1}^{n} M_k\right) = \sum_{k=1}^{n} \mu(M_k). \tag{9}$$

For an additive function $\mu(M)$,

$$\mu(M' \setminus M) = \mu(M') - \mu(M), \tag{10}$$

if $M \in \mathfrak{K}$, $M' \in \mathfrak{K}$, and $M \subset M'$. Indeed, $M' = M \cup (M' \setminus M)$, and by virtue of (9),

$$\mu(M') = \mu(M) + \mu(M' \setminus M).$$

A particularly important case occurs when an additive function $\mu(M)$ is *non-negative*, i.e. $\mu(M) \geqslant 0$ for all $M \in \mathfrak{K}$.

A non-negative additive function $\mu(M)$ is monotone, i.e.

$$\mu(M') \geqslant \mu(M)$$

for $M' \supset M$. This follows from (10) since $\mu(M' \setminus M) \geqslant 0$.

An additive set function $\mu(M)$ is said to be *σ-additive* or *countably additive* if

$$\mu(M) = \sum_{k=1}^{\infty} \mu(N_k), \tag{11}$$

where the set $M \in \mathfrak{K}$ is the union of disjoint sets N_k of $\mathfrak{K}$, and the series (11) is absolutely convergent in Z.

In the fundamental case when an additive function $\mu(M)$ is non-negative, we supplement (11) by two additional equivalent forms, namely:

1) if $M_n \in \mathfrak{K}$, $M_n \subset M_{n+1}$, $M \in \mathfrak{K}$, and if $M = \bigcup_{n=1}^{\infty} M_n$, then

$$\mu(M) = \lim \mu(M_n); \tag{11'}$$

2) if $N_k \in \mathfrak{R}$, $N_k \supset N_{k+1}$, and if $\bigcap_{k=1}^{\infty} N_k = \emptyset$, then

$$\lim \mu(N_k) = 0. \tag{11''}$$

It is possible to establish a one-to-one correspondence between representations of the set M in the form of a countable sum of disjoint sets N_k, and representations of the set M in the form of a countable sum of monotone increasing sets M_n by letting

$$M_n = \bigcup_{k=1}^{n} N_k, \quad N_k = M_k \setminus M_{k-1}, \quad M_0 = \emptyset, \ k = 1, 2, \ldots$$

Then

$$\mu(M_n) = \sum_{k=1}^{n} \mu(N_k)$$

and the conditions (11) and (11′) imply one another. For the proof of the equivalence of (11′) and (11″) it is sufficient to let $N_k = M \setminus M_k$, since

$$\lim \mu(N_k) = \mu(M) - \lim \mu(M_k)$$

implies that (11′) and (11″) also follow from one another.

It is natural to consider σ-additive set functions on Borel fields of sets. A non-negative σ-additive function $\mu(M)$ defined on a Borel field of sets $\mathfrak{B}(M \in \mathfrak{B})$ will be called a *measure*.

A positive Lebesgue operation $L(F)$ generates some measure $\mu(M)$, namely, we have:

LEMMA 2.5.4. *Let a positive Lebesgue operation $L(F)$ be defined on a Baire *-field (or field) of functions $\mathfrak{F}$, and let $\mathfrak{B}$ be the collection of sets whose characteristic functions belong to $\mathfrak{F}$. Then $\mathfrak{B}$ is a Borel field of sets, and the set function*

$$\mu(M) = L(\chi_M), \qquad M \in \mathfrak{B}, \tag{12}$$

is a measure with domain $\mathfrak{B}$.

In fact, the characteristic functions χ_M of the sets M in $\mathfrak{B}$ form a sublattice of $\mathfrak{F}$ which is σ-closed, like the lattice of functions $\mathfrak{F}$. Moreover, if $M' \supset M$, then whenever $\chi_{M'}$ and χ_M belong to this sublattice, the function $\chi_{M' \setminus M} = \chi_{M'} - \chi_M$ also belongs to this sublattice. Thus $\mathfrak{B}$ is a Borel field, and the function $\mu(M)$ defined on $\mathfrak{B}$ by (12) is non-negative and additive, since the operation $L(F)$ is positive and linear. Hence,

$$\mu(M) = L(\chi_M) = L(\chi_{M_1} + \chi_{M_2}) = L(\chi_{M_1}) + L(\chi_{M_2}) = \mu(M_1) + \mu(M_2),$$

if $M = M_1 \cup M_2$ and if the sets M_1 and M_2 in $\mathfrak{B}$ are disjoint. Finally, the σ-additivity of the (non-negative) function $\mu(M)$ follows from (6–7) regarding the continuity of the operation $L(F)$:

$$\operatorname{Lim} \mu(N_k) = \lim L(\chi_k) = 0,$$

where $\{N_k\}$ is a decreasing sequence in $\mathfrak{B}$ with an empty intersection, and the $\chi_k(\tau)$ are the characteristic functions of the sets N_k.

If the field $\mathfrak{F}$ on which the positive Lebesque operation $L(F)$ is defined is regular, then the measure $\mu(M)$ generated by the operation $L(F)$ is defined on the regular Borel field $\mathfrak{B} = \mathfrak{M}(\mathfrak{F})$ (Theorem 2.1.1). Here the field coincides with the closure with respect to uniform convergence of the ∗-field $(\mathfrak{E}(\mathfrak{B})/\mathfrak{J})$, where $\mathfrak{J} = \mathfrak{F}^+$ (Theorem 2.3.5), i.e. $\mathfrak{F} = \mathfrak{J}(\mathfrak{B})$.

Let $\chi_k(\tau)$ be the characteristic functions of the disjoint sets N_k in $\mathfrak{B}$. For functions

$$\varphi(\tau) = \sum \alpha_k \chi_k(\tau)$$

of the ∗-field $(\mathfrak{E}(\mathfrak{B})/\mathfrak{J})$, the operation $L(F)$ can be represented in the form

$$L(\varphi) = \sum \alpha_k \mu(N_k), \tag{13}$$

where the series $\sum |\alpha_k| \mu(N_k)$ converges in Z. Indeed, this is obvious for functions $\varphi(\tau)$ which have a finite number of values; for non-negative $\varphi(\tau)$ this follows from the continuity of the operation $L(F)$, and it still holds for any $\varphi(\tau) \in (\mathfrak{E}(\mathfrak{B})/\mathfrak{J})$ since $\varphi(\tau) = \varphi_1(\tau) - \varphi_2(\tau) + i\varphi_3(\tau) - i\varphi_4(\tau)$, where $\varphi_k(\tau) \geqslant 0$. Thus $L(F)$ induces a positive linear operation $L(\varphi)$ on the ∗-field $(\mathfrak{E}(\mathfrak{B})/\mathfrak{J})$, which is defined by (13), and according to Theorem 2.5.1, $L(\varphi)$ in turn uniquely defines an operation $L(F)$ which is a positive linear extension to the closure $\mathfrak{F}$ (with respect to uniform convergence) of the ∗-field $(\mathfrak{E}(\mathfrak{B})/\mathfrak{J})$.

Thus, in the case of a positive Lebesgue operation defined on a regular Baire ∗-field (or field) $\mathfrak{F}$, we have indicated an analytical process which recovers the operation $L(F)$ on $\mathfrak{F}$ from the measure $\mu(M)$ generated by $L(F)$. In the following section we show that this process can be applied to an arbitrary measure $\mu(M)$ on a regular Borel field of sets $\mathfrak{B}$, and it generates positive Lebesgue operations which will be called *integrals with respect to* $\mu(M)$.

§ 6. The Integral

6.1. The definition of integral. In this section we will assume, without explicitly stating it, that the measure $\mu(M)$ (with values in a complete vector ∗-order Z) is defined on a *regular Borel field* $\mathfrak{B}$. We will introduce an operation $I(F)$ with respect to such a measure in the following manner.

Let $\chi_k(\tau)$ be the characteristic functions of the sets N_k of a $(\mathfrak{B})$-partition which is admissible for the *elementary* $(\mathfrak{B})$-function

$$\varphi(\tau) = \sum_k \alpha_k \chi_k(\tau), \tag{1}$$

and let $\mathfrak{E}_\mu^+(\mathfrak{B})$ be the collection of all non-negative elementary $(\mathfrak{B})$-functions

(1) for which the series

$$\sum_k \alpha_k \mu(N_k)$$

converges in Z. The collection $\mathfrak{E}_\mu^+(\mathfrak{B}) = \mathfrak{I}$ is obviously a majorant system, and the family of functions

$$\tilde{\mathfrak{E}}_\mu(\mathfrak{B}) = \tilde{\mathfrak{E}}_\mu = (\tilde{\mathfrak{E}}(\mathfrak{B})/\mathfrak{I}) \tag{2}$$

is characterized by the fact that for functions $\varphi(\tau)$ of this family, the series $\sum_k |\alpha_k| \mu(N_k)$ converges, where the N_k are sets of admissible partitions in the representation (1) of the function $\varphi(\tau)$ in $\tilde{\mathfrak{E}}(\mathfrak{B})$.

For a function $\varphi(\tau) \in \tilde{\mathfrak{E}}_\mu(\mathfrak{B})$, we let (cf. (2.5.13))

$$I(\varphi) = \sum_k \alpha_k \mu(N_k), \tag{3}$$

where, according to what has been said, the series on the right *converges absolutely* in Z.

$I(\varphi)$ is independent of the choice of the representation (1). If another representation

$$\varphi(\tau) = \sum_r \alpha_r' \chi_r'(\tau),$$

has been given, which corresponds to an admissible $(\mathfrak{B})$-partition $T = \bigcup_r N_r'$, and if $T = \bigcup_{k,r}(N_k \cap N_r')$ is the product of the two $(\mathfrak{B})$-partitions, then since the measure $\mu(M)$ is σ-additive,

$$\sum_k \alpha_k \mu(N_k) = \sum_k \sum_r \alpha_k \mu(N_k \cap N_r') = \sum_r \sum_k \alpha_r' \mu(N_k \cap N_r') = \sum_r \alpha_r' \mu(N_r').$$

Thus, *the operation $I(\varphi)$ is uniquely defined on the regular *-field of functions* $\tilde{\mathfrak{E}}_\mu(\mathfrak{B})$ (cf., Theorem 2.3.1 and the remark after Theorem 2.2.1). *It is linear and positive.* Let $\varphi(\tau)$ and $\psi(\tau)$ belong to $\tilde{\mathfrak{E}}_\mu(\mathfrak{B})$ and let

$$\varphi(\tau) = \sum \alpha_k \chi_k(\tau), \qquad \psi(\tau) = \sum \beta_k \chi_k(\tau)$$

be their representations by means of a *common* partition $\{N_k\}$. Then

$$\varphi(\tau) + \psi(\tau) = \sum (\alpha_k + \beta_k) \chi_k(\tau), \qquad \alpha\varphi(\tau) = \sum \alpha\alpha_k \chi_k(\tau)$$

and hence

$$I(\varphi + \psi) = \sum_k (\alpha_k + \beta_k) \mu(N_k) =$$
$$= \sum_k \alpha_k \mu(N_k) + \sum \beta_k \mu(N_k) = I(\varphi) + I(\psi),$$
$$I(\alpha\varphi) = \sum_k \alpha\alpha_k \mu(N_k) = \alpha I(\varphi),$$

i.e. the operation $I(\varphi)$ is linear. Finally, if $\varphi(\tau) \geqslant 0$, i.e. if $\alpha_k \geqslant 0$, then $I\varphi \geqslant 0$.

By Theorem 2.5.1, *the operation $I(\varphi)$ on $\tilde{\mathfrak{E}}_\mu(\mathfrak{B})$ admits a unique positive linear extension $I(F)$ on the closure $\mathfrak{G}_\mu(\mathfrak{B})$ (with respect to uniform convergence)* *of the family* $\mathfrak{E}_\mu(\mathfrak{B})$. By virtue of (2.3.11), and (2), we have

$$\mathfrak{G}_\mu(\mathfrak{B}) = \mathfrak{J}(\mathfrak{B}) = (\tilde{\mathfrak{G}}(\mathfrak{B})/\mathfrak{J}), \tag{4}$$

where $\mathfrak{J} = \mathfrak{E}_\mu^+(\mathfrak{B})$, and hence (Theorem 2.3.1) $\mathfrak{G}_\mu(\mathfrak{B})$ is a Baire *-field of functions. The positive linear operation $I(F)$ on $\mathfrak{G}_\mu(\mathfrak{B})$ is by definition the *integral with respect to the measure* $\mu(M)$. We will reserve the same name for the operation which an integral $I(F)$ on $\mathfrak{G}_\mu(\mathfrak{B})$ induces on any of the series of similar (Sec. 2.3.4) Baire *-fields $\mathfrak{F}$ which satisfy the condition

$$\mathfrak{G}(\mathfrak{B}) \subset \mathfrak{F} \subset \mathfrak{G}_\mu(\mathfrak{B}). \tag{5}$$

In all of these cases we will again speak of the integral $I(F)$ on $\mathfrak{F}$.

Since $\mathfrak{F} = \mathfrak{J}(\mathfrak{B})$, where $\mathfrak{J} = \mathfrak{F}^+$ (Theorem 2.3.5), the integral $I(F)$ on $\mathfrak{F}$ can also be obtained by the closure (with respect to uniform convergence) of the operation $I(\varphi)$ on the family $(\mathfrak{E}(\mathfrak{B})/\mathfrak{J}) \subset \mathfrak{E}_\mu(\mathfrak{B})(\mathfrak{J} = \mathfrak{F}^+)$. For $F(\tau) \in \mathfrak{G}_\mu(\mathfrak{B})$, we will write

$$I(F) = \int_T F(\tau)\, \mu(M_\tau).$$

With the help of the notion of the integral, the results at the end of Sec. 2.5.2 regarding a positive Lebesgue operation can be formulated as the following theorem:

THEOREM 2.6.1. *A positive Lebesgue operation* $L(F)$, *defined on a regular Baire *-field of functions* $\mathfrak{F}$, *coincides with the integral* $I(F)$ *on* $\mathfrak{F}$ *with respect to the measure* $\mu(M)$ *which is generated by the operation* $L(F)$ *on the field* $\mathfrak{B} = \mathfrak{M}(\mathfrak{F})$.

REMARK. If the integral $I(F)$ is restricted to $\mathfrak{G}(\mathfrak{B})$, then its definition is significantly simplified, since now the operation $I(\varphi)$ is defined on the *-field $\mathfrak{E}(\mathfrak{B})$ of the elementary $(\mathfrak{B})$-functions that assume a finite number of values, and thus the sums in (1) and (3) are finite. In this form, the definition can be extended to the case when, in place of a measure $\mu(M)$ on $\mathfrak{B}$, we have a non-negative additive function $\mu(M)$ on a field of sets $\mathfrak{K}$ (see Sec. 2.5.2). Namely, we form the finite sum (3) in the representation (1) of a function $\varphi(\tau) \in \mathfrak{E}(\mathfrak{K})$ with respect to an admissible $(\mathfrak{K})$-partition. In accord with Theorem 2.5.1, we extend the operation $I(\varphi)$ on $\mathfrak{E}(\mathfrak{K})$ to a positive linear operation $I(F)$ on the closure $\mathfrak{G}(\mathfrak{K})$ (with respect to uniform convergence) of the *-field $\mathfrak{E}(\mathfrak{K})$. The operation $I(F)$ on $\mathfrak{G}(\mathfrak{K})$ will also be called an integral.

6.2. The indefinite integral. Let $F(\tau) \in \mathfrak{G}_\mu(\mathfrak{B})$, let $M \in \mathfrak{B}$, and let $\chi(\tau)$ be the characteristic function of M. Then the product $\chi(\tau)F(\tau)$ belongs to $\mathfrak{G}_\mu(\mathfrak{B})$ (Theorem 2.3.6) and, according to the definition,

$$\int_M F(\tau)\, \mu(M_\tau) = \int_T \chi(\tau)\, F(\tau)\, \mu(M_\tau) = I(\chi F).$$

Since the positive linear operation $I(F)$ is monotone, it follows that (see (2.5.2) and (2.5.2')):

1) *If* $F(\tau)$ *and* $G(\tau)$ *are real functions in* $\mathfrak{G}_\mu(\mathfrak{B})$, *and if* $G(\tau) \geqslant F(\tau)$ *on the set* M, *then*

$$\int_M G(\tau)\,\mu(\underset{\tau}{M}) \geqslant \int_M F(\tau)\,\mu(\underset{\tau}{M}). \tag{6}$$

In particular, if $\alpha \leqslant F(\tau) \leqslant \beta$ *on* M, *then*

$$\alpha\mu(M) \leqslant \int_M F(\tau)\,\mu(\underset{\tau}{M}) \leqslant \beta\mu(M). \tag{7}$$

2) *The integral* $\int_M |F(\tau)|\,\mu(\underset{\tau}{M})$ *is an absolute majorant for* $\int_M F(\tau)\,\mu(\underset{\tau}{M})$ *in* Z.

For fixed $F(\tau) \in \mathfrak{G}_\mu(\mathfrak{B})$, and for variable $M \in \mathfrak{B}$, the set function

$$\nu(M) = \int_M F(\tau)\,\mu(\underset{\tau}{M})$$

is called the *indefinite integral generated by the function* $F(\tau)$. In view of the linearity of the integral,

$$\nu(M) = \nu(M_1) + \nu(M_2),$$

if the sets M_1 and M_2 in $\mathfrak{B}$ are disjoint and if $M = M_1 \cup M_2$. In other words, *a definite integral is an additive set function on a field* $\mathfrak{B}$.

Let $\mu(M)$ be a non-negative additive set function on a field $\mathfrak{R}$, and let $\{M_n\}$ be a decreasing sequence of sets in $\mathfrak{R}$. A set function $\nu(M)$ on $\mathfrak{R}$ is said to be *absolutely continuous with respect to the measure* $\mu(M)$, if whenever

$$\lim \mu(M_n) = 0,$$

$\nu(M_n)$ is also infinitely small in Z.

If, instead of additive functions, we consider the measures $\mu(M)$ and $\nu(M)$ to have been defined on a Borel field $\mathfrak{B}_\mu (= \mathfrak{B}_\nu)$, then it is easy to show that $\nu(M)$ *is absolutely continuous with respect to* $\mu(M)$ *if and only if* $\mu(N) = 0$ *implies that* $\nu(N) = 0$. Indeed, if the measure $\nu(M)$ is absolutely continuous with respect to $\mu(M)$, and if $\mu(N) = 0$, then by setting $M_n = N, n = 1, 2, \ldots,$ we obtain $\lim_n \nu(M_n) = \nu(N) = 0$, since $\lim_n \mu(M_n) = \mu(N) = 0$. Conversely, if $\mu(N) = 0$ implies that $\nu(N) = 0$, and if $\{M_n\}$ is a decreasing sequence of sets in $\mathfrak{B}_\mu$ with $\lim_n \mu(M_n) = 0$, then by setting $M_0 = \bigcap_n M_n$ we obtain

$$M_0 \in \mathfrak{B}_\mu = \mathfrak{B}_\nu, \quad 0 \leqslant \lim_n \mu(M_n \setminus M_0) = \lim_n \mu(M_n) - \mu(M_0) =$$
$$= -\mu(M_0), \quad \mu(M_0) = 0,$$

and thus $\nu(M_0) = 0$ and

$$\lim_n \nu(M_n) = \lim_n \nu(M_n \setminus M_0) + \nu(M_0) = 0,$$

since both of the last terms are zero.

THEOREM 2.6.2. *The indefinite integral $\nu(M)$ generated by a function $F(\tau)\in\mathfrak{G}_\mu(\mathfrak{B})$ is absolutely continuous with respect to the measure $\mu(M)$.*

In fact, an arbitrary function $F(\tau)$ in $\mathfrak{G}_\mu(\mathfrak{B})$ can be represented in the form

$$F(\tau)=G(\tau)+\varphi(\tau),\quad G(\tau)\in\mathfrak{G}(\mathfrak{B}),\quad \varphi(\tau)\in\tilde{\mathfrak{E}}_\mu(\mathfrak{B}),$$

and hence it is sufficient to show that in the decomposition

$$\nu(M)=\int_M G(\tau)\,\underset{\tau}{\mu}(M)+\int_M \varphi(\tau)\,\underset{\tau}{\mu}(M)=\nu_1(M)+\nu_2(M)$$

each of the terms is absolutely continuous with respect to $\mu(M)$.

1) Let $G(\tau)$ be an arbitrary function in $\mathfrak{G}(\mathfrak{B})$ and let $\gamma=\sup\limits_\tau|G(\tau)|$. Then

$$\int_M |G(\tau)|\,\underset{\tau}{\mu}(M)\leqslant\gamma\mu(M)$$

and $\gamma\mu(M)$ is an absolute majorant for the first term $\nu_1(M)$ which, like $\mu(M)$, is thus infinitely small.

2) For the second term $\nu_2(M)$, we can assume without loss of generality that the function $\varphi(\tau)$ in $\tilde{\mathfrak{E}}_\mu(\mathfrak{B})$ is non-negative, i.e. that $\varphi(\tau)\in\tilde{\mathfrak{E}}^+_\mu(\mathfrak{B})$, since for any function $\varphi(\tau)\in\tilde{\mathfrak{E}}_\mu(\mathfrak{B})$, we have $\varphi(\tau)=\varphi_1(\tau)-\varphi_2(\tau)+i\varphi_3(\tau)-i\varphi_4(\tau)$, where $\varphi_k(\tau)\in\tilde{\mathfrak{E}}^+_\mu(\mathfrak{B})$.

From the representation

$$\int_M \varphi(\tau)\,\underset{\tau}{\mu}(M)=I(\varphi\chi_M)=\sum_{k=1}^{\infty}\alpha_k\mu(M\cap N_k),$$

where $0\leqslant\varphi(\tau)\in\tilde{\mathfrak{E}}_\mu(\mathfrak{B})$, it follows that

$$\int_M \varphi(\tau)\,\underset{\tau}{\mu}(M)\leqslant\sum_{k=1}^{n}\alpha_k\mu(M\cap N_k)+\sum_{k=n+1}^{\infty}\alpha_k\mu(N_k).$$

In this inequality, let M range through a sequence $\{M_n\}\in\mathfrak{B}$ for which $\lim\limits_n\mu(M_n)=0$. Then certainly $\lim\limits_n\mu(M_nN_k)=0$, and hence

$$0\leqslant\lim\int_{M_n}\varphi(\tau)\,\underset{\tau}{\mu}(M)=\inf_n\int_{M_n}\varphi(\tau)\,\underset{\tau}{\mu}(M)\leqslant\sum_{k=n+1}^{\infty}\alpha_k\mu(N_k).$$

By virtue of the convergence of the series $\sum\limits_{k=1}^{\infty}\alpha_k\mu(N_k)$ (which represents $I(\varphi)$), the series on the right-hand side is infinitely small, so that $\lim\nu_2(M_n)=0$, i.e. the function $\nu_2(M)$ is absolutely continuous with respect to $\mu(M)$.

THEOREM 2.6.3. *The indefinite integral $\nu(M)$ generated by a function $F(\tau)\in\mathfrak{G}_\mu(\mathfrak{B})$ is a σ-additive set function on the Borel field $\mathfrak{B}$.*

It is sufficient to prove the theorem for a non-negative function $F(\tau)\in\mathfrak{G}_\mu(\mathfrak{B})$, when the additive function $\nu(M)$ is non-negative. In this case the condition that the function $\nu(M)$ be σ-additive can be written in the form:

$$\lim \nu(N_k) = 0, \tag{8}$$

if

$$N_{k+1} \subset N_k \in \mathfrak{B}, \quad \bigcap_k N_k = \varnothing. \tag{9}$$

But, since the measure $\mu(M)$ is countably additive, it follows ftom (9) that $\lim \mu(N_k) = 0$, and that (8) holds since the function $\nu(M)$ is absolutely continuous with respect to $\mu(M)$.

REMARK. If the function $F(\tau)$ in $\mathfrak{G}_\mu(\mathfrak{B})$ generating the indefinite integral (7) is non-negative, then the indefinite integral $\nu(M)$ is non-negative and is a measure on $\mathfrak{B}$. In the general case, $F(\tau) = F_1(\tau) - F_2(\tau) + iF_3(\tau) - iF_4(\tau)$, where the functions $F_k(\tau)$ in $\mathfrak{G}_\mu(\mathfrak{B})$ are non-negative, and hence

$$\nu(M) = \nu_1(M) - \nu_2(M) + i\nu_3(M) - i\nu_4(M),$$

where the indefinite integrals $\nu_k(M)$, which are generated by the functions $F_k(\tau)$, are measures on the Borel field $\mathfrak{B}$.

In the case of an integral $I(F)$ with respect to a non-negative additive function $\mu(M)$ on a field $\mathfrak{R}$, it is easy to see that the product $\chi(\tau)F(\tau)$ belongs to the domain $\mathfrak{G}(\mathfrak{R})$ of the integral $I(F)$ if $F(\tau) \in \mathfrak{G}(\mathfrak{R})$ and if $\chi(\tau)$ is the characteristic function of the set $M \in \mathfrak{R}$. Therefore the definition (6) holds, and for fixed $F(\tau) \in \mathfrak{G}(\mathfrak{R})$, and variable $M \in \mathfrak{R}$, (7) again defines an indefinite integral on the field $\mathfrak{R}$. Theorem 2.6.2 (with $\mathfrak{G}_\mu(\mathfrak{B})$ replaced by $\mathfrak{G}(\mathfrak{R})$) holds in this case, since in the proof we now have $F(\tau) = G(\tau)$, and part 2) is eliminated. Theorem 2.6.3 also holds (with $\mathfrak{B}$ replaced by $\mathfrak{R}$ and $\mathfrak{G}_\mu(\mathfrak{B})$ replaced by $\mathfrak{G}(\mathfrak{R})$) if it is required that the function $\mu(M)$ be countably additive on $\mathfrak{R}$.

6.3 The continuity of the integral $I(F)$ on $\mathfrak{G}_\mu(\mathfrak{B})$. In view of the greater simplicity of the definition of the integral $I(F)$ on the Baire $*$-field $\mathfrak{G}(\mathfrak{B})$ of bounded functions, it is sometimes useful to approximate the integral of a function in $\mathfrak{G}_\mu(\mathfrak{B})$ by integrals of functions in $\mathfrak{G}(\mathfrak{B})$. With this aim, we introduce a concept which turns out to be useful in what follows.

A sequence $\{M_n\}$:

$$M_n \in \mathfrak{B}, \quad M_n \subset M_{n+1}, \quad T = \bigcup_{n=1}^{\infty} M_n$$

is said to be *admissible for a* ($\mathfrak{B}$)-*measurable function* $F(\tau)$ if $F(\tau)$ is bounded on each of the sets M_n. Along with the sequence $\{M_n\}$, for a function $F(\tau) \in \tilde{\mathfrak{G}}(\mathfrak{B})$, we will also say that the sequence $\{\vartheta_n(\tau)\}$ of characteristic functions of the sets M_n is *admissible*. For a function $F(\tau) \in \tilde{\mathfrak{G}}(\mathfrak{B})$, and an admissible sequence $\{\vartheta_n(\tau)\}$ for $F(\tau)$, we have

$$\vartheta_n(\tau) F(\tau) \in \mathfrak{G}(\mathfrak{B}), \quad \lim \vartheta_n(\tau) F(\tau) = F(\tau).$$

LEMMA 2.6.1. *A non-negative function $F(\tau) \in \tilde{\mathfrak{G}}(\mathfrak{B})$ belongs to $\mathfrak{G}_\mu(\mathfrak{B})$ if and only if the set $\{I(\vartheta_n F)\}$ is bounded in Z, where $\{\vartheta_n(\tau)\}$ is an admissible sequence for $F(\tau)$.*

In fact, if $F(\tau) \in \mathfrak{G}_\mu(\mathfrak{B})$, and if $F(\tau) \geqslant 0$, then $I(\vartheta_n F) \leqslant I(F)$, since the operator $I(F)$ is monotone. Conversely, let

$$I(\vartheta_n F) \leqslant \eta \in Z \tag{10}$$

and let $F(\tau) = \varphi(\tau)$ be a non-negative function in $\mathfrak{E}(\mathfrak{B})$. If $\varphi(\tau) = \sum_{k=1}^{\infty} \alpha_k \chi_k(\tau)$ is the representation (1) of $\varphi(\tau)$, then it follows from (10) that

$$\sum_{k=1}^{\infty} \alpha_k \mu(N_k \cap M_n) = I(\vartheta_n \varphi) \leqslant \eta \tag{11}$$

and

$$\sum_{k=1}^{m} \alpha_k \mu(N_k \cap M_n) \leqslant \eta.$$

Whence, for $n \to \infty$, we have

$$\sum_{k=1}^{m} \alpha_k \mu(N_k) \leqslant \eta,$$

since $\lim_{n\to\infty} \mu(N_k \cap M_n) = \mu(N_k)$. Therefore, the series $\sum_{k=1}^{\infty} \alpha_k \mu(N_k)$ converges, i.e. $\varphi(\tau) \in \tilde{\mathfrak{E}}_\mu^+(\mathfrak{B})$. For an arbitrary non-negative $(\mathfrak{B})$-measurable function, and for $\varepsilon > 0$, there exists a function $\psi(\tau) \in \tilde{\mathfrak{E}}(\mathfrak{B})$ such that

$$0 \leqslant F(\tau) \leqslant \psi(\tau) < F(\tau) + \varepsilon \tag{12}$$

(cf., (2.3.7)), and hence

$$I(\vartheta_n \psi) \leqslant I(\vartheta_n F) + \varepsilon\mu(M_n) < \eta + \varepsilon\mu(T) = \eta',$$

when (10) holds. According to what has been proved, $\psi(\tau) \in \mathfrak{E}_\mu^+(\mathfrak{B})$, and in view of (12), $F(\tau) \in (\tilde{\mathfrak{G}}(\mathfrak{B})/\mathfrak{J})$, with $\mathfrak{J} = \tilde{\mathfrak{E}}_\mu^+(\mathfrak{B})$, i.e. $F(\tau) \in \mathfrak{G}_\mu(\mathfrak{B})$ (cf. (4)).

The following theorem is true both for $\mathfrak{F} = \mathfrak{G}(\mathfrak{B})$ and for $\mathfrak{F} = \mathfrak{G}_\mu(\mathfrak{B})$.

THEOREM 2.6.4. *The integral $I(F)$ on $\mathfrak{F}$ with respect to a measure $\mu(M)$ is a positive Lebesgue operation.*

The integral $I(F)$ has been defined on a Baire $*$-field of functions $\mathfrak{F}$, and for the proof of the theorem it is necessary to verify that it is continuous, i.e. to show that for every sequence $\{G_n(\tau)\}$ in $\mathfrak{F}$ which satisfies

$$G_k(\tau) \geqslant G_{k+1}(\tau), \quad \lim_{k\to\infty} G_k(\tau) = 0, \tag{13}$$

we have

$$\lim_{k\to\infty} I(G_k) = 0. \tag{14}$$

In the case $\mathfrak{F} = \mathfrak{G}(\mathfrak{B})$ ($\mathfrak{B}$ is the domain of the measure $\mu(M)$) which we consider first we have for any positive ε, the estimate

$$0 \leqslant I(G_k) \leqslant \gamma\mu(N_k) + \varepsilon\mu(T), \tag{15}$$

where $\gamma = \sup_{\tau} G_1(\tau)$ and $N_k = T(G_k > \varepsilon)$. Indeed,

$$I(G_k) = \int_{N_k} G_k(\tau)\,\mu(M_\tau) + \int_{T\setminus N_k} G_k(\tau)\,\mu(M_\tau) \leqslant$$
$$\leqslant \sup_{\tau} G_k(\tau)\,\mu(N_k) + \varepsilon\mu(T \setminus N_k) \leqslant \gamma\mu(N_k) + \varepsilon\mu(T).$$

On the other hand, it follows from (13) that

$$N_k \supset N_{k+1}, \quad \bigcap_k N_k = \varnothing,$$

and hence $\lim \mu(N_k) = 0$ by virtue of the countable additivity of the measure $\mu(M)$. Therefore, the estimate (15) gives

$$0 \leqslant \lim I(G_k) = \inf_k \{I(G_k)\} \leqslant \varepsilon\mu(T),$$

whence (14) follows for the case $\mathfrak{F} = \mathfrak{G}(\mathfrak{B})$, since $\varepsilon > 0$ was arbitrary. In the general case, $\mathfrak{F} = \mathfrak{G}_\mu(\mathfrak{B})$, we choose for the function $G_1(\tau) \in \mathfrak{G}_\mu(\mathfrak{B})$ an admissible sequence $\{\vartheta_n(\tau)\}$ (respectively, $\{M_n\}$), and by using it we estimate the integral $I(G_k)$ $(G_k(\tau) \leqslant G_1(\tau))$. We have

$$0 \leqslant I(G_k) = I(\vartheta_n G_k) + I((1 - \vartheta_n) G_k) \leqslant I(\vartheta_n G_k) + I((1 - \vartheta_n) G_1).$$

Here, the functions $\vartheta_n(\tau) G_k(\tau) = G_k'(\tau) \in \mathfrak{G}(\mathfrak{B})$ satisfy conditions similar to (13), and hence, by what has already been proved, $\lim_{k\to\infty} I(\vartheta_n G_k) = 0$, and therefore

$$0 \leqslant \lim_{k\to\infty} I(G_k) \leqslant I((1 - \vartheta_n) G_1) = \int_{T\setminus M_n} G_1(\tau)\,\mu(M_\tau). \tag{16}$$

Since the indefinite integral is countably additive, the right-hand side of (16) converges to zero since

$$T \setminus M_n \supset T \setminus M_{n+1}, \quad \bigcap_{n=1}^{\infty} (T \setminus M_n) = \varnothing.$$

Therefore, (16) gives (14) for $\mathfrak{F} = \tilde{\mathfrak{G}}_\mu(\mathfrak{B})$.

COROLLARY. *If $\{\vartheta_n(\tau)\}$ is an admissible sequence for a function $F(\tau) \in \mathfrak{G}_\mu(\mathfrak{B})$, then*

$$\lim_{n\to\infty} I(\vartheta_n F) = I(F).$$

REMARK. Theorems 2.6.1 and 2.6.4 establish a correspondence, unique in one direction, between positive Lebesgue operations on *regular* Baire fields of functions and measures on *regular* Borel fields of sets.

The following lemma and theorem supplement Theorem 2.6.4.

LEMMA 2.6.2. *Let a sequence $\{F_n(\tau)\}$ in $\mathfrak{G}_\mu(\mathfrak{B})$ converge to the function $F(\tau)$ and let*

$$F_n(\tau) \geqslant 0, \quad I(F_n) \leqslant \eta \in Z. \tag{17}$$

Then $F(\tau)$ belongs to $\mathfrak{G}_\mu(\mathfrak{B})$ and

$$I(F) \leqslant \eta. \tag{18}$$

In fact, the functions $F(\tau)$ and $\Phi(\tau) = \sup_n \{F_n(\tau)\}$ are ($\mathfrak{B}$)-measurable, and a sequence $\{\vartheta_k(\tau)\}$ which is admissible for $\Phi(\tau)$ is obviously admissible for $F(\tau)$ and for all the functions $F_n(\tau)$. Therefore the functions $\vartheta_k(\tau)F_n(\tau)$ and their limits (for $n\to\infty$) $\vartheta_k(\tau)F(\tau)$, belong to $\mathfrak{G}(\mathfrak{B})$, so that by

$$\vartheta_k(\tau)F_n(\tau) \leqslant \vartheta_k(\tau)\Phi(\tau) \in \mathfrak{G}(\mathfrak{B}) \subset \mathfrak{G}_\mu(\mathfrak{B})$$

and by Theorem 2.6.4, we have

$$\lim_{n\to\infty} I(\vartheta_k F_n) = I(\vartheta_k F).$$

Therefore $I(\vartheta_k F_n) \leqslant I(F_n) \leqslant \eta$ (see (17)) implies that

$$I(\vartheta_k F) \leqslant \eta \in Z,$$

which, according to Lemma 2.6.1, is the criterion that $F(\tau)$ belong to $\mathfrak{G}_\mu(\mathfrak{B})$. By virtue of the continuity of the integral (Theorem 2.6.4) for $k\to\infty$, we have (18).

THEOREM 2.6.5. *For a non-decreasing sequence $\{F_n(\tau)\}$ in $\mathfrak{G}_\mu(\mathfrak{B})$, let the limits* $\lim F_n(\tau)$ *and* $\lim\limits_{n\to\infty} I(F_n)$ *exist. Then the limit* $F(\tau) = \lim F_n(\tau)$ *belongs to $\mathfrak{G}_\mu(\mathfrak{B})$, and*

$$\lim I(F_n) = I(F). \tag{19}$$

Indeed, we apply Lemma 2.6.2 to the sequence $\{F_n'(\tau)\} = \{F_n(\tau) - F_1(\tau)\}$, and hence the limit $F'(\tau) = \lim F_n'(\tau) = F(\tau) - F_1(\tau)$ belongs to $\mathfrak{G}_\mu(\mathfrak{B})$, and when $F'(\tau)$ belongs to $\mathfrak{G}_\mu(\mathfrak{B})$, so does the function $F(\tau) = F'(\tau) + F_1(\tau)$.

Equation (19) now follows from the continuity of the integral, i.e. from Theorem 2.6.4.

6.4. Sets of measure zero. Sets N in the domain $\mathfrak{B}$ of a measure $\mu(M)$ and which satisfy the condition $\mu(N) = 0$ (*sets of measure zero with respect to* $\mu(M)$) obviously form a Borel subfield $\mathfrak{B}_0$ of the field $\mathfrak{B}$. For a set $N \in \mathfrak{B}_0$, and for any function $F(\tau)$ in $\mathfrak{G}_\mu(\mathfrak{B})$,

$$\int_N F(\tau)\,\mu(\underset{\tau}{M}) = 0,$$

as this follows from the absolute continuity of the indefinite integral. Therefore

$$\int_M F(\tau)\,\mu(\underset{\tau}{M}) = \int_M G(\tau)\,\mu(\underset{\tau}{M}), \tag{20}$$

when the values of the functions $F(\tau)$ and $G(\tau)$ coincide everywhere on the set $M \in \mathfrak{B}$ except perhaps on a set $N \in \mathfrak{B}_0$. In particular, if $G(\tau) = 0$, then

$$\int_M F(\tau)\,\mu(M_\tau) = 0, \tag{21}$$

i.e. the indefinite integral is zero when $F(\tau)$ vanishes everywhere except perhaps on a set $N \in \mathfrak{B}_0$.

Conversely, *if* (21) *holds for a non-negative function* $F(\tau) \in \mathfrak{G}_\mu(\mathfrak{B})$, *then* $F(\tau) = 0$ *on* M *except perhaps on a set* N *of measure zero.*

Indeed, let $N = M(F > 0)$, and let $N_k = M\left(F > \frac{1}{k}\right)$. Then $N_k \subset N_{k+1}$, $N = \bigcup_{k=1}^{\infty} N_k$, and

$$0 \leqslant \frac{\mu(N_k)}{k} \leqslant \int_{N_k} F(\tau)\,\mu(M_\tau) \leqslant \int_M F(\tau)\,\mu(M_\tau) = 0.$$

Thus, $\mu(N_k) = 0$, and hence

$$\mu(N) = \lim_{k \to \infty} \mu(N_k) = 0.$$

From this property of the integral we have:

THEOREM 2.6.6. *Two functions,* $F(\tau)$ *and* $G(\tau)$, *generate the same indefinite integral* $\nu(M)$ *if and only if they coincide to within values on a set of measure zero with respect to* $\mu(M)$.

In fact, the sufficiency of the condition is already known, and if (20) holds for any $M \in \mathfrak{B}$, and if at the same time one of the sets

$$M = T(F > G), \qquad M' = T(G > F)$$

has *positive measure*, then one of the integrals

$$\int_M (F(\tau) - G(\tau))\,\mu(M_\tau), \qquad \int_{M'} (G(\tau) - F(\tau))\,\mu(M_\tau)$$

is different from zero, which contradicts (20).

For $M = T$, (20) becomes $I(F) = I(G)$, which leads to the possibility of extending somewhat the concept of the integral $I(F)$. Namely, we permit functions $F(\tau)$ which are not necessarily defined everywhere on the set T, but are defined except perhaps on some sets of measure zero, or in other words, they are defined *almost everywhere with respect to* $\mu(M)$. I et such a function $F(\tau)$ be defined on $T \setminus N$, where $\mu(N) = 0$, and suppose its definition is completed on N to a function $F'(\tau) \in \mathfrak{G}_\mu(\mathfrak{B})$. By letting

$$I(F) = I(F'), \tag{22}$$

it is easy to see that $I(F)$ is independent of the values which the function

$F'(\tau)$ takes on the set N. Therefore, (22) can serve as the definition of the integral $I(F)$ for functions $F(\tau)$ which are defined almost everywhere on T (with respect to $\mu(M)$).

In a corresponding manner, by neglecting sets of measure zero it is possible to give a wider meaning to the notion of Baire ∗-fields of functions and Lebesgue operations. We will not go into the details of this; however, we note the case of monotone sequences of functions.

THEOREM 2.6.7. *A non-decreasing sequence* $\{F_n(\tau)\}$ *in* $\mathfrak{G}_\mu(\mathfrak{B})$ *converges almost everywhere (with respect to* $\mu(M)$*) to a function* $F(\tau) \in \mathfrak{G}_\mu(\mathfrak{B})$ *if and only if the set* $\{I(F_n)\}$ *is bounded in Z and lim* $I(F_n) = I(F)$.

NECESSITY. Like the sequence $\{F_n(\tau)\}$, the sequence $I(F_n)$ is also non-decreasing, and since $F_n(\tau) \leqslant F(\tau)$, we have $I(F_n) \leqslant I(F)$. Therefore $\lim I(F_n)$ exists. The equation $I(F) = \lim I(F_n)$ is now guaranteed by Theorem 2.6.5.

SUFFICIENCY. Since the sequence $\{I(F_n)\}$ is bounded,

$$I(F_n - F_1) \leqslant \eta \in Z. \tag{17'}$$

If

$$M_{n,k} = T(F_n - F_1 > k), \qquad M_k = \bigcup_{n=1}^{\infty} M_{n,k},$$

then

$$M_{n,k} \subset M_{n+1,k}, \qquad k\mu(M_{n,k}) \leqslant I(F_n - F_1) \leqslant \eta, \qquad M_k \supset M_{k+1}$$

and the intersection $N = \bigcap_{k=1}^{\infty} M_k$ coincides with the set on which the sequence $\{F_n(\tau)\}$ diverges.

Since the measure $\mu(M)$ is countably additive,

$$\mu(M_k) = \lim_{n\to\infty} \mu(M_{n,k}) \leqslant \frac{\eta}{k}, \tag{23}$$

and hence $\mu(N) = 0$, i.e. $\{F_n(\tau)\}$ converges almost everywhere (with respect to $\mu(M)$) on T. Finally, the application of Theorem 2.6.5 to $\lim F_n(\tau)$ on $T \setminus N$ shows that the function $F(\tau)$, which was obtained by an arbitrary extension of the definition of $\lim F_n(\tau)$ to the set N, belongs to $\mathfrak{G}_\mu(\mathfrak{B})$. Moreover, by virtue of the same theorem, $\lim I(F_n) = I(F)$.

REMARK. The functions $F_n(\tau)$ could have been assumed to be defined almost everywhere, and the inequality $F_n(\tau) \leqslant F_{n+1}(\tau)$ could then be understood in the corresponding sense.

Two functions, $F(\tau)$ and $G(\tau)$, in $\tilde{\mathfrak{G}}(\mathfrak{B})$ are said to be *equivalent with respect to a measure* $\mu(M)$ on $\mathfrak{B}$, and we write

$$F(\tau) \equiv G(\tau)(\mu), \tag{24}$$

if $F(\tau) = G(\tau)$ almost everywhere on T with respect to $\mu(M)$.

The congruence (24) defines a linear equivalence relation in the vector $*$-system of each Baire $*$-field of functions $\mathfrak{F}$ which is similar to $\mathfrak{G}(\mathfrak{B})$ $(\mathfrak{G}(\mathfrak{B}) \subset \mathfrak{F} \subset \tilde{\mathfrak{G}}(\mathfrak{B}))$. The collection $\mathring{\mathfrak{F}}$ of types $\mathring{F}$ with respect to the equivalence (24) (end of Sec. 1.5.1) forms a vector $*$-system, and in the lattice $\mathfrak{F}_r$ of real functions in $\mathfrak{F}$, (24) permits us to define a proper order relation, $\mathring{G} \geqslant \mathring{F}$ if $G(\tau) \geqslant F(\tau)$ almost everywhere with respect to $\mu(M)$, where $F(\tau)$ and $G(\tau)$ are representatives of the types $\mathring{G}$ and $\mathring{F}$. This order relation converts the set of Hermitian elements in $\mathring{\mathfrak{F}}$ (of types of real functions in $\mathfrak{F}$) into a lattice, and hence $\mathring{\mathfrak{F}}$ is a vector $*$-lattice. We will show that it is σ-closed.

In fact, let

$$\mathring{F}_n \in \mathring{\mathfrak{F}}, \quad \mathring{F}_n \leqslant \mathring{F}_{n+1} \leqslant \mathring{\Phi} \in \mathring{F}.$$

Then functions $F_n(\tau) \in \mathfrak{F}$ and $\Phi(\tau) \in \mathfrak{F}$ can be found for which

$$F_n(\tau) \leqslant F_{n+1}(\tau), \qquad F_n(\tau) \leqslant \Phi(\tau)$$

almost everywhere. Since sets of measure zero form a Borel field, the last relations are simultaneously satisfied almost everywhere in T with respect to the measure $\mu(M)$. Consequently, the limit

$$\lim F_n(\tau) = F(\tau)$$

exists almost everywhere on T. Without loss of generality, it is possible to assume that this limit exists for all $\tau \in T$, so that $F(\tau) \in \mathfrak{F}$, and

$$\lim \mathring{F}_n = \mathring{F} \in \mathring{\mathfrak{F}}. \tag{25}$$

For $\mathfrak{F} \subset \tilde{\mathfrak{G}}_\mu(\mathfrak{B})$, if we let

$$\mathring{I}(\mathring{F}) = I(F),$$

where $F(\tau)$ is a representative of the type $\mathring{F}$, we uniquely define on $\mathring{\mathfrak{F}}$ a positive linear operation $\mathring{I}(\mathring{F})$ which is continuous in the sense that, by (25),

$$\lim \mathring{I}(\mathring{F}_n) = \mathring{I}(\mathring{F}). \tag{26}$$

In reference to the functions $F(\tau) = \chi_M(\tau)$, where $M \in \mathfrak{B}$, we will let

$$\mathring{M} = \mathring{\chi}_M, \quad \mathring{\mu}(\mathring{M}) = \mathring{I}(\mathring{\chi}_M) = \mu(M) \tag{27}$$

and we will say that $\mathring{M}$ is the *type of the set M with respect to μ*. A regular (Borel) field of sets $\mathfrak{B}$ is a σ-closed Boolean algebra; the collection $\mathring{\mathfrak{B}}$ of types $\mathring{M}$ of sets $M \in \mathfrak{B}$ is also a σ-closed Boolean algebra which is a homomorphic image of the Boolean algebra $\mathfrak{B}$. The function $\mathring{\mu}(\mathring{M})$, defined by (27) in the Boolean algebra $\mathring{\mathfrak{B}}$, is:

1) additive, i.e.

$$\mathring{\mu}(\mathring{M}_1 \cup \mathring{M}_2) = \mathring{\mu}(\mathring{M}_1) + \mathring{\mu}(\mathring{M}_2),$$

if $\mathring{M}_1 \cap \mathring{M}_2 = \varnothing$, or in the general case

$$\mathring{\mu}(\mathring{M}_1 \cup \mathring{M}_2) + \mathring{\mu}(\mathring{M}_1 \cap \mathring{M}_2) = \mathring{\mu}(\mathring{M}_1) + \mathring{\mu}(\mathring{M}_2);$$

2) positive, i.e. $\mathring{\mu}(\mathring{M}) > 0$ for $\mathring{M} \neq 0$ (indeed, the equation $\mathring{\mu}(\mathring{M}) = 0$ is equivalent to $\mu(M) = 0$ for a representative M of the type $\mathring{M}$, and hence $\mathring{M} = 0$ if $\mathring{\mu}(\mathring{M}) = 0$);

3) continuous, in the sense that

$$\lim \mathring{\mu}(\mathring{M}_n) = \mathring{\mu}(\mathring{M}), \tag{28}$$

if $\mathring{M}_n \leqslant \mathring{M}_{n+1}$ and if $\mathring{M} = \sup\{\mathring{M}_n\}$ (in fact, (28) is a special case of (26)).

We note that the functions $F(\tau)$ and $G(\tau)$ in (24) can be assumed to be defined almost everywhere on T (with respect to $\mu(M)$), and only their extension to all of T would belong to $\tilde{\mathfrak{G}}(\mathfrak{B})$. Correspondingly, if the families $\tilde{\mathfrak{G}}(\mathfrak{B})$, $\mathfrak{G}(\mathfrak{B})$, $\mathfrak{F}$, and $\mathfrak{G}_\mu(\mathfrak{B})$ are extended by admitting functions which are defined almost everywhere on T, then this modification has no effect on the collection of types $\mathring{\mathfrak{F}}$.

6.5. The transformation of a measure and an integral. Let a measure $\mu(M)$ be defined on a regular Borel field $\mathfrak{B} \subset \mathfrak{P}(T)$, and let

$$\tau' = \Phi(\tau) \tag{29}$$

be a mapping of the set T into some set T'. To any set $M' \subset T'$ there corresponds the (complete) inverse image $M = \Phi^{-1}(M')$, i.e. the collection of all elements τ in T whose images under the mapping (29) belong to M'.

The system $\mathfrak{B}_{T'}$ of subsets of a set T' whose complete inverse images belong to the Borel field $\mathfrak{B}$ is again a regular Borel field, and the formula

$$\mu'(M') = \mu(M),$$

where M is the complete inverse image of the set $M' \in \mathfrak{B}_{T'}$, defines a measure $\mu'(M')$ on $\mathfrak{B}_{T'}$.

In fact, the inverse image of T' equals T, and since $T \in \mathfrak{B}$, we have $T' \in \mathfrak{B}_{T'}$. If M_k and M are, respectively, the inverse images of the sets M'_k and M' in $\mathfrak{B}_{T'}$, then $\bigcup_k M_k$ and $T \setminus M$ are the inverse images of the sets $\bigcup_k M'_k$ and $T' \setminus M'$. Hence, whenever $\mathfrak{B}$ is a regular Borel field, $\mathfrak{B}_{T'}$ is also a regular Borel field. Moreover, if the sets M'_k in $\mathfrak{B}_{T'}$ are disjoint, if $M' = \bigcup_k M'_k$, and if $M = \bigcup_k M_k$ is the inverse image of the set M', then the sets M_k also are disjoint, and the equation

$$\mu(M) = \sum_{k=1}^{\infty} \mu(M_k),$$

which holds for the measure $\mu(M)$, is equivalent to

$$\mu'(M') = \sum_{k=1}^{\infty} \mu'(M'_k).$$

Thus, the non-negative function $\mu'(M')$ is σ-additive, i.e. $\mu'(M')$ is a measure on $\mathfrak{B}_{T'}$.

Now let $\mathfrak{B}'$ be any regular Borel subfield of the field $\mathfrak{B}_{T'}$. The measure $\mu'(M')$ induces a measure on $\mathfrak{B}'$. Each measure obtained from the measure $\mu(M)$ in this manner will be called a *transformation of the measure* $\mu(M)$ *under the mapping* (29), and we use the notation $\mu'(M')$, which was already used for the "maximal" of these transformed measures.

We consider an integral on $\mathfrak{G}_{\mu'}(\mathfrak{B}')$ with respect to the measure $\mu'(M')$ on $\mathfrak{B}'$:

$$I'(F') = \int_{T'} F'(\tau')\,\mu'(M'_{\tau'}),$$

and we also consider the system $\tilde{\mathfrak{G}}(\mathfrak{B}')$ of $(\mathfrak{B}')$-measurable functions.

LEMMA 2.6.3. *If a function* $F'(\tau')$ *belongs to* $\tilde{\mathfrak{G}}(\mathfrak{B}')$, *or to* $\mathfrak{G}_{\mu'}(\mathfrak{B}')$, *then*

$$F(\tau) = F'(\Phi(\tau)) \tag{30}$$

belongs to $\tilde{\mathfrak{G}}(\mathfrak{B})$ *or* $\mathfrak{G}_{\mu}(\mathfrak{B})$, *respectively.*

Indeed, let $\chi(\tau) = \chi'(\Phi(\tau))$ be the characteristic function of the inverse image M of the set $M' \in \mathfrak{B}'$ whose characteristic function is $\chi'(\tau)$. The mapping (30) transforms the representation $\varphi'(\tau) = \sum \alpha_k \chi'_k(\tau')$ of $\varphi'(\tau') \in \tilde{\mathfrak{E}}(\mathfrak{B}')$, with respect to the $(\mathfrak{B}')$-partition $T' = \cup N'_k$, into the representation

$$\varphi(\tau) = \varphi'(\Phi(\tau)) = \sum \alpha_k \chi_k(\tau)$$

of the function $\varphi(\tau) \in \tilde{\mathfrak{E}}(\mathfrak{B})$ with respect to the $(\mathfrak{B})$-partition $T = \cup N_k$, where the N_k are inverse images of the sets N'_k. If, in addition, $\varphi'(\tau') \in \tilde{\mathfrak{E}}_{\mu'}(\mathfrak{B}')$, then since $\mu'(N'_k) = \mu(N_k)$, the series $\sum_k |\alpha_k| \mu(N_k)$ converges, i.e. $\varphi(\tau)$ belongs to $\tilde{\mathfrak{E}}_{\mu}(\mathfrak{B})$. By forming the closures of $\tilde{\mathfrak{E}}(\mathfrak{B}')$ and $\tilde{\mathfrak{E}}_{\mu'}(\mathfrak{B}')$ with respect to uniform convergence, i.e. by forming the families of functions $\tilde{\mathfrak{G}}(\mathfrak{B}')$ and $\mathfrak{G}_{\mu'}(\mathfrak{B}')$, we see that the mapping (30) transforms them into $\tilde{\mathfrak{G}}(\mathfrak{B})$ and $\mathfrak{G}_{\mu}(\mathfrak{B})$, respectively.

If $\varphi'(\tau') \in \tilde{\mathfrak{E}}_{\mu'}(\mathfrak{B}')$, then

$$\sum_k \alpha_k \mu'(N'_k) = \sum_k \alpha_k \mu(N_k),$$

i.e. $I'(\varphi') = I(\varphi)$, where $\varphi(\tau) = \varphi'(\Phi(\tau))$. On passing from $\tilde{\mathfrak{E}}_{\mu'}(\mathfrak{B}')$ to $\mathfrak{G}_{\mu'}(\mathfrak{B}')$, we obtain the equation $I'(F') = I(F)$ for any function $F'(\tau') \in \mathfrak{G}_{\mu'}(\mathfrak{B}')$, and for its image $F(\tau)$ under the mapping (30). Thus, we have:

THEOREM 2.6.8. *Let the functions* $F'(\tau')$ *and* $F(\tau)$ *be connected by* (30), *and let* $F'(\tau')$ *belong to* $\mathfrak{G}_{\mu'}(\mathfrak{B}')$. *Then* $F(\tau)$ *belongs to* $\mathfrak{G}_{\mu}(\mathfrak{B})$ *and* $I'(F') = I(F)$, *i.e.*

$$\int_{T'} F'(\tau')\mu'(M'_{\tau'}) = \int_{T} F'(\Phi(\tau))\mu(M_{\tau}). \tag{31}$$

If a regular Borel field $\mathfrak{B}' \subset \mathfrak{P}(T')$ has been defined on the set T', then the mapping (29) of T into T' is said to be $(\mathfrak{B} \to \mathfrak{B}')$-*measurable* if the inverse image of the set $M' \in \mathfrak{B}'$ is $(\mathfrak{B})$-measurable. By letting

$$\mu'(M') = \mu(M), \qquad M' \in \mathfrak{B}', \qquad M = \Phi^{-1}(M'), \tag{32}$$

for any $(\mathfrak{B} \to \mathfrak{B}')$-measurable mapping, we obtain a measure $\mu'(M')$ on $\mathfrak{B}'$ which is one of the transformations of the measure $\mu(M)$ under the mapping (29).

We now assume that a measure $\mu'(M')$ has been defined on a Borel field $\mathfrak{B}' \in \mathfrak{P}(T')$, that (29) is a one-to-one mapping of T onto T', that the mapping Φ is $(\mathfrak{B} \to \mathfrak{B}')$-measurable, that the inverse mapping Φ^{-1} is $(\mathfrak{B}' \to \mathfrak{B})$-measurable, and moreover, that (32) is satisfied. Under all of these conditions we will say that the mapping Φ is an *isomorphism of the measures* $\mu(M)$ *and* $\mu'(M')$, and we will say that the measures are *isomorphic*.

By neglecting sets of measure zero, the concept of the isomorphism of two measures can be given a more general meaning. For example, we can consider that the mappings Φ and Φ^{-1} have been defined after removing sets of measure zero from T and T' (in T with respect to $\mu(M)$ and in T' with respect to $\mu'(M')$). Another possibility is the transition to the Boolean algebras $\mathring{\mathfrak{B}}$ and $\mathring{\mathfrak{B}}'$, and to the continuous additive functions $\mathring{\mu}(\mathring{M})$ and $\mathring{\mu}'(\mathring{M}')$ on $\mathring{\mathfrak{B}}$ and $\mathring{\mathfrak{B}}'$ which are generated by the measures $\mu(M)$ and $\mu'(M')$. Namely, we say that two measures $\mu(M)$ and $\mu'(M')$ are *isomorphic* (in this new sense) if an isomorphism can be established between the Boolean algebras $\mathring{\mathfrak{B}}$ and $\mathring{\mathfrak{B}}'$ such that $\mathring{\mu}(\mathring{M}) = \mathring{\mu}'(\mathring{M}')$ when M and M' correspond to one another.

6.6. A generalization of the concept of integral. Let $\mu_1(M)$, $\mu_2(M)$, $\mu_3(M)$, and $\mu_4(M)$ be measures (or non-negative additive functions), defined on the same regular Borel field of sets $\mathfrak{B}$ (or field $\mathfrak{R}$), with values in a complete vector *-order Z. These measures (functions) generate the function

$$\mu(M) = \mu_1(M) - \mu_2(M) + i\mu_3(M) - i\mu_4(M),$$

which is defined, and which is σ-additive, on the Borel field $\mathfrak{B}$ (additive on the field $\mathfrak{R}$). The integral $I(F)$ with respect to $\mu(M)$ is defined by the formula

$$I(F) = I_1(F) - I_2(F) + iI_3(F) - iI_4(F),$$

where the $I_k(F)(k = 1, 2, 3, 4)$ are integrals with respect to the measures (functions) $\mu_k(M)$.

The integral $I(F)$ with respect to $\mu(M)$ is defined on the intersection

$$\mathfrak{G}_\mu(\mathfrak{B}) = \bigcap_{k=1}^{4} \mathfrak{G}_{\mu_k}(\mathfrak{B}),$$

which is again a Baire $*$-field of functions and which contains $\mathfrak{G}(\mathfrak{B})$. If the $I_k(F)$ are integrals on $\mathfrak{G}(\mathfrak{R})$, then the integral $I(F)$ is also defined on $\mathfrak{G}(\mathfrak{R})$.

§ 7. Upper and Lower Operations

7.1. The operation $\bar{L}(F)$. *Any positive linear operation $L(F)$ which is defined on a field or $*$-field of functions $\mathfrak{F}$, and whose range is in a complete vector order or $*$-order Z, can be extended to a positive Lebesgue operation if it satisfies the obvious necessary condition* (2.5.6–7), *and if, in addition, it satisfies one of the conditions* I *or* II *of Section* 1.9.2.

The proof of this important assertion will not be completed until § 9 (Theorems 2.9.1 and 2.9.2).

In the intermediate stage of the extension of the operation $L(F)$, and namely for a σ-extension of its domain, the statement above loses its vector character. Therefore, we will first assume that *the range of the operation $L(F)$ is not a field, but a semifield of functions $\mathfrak{F}$.*

An operation $L(F)$, defined on a semifield of (real) functions $\mathfrak{F} \subset \tilde{\mathfrak{G}}_r(T)$, is by definition *linear* if

$$L(F+G) = L(F) + L(G), \quad L(\alpha F) = \alpha L(F), \qquad \alpha \geqslant 0,$$

for functions $F(\tau)$ and $G(\tau)$ in $\mathfrak{F}$. It is *monotone* if

$$L(F) \geqslant L(G)$$

for $F(\tau) \geqslant G(\tau)$.

We let $[\mathfrak{F}]$ denote the collection of functions $F(\tau)$ in $\tilde{\mathfrak{G}}_r(T)$ which satisfy an inequality of the form

$$G'(\tau) \leqslant F(\tau) \leqslant F'(\tau) \tag{1}$$

for some $G'(\tau)$ and $F'(\tau)$ in $\mathfrak{F}$.

Like $\mathfrak{F}$, $[\mathfrak{F}]$ is also a field or semifield of functions, and moreover, $[\mathfrak{F}] \supset \mathfrak{F}$. For example, $[\mathfrak{F}]$ coincides with $\mathfrak{G}_r(T)$ if $\mathfrak{F}$ is any regular field consisting of bounded functions.

For a monotone linear operation $L(F)$ on a semifield $\mathfrak{F}$, the formula

$$\hat{L}(F) = \inf L(F'), \qquad F(\tau) \leqslant F'(\tau) \in \mathfrak{F} \tag{2}$$

defines an operation $\hat{L}(F)$ on the family of function $[\mathfrak{F}]$, since for any $F(\tau) \in [\mathfrak{F}]$, the lower bound in (2) exists. In fact,

1) by virtue of (1), the set $\{L(F')\}$ is not empty and is bounded below, since by the monotonicity of $L(F)$, we have $L(F') \geqslant L(G')$ for a fixed function $G'(\tau)$ in $\mathfrak{F}$;

2) the set $\{L(F')\}$ is directed below, since if

$$F(\tau) \leqslant F_1'(\tau) \in \mathfrak{F}, \quad F(\tau) \leqslant F_2'(\tau) \in \mathfrak{F},$$

then the function $F'(\tau) = \inf\{F_1', F_2'\} \in \mathfrak{F}$ satisfies the condition $F(\tau) \leqslant F'(\tau)$, and since $L(F)$ is monotone,

$$L(F') \leqslant L(F_1'), \quad L(F') \leqslant L(F_2').$$

From the monotonicity of $L(F)$, and the definition of the operation $\widehat{L}(F)$, it follows that $\widehat{L}(F)$ is monotone, i.e.

$$\widehat{L}(F) \geqslant \widehat{L}(G) \quad \text{for} \quad F(\tau) \geqslant G(\tau).$$

The operation $\widehat{L}(F)$ *is an extension of the operation* $L(F)$, *and for any* $F(\tau)$ *and* $G(\tau)$ *in* $[\mathfrak{F}]$, *it satisfies the inequality*

$$\widehat{L}(F+G) = \widehat{L}(\Phi + \Psi) \leqslant \widehat{L}(\Phi) + \widehat{L}(\Psi) \leqslant \widehat{L}(F) + \widehat{L}(G), \tag{3}$$

where $\Phi(\tau) = \sup\{F, G\}$, $\Psi(\tau) = \inf\{F, G\}$. *It is also positive homogeneous*:

$$\widehat{L}(\alpha F) = \alpha \widehat{L}(F), \quad \alpha \geqslant 0. \tag{4}$$

Indeed, if $F(\tau) \in \mathfrak{F}$, then $L(F)$ is the smallest element of the set $\{L(F')\}$, and $\widehat{L}(F) = L(F)$. In particular, $\widehat{L}(0) = L(0) = 0$. Now let

$$F(\tau) \leqslant F'(\tau) \in \mathfrak{F}, \quad G(\tau) \leqslant G'(\tau) \in \mathfrak{F}, \quad \Phi(\tau) \leqslant \Phi''(\tau) \in \mathfrak{F},$$
$$\Psi(\tau) \leqslant \Psi''(\tau) \in \mathfrak{F}, \quad \Phi' = \sup\{F', G'\}, \quad \Psi' = \inf\{F', G'\}.$$

Then

$$F(\tau) + G(\tau) = \Phi(\tau) + \Psi(\tau) \leqslant \Phi''(\tau) + \Psi''(\tau) \in \mathfrak{F}$$

and hence

$$\widehat{L}(F+G) \leqslant L(\Phi'' + \Psi'') = L(\Phi'') + L(\Psi''). \tag{5}$$

Since $\Phi'(\tau) \geqslant \Phi(\tau)$, $\Psi'(\tau) \geqslant \Psi(\tau)$, and $\Phi'(\tau) + \Psi'(\tau) = F'(\tau) + G'(\tau)$, we have

$$\widehat{L}(\Phi) + \widehat{L}(\Psi) \leqslant L(\Phi') + L(\Psi') = L(F') + L(G'). \tag{6}$$

In the right-hand sides of (5) and (6) it is possible to pass to the lower bounds of $L(\Phi'')$, $L(\Psi'')$, $L(F')$, and $L(G')$, which gives

$$\widehat{L}(F+G) \leqslant \widehat{L}(\Phi) + \widehat{L}(\Psi)$$

and

$$\widehat{L}(\Phi) + \widehat{L}(\Psi) \leqslant \widehat{L}(F) + \widehat{L}(G),$$

i.e. (3). Finally, (4) follows from definition (2) of the operation $\widehat{L}(F)$, and from (1.7.2′) for the lower bound.

In particular, if $\mathfrak{F}$ is a field of functions, then $[\mathfrak{F}]$ is also a field of functions, and whenever $F(\tau)$ belongs to $[\mathfrak{F}]$, the function $-F(\tau)$ also belongs to $[\mathfrak{F}]$. Therefore, in this case (3) applies for $F(\tau)\in[\mathfrak{F}]$ and $G(\tau) = -F(\tau)$, and this gives

$$0=\hat{L}(0)\leqslant\hat{L}(F)+\hat{L}(-F)$$

or

$$\check{L}(F)\leqslant\hat{L}(F), \tag{7}$$

where by definition,

$$\check{L}(F)=-\hat{L}(-F). \tag{8}$$

If $F(\tau)\in\mathfrak{F}$, where $\mathfrak{F}$, as above, is a field, then $-F(\tau)\in\mathfrak{F}$, and hence

$$\check{L}(F)=-\hat{L}(-F)=-L(-F)=L(F),$$

i.e. in the case under consideration the *operation* $\check{L}(F)$ *is also an extension of the operation* $L(F)$.

In the case when $\mathfrak{F}$ is a semifield, we will also define the operation $\check{L}(F)$. With this aim, we let $-\mathfrak{F}$ denote the collection of functions $-F(\tau)$, where $F(\tau)\in\mathfrak{F}$. The collection $-\mathfrak{F}$ is also a semifield of functions, and if $F(\tau)\in[-\mathfrak{F}]$, then $-F(\tau)\in[\mathfrak{F}]$. We define the operation $\check{L}(F)$ on the semifield $[-\mathfrak{F}]$ by (8).

The operation $\check{L}(F)$ *is monotone, positive homogeneous, and satisfies the inequality*

$$\check{L}(F+G)\geqslant\check{L}(\Phi)+\check{L}(\Psi)\geqslant\check{L}(F)+\check{L}(G), \tag{3'}$$

where the functions $\Phi(\tau)$ *and* $\Psi(\tau)$ *are defined in the same way as in* (3).

Indeed, if $F(\tau)\geqslant G(\tau)$ for functions $F(\tau)$ and $G(\tau)$ in $[-\mathfrak{F}]$, then $-F(\tau)$ and $-G(\tau)$ belong to $[\mathfrak{F}]$, $-G(\tau)\geqslant-F(\tau)$. $\hat{L}(-G)\geqslant\hat{L}(-F)$ in view of the monotonicity of $\hat{L}(F)$, and hence $\check{L}(F)\geqslant\check{L}(G)$. For $\alpha\geqslant0$ and $F(\tau)\in[-\mathfrak{F}]$, we have by virtue of (4),

$$\check{L}(\alpha F)=-\hat{L}(-\alpha F)=-\alpha\hat{L}(-F)=\alpha\check{L}(F), \tag{4'}$$

i.e. the operation $\check{L}(F)$ is positive homogeneous. Finally, (3′) is obtained by application of (3) to the functions $-F(\tau)$ and $-G(\tau)$ in $\mathfrak{F}$. Since sup $(-F, -G) = -\inf\{F, G\} = -\Psi(\tau)$, and $\inf\{-F, -G\} = -\Phi(\tau)$, then

$$\hat{L}(-(F+G))\leqslant\hat{L}(-\Psi)+\hat{L}(-\Phi)\leqslant\hat{L}(-F)+\hat{L}(-G),$$

whence after the multiplication by -1, we obtain (3′).

We note that (7) applies only for functions $F(\tau)$ in the intersection $\mathfrak{F}_0 = [\mathfrak{F}]\cap[-\mathfrak{F}]$.* For $F(\tau)\in\mathfrak{F}_0$, the functions $F(\tau)$ and $G(\tau) = -F(\tau)$ belong to $[\mathfrak{F}]$, and hence

* This intersection can have the function $F(\tau) = 0$ as its unique element.

$$\widehat{L}(F)+\widehat{L}(-F)\geqslant 0,$$

which is equivalent to (7).

We let $\overline{\mathfrak{F}}$ denote the collection of those functions $F(\tau)$ of the family $\mathfrak{F}_0=[\mathfrak{F}]\cap[-\mathfrak{F}]$ for which

$$\breve{L}(F)=\widehat{L}(F), \tag{9}$$

and we define an operation on $\overline{\mathfrak{F}}$ by

$$\overline{L}(F)=\widehat{L}(F)=\breve{L}(F), \qquad F(\tau)\in\overline{\mathfrak{F}}.$$

THEOREM 2.7.1. *The collection $\overline{\mathfrak{F}}$ is a field of functions, and $\overline{L}(F)$ is a positive linear operation on $\overline{\mathfrak{F}}$. If $\mathfrak{F}$ is already a field of functions, then $\overline{L}(F)$ is an extension of the operation $L(F)$.*

In fact, let $F(\tau)$ and $G(\tau)$ belong to $\overline{\mathfrak{F}}$. Then the functions

$$F(\tau)+G(\tau),\ \Phi(\tau)=\sup\{F,G\},\quad \Psi(\tau)=\inf\{F,G\} \tag{10}$$

are contained in $\mathfrak{F}_0=[\mathfrak{F}]\cap[-\mathfrak{F}]$, since $\mathfrak{F}_0$ is a semifield of functions and since $\overline{\mathfrak{F}}\subset\mathfrak{F}_0$. Therefore, we can apply both (3) and (3′) to the functions (10), and this gives

$$\begin{aligned}\widehat{L}(F+G)\leqslant\widehat{L}(\Phi)+\widehat{L}(\Psi)\leqslant\overline{L}(F)+\overline{L}(G)\leqslant\\ \leqslant\breve{L}(\Phi)+\breve{L}(\Psi)\leqslant\breve{L}(F+G)\end{aligned} \tag{11}$$

and, in particular,

$$\widehat{L}(F+G)\leqslant\breve{L}(F+G),\qquad \widehat{L}(\Phi)+\widehat{L}(\Psi)\leqslant\breve{L}(\Phi)+\breve{L}(\Psi).$$

Whence it follows from (7) that

$$\begin{aligned}\widehat{L}(F+G)=\breve{L}(F+G)=\overline{L}(F+G),\\ \widehat{L}(\Phi)=\breve{L}(\Phi)=\overline{L}(\Phi),\ \widehat{L}(\Psi)=\breve{L}(\Psi)=\overline{L}(\Psi)\end{aligned}$$

and thus $F(\tau)+G(\tau)\in\overline{\mathfrak{F}}$, $\Phi(\tau)\in\overline{\mathfrak{F}}$, and $\Psi(\tau)\in\overline{\mathfrak{F}}$. Moreover, it now follows from (11) that

$$\overline{L}(F+G)=\overline{L}(F)+\overline{L}(G), \tag{12}$$

and (4) and (4′) show that for $F(\tau)\in\overline{\mathfrak{F}}$ and $\alpha\geqslant 0$,

$$\widehat{L}(\alpha F)=\alpha\widehat{L}(F)=\alpha\breve{L}(F)=\breve{L}(\alpha F),$$

i.e. $\alpha F\in\overline{\mathfrak{F}}$ and

$$\overline{L}(\alpha F)=\alpha\overline{L}(F). \tag{13}$$

Finally, the condition (9) ($F(\tau)$ belongs to $\overline{\mathfrak{F}}$), i.e. the condition

$$\widehat{L}(F)+\widehat{L}(-F)=0 \quad (F(\tau)\in[\mathfrak{F}]\cap[-\mathfrak{F}]),$$

shows that whenever $F(\tau)$ belongs to $\bar{\mathfrak{F}}$, $-F(\tau)$ also belongs to $\bar{\mathfrak{F}}$, and

$$\bar{L}(-F) = -\bar{L}(F). \tag{14}$$

Thus, in addition to $F(\tau)$ and $G(\tau)$, the family $\bar{\mathfrak{F}}$ contains the functions (10) and $\alpha F(\tau)$ for any real α, i.e. $\bar{\mathfrak{F}}$ is a field of functions. By virtue of (12), (13), (14), and the monotonicity of the operation $\hat{L}(F)$, the operation $\bar{L}(F)$ is a positive linear operation on $\bar{\mathfrak{F}}$. If $\mathfrak{F}$ is a field of functions, then not only $\hat{L}(F)$ but also $\check{L}(F)$ (and this means $\bar{L}(F)$) is an extension of the operation $L(F)$.

In the case when $L(F)$ is a positive linear operation on a field of functions $\mathfrak{F}$, we will call $\hat{L}(F)$ and $\check{L}(F)$, respectively, the *upper* and *lower operations* with respect to $L(F)$. In particular, if $L(F) = I_r(F)$, where the operation $I_r(F)$ is the real integral induced by the integral $I(F)$ (see the remark in Sec. 2.6.1) on the regnlar field $\mathfrak{G}_r(\mathfrak{R})$ of real functions in $\mathfrak{G}(\mathfrak{R})$, then the operations $\hat{I}_r(F)$ and $\check{I}_r(F)$ are by definition the *upper* and *lower integral*, respectively.

7.2. The completion of an operation $L(F)$. An operation $\bar{L}(F)$ is by definition the *completion* of a positive linear operation $L(F)$ if the latter has been defined on a field of functions $\mathfrak{F}$. Such an operation $L(F)$ is said to be *complete* if it coincides with its completion $\bar{L}(F)$.

THEOREM 2.7.2. *The completion $\bar{L}(F)$ is the smallest extension of the operation $L(F)$ to a complete operation.*

We show first that $\bar{L}(F)$ *is a complete operation.* For any function $F(\tau) \in [\bar{\mathfrak{F}}] = [\mathfrak{F}]$* we have

$$\hat{\bar{L}}(F) = \hat{L}(F). \tag{15}$$

Indeed, if $F(\tau) \leqslant F'(\tau) \in \bar{\mathfrak{F}}$, then

$$\hat{L}(F) \leqslant \hat{L}(F') = \bar{L}(F')$$

and passing to the lower bound in the right-hand side gives

$$\hat{L}(F) \leqslant \hat{\bar{L}}(F). \tag{16}$$

On the other hand, from the definition of the upper operations $\hat{L}(F)$ and $\hat{\bar{L}}(F)$, and the inclusion $\bar{\mathfrak{F}} \supset \mathfrak{F}$, the inequality $\hat{\bar{L}}(F) \leqslant \hat{L}(F)$ follows, which together with (16) proves (15). Equation (15) is equivalent to the equation

$$\check{\bar{L}}(F) = \check{L}(F), \qquad F(\tau) \in [\mathfrak{F}],$$

which together with (15) shows that $\bar{\bar{\mathfrak{F}}} = \bar{\mathfrak{F}}$ and $\bar{\bar{L}}(F) = \bar{L}(F)$, i.e. that the operation $\bar{L}(F)$ is complete.

---

* From $\bar{\mathfrak{F}} \subset [\mathfrak{F}]$ it follows that $[\bar{\mathfrak{F}}] \subset [\mathfrak{F}]$, and $[\bar{\mathfrak{F}}] \supset [\mathfrak{F}]$ follows from $\bar{\mathfrak{F}} \supset \mathfrak{F}$.

The minimality of the extension $\bar{L}(F)$. Let a positive linear operation $L'(F)$ on the field of functions $\mathfrak{F}'$ be an arbitrary complete extension of the operation $L(F)$. Then, since $\mathfrak{F}' \supset \mathfrak{F}$, we have

$$\hat{L}'(F) \leqslant \hat{L}(F), \tag{17}$$

and hence

$$\check{L}'(F) \geqslant \check{L}(F) \tag{17'}$$

for $F(\tau) \in [\mathfrak{F}] \subset [\mathfrak{F}']$. For functions $F(\tau) \in \bar{\mathfrak{F}}$, in view of (17) and (17′),

$$\hat{L}'(F) \leqslant \bar{L}(F) \leqslant \check{L}'(F),$$

since $\hat{L}(F) = \check{L}(F) = \bar{L}(F)$. Therefore, for $F(\tau) \in \bar{\mathfrak{F}}$,

$$\bar{L}(F) = \hat{L}'(F) = \check{L}'(F) = \bar{L}'(F) = L'(F)$$

and $F(\tau) \in \bar{\mathfrak{F}}' = \mathfrak{F}'$, i.e. $\bar{\mathfrak{F}} \subset \mathfrak{F}'$ and $\bar{L}(F) = L'(F)$.

Thus, $L'(F)$ is an extension of the operation $\bar{L}(F)$, i.e. it is the smallest complete extension of the operation $L(F)$.

The domain $\mathfrak{F}$ of a complete operation $L(F)$ is closed in the following sense.

Let $G(\tau) \in \mathfrak{F}'$, *and let a sequence* $\{F_n(\tau)\}$ *in* $\mathfrak{F}$ *converge uniformly with respect to* $G(\tau)$ *to a function* $F(\tau)$, *i.e.*

$$|F_n(\tau) - F(\tau)| \leqslant \varepsilon_n G(\tau), \qquad \lim_n \varepsilon_n = 0. \tag{18}$$

Then the limit $F(\tau)$ *belongs to* $\mathfrak{F}$.

In fact, (18) is equivalent to

$$F_n(\tau) - \varepsilon_n G(\tau) \leqslant F(\tau) \leqslant F_n(\tau) + \varepsilon_n G(\tau),$$

whence it follows that $F(\tau) \in \mathfrak{F}$ and that $\varepsilon_n L(G)$ is an absolute majorant of both of the differences $L(F_n) - \hat{L}(F)$ and $L(F_n) - \check{L}(F)$ (in view of the monotonicity of the operations $\hat{L}(F)$ and $\check{L}(F)$).

Therefore

$$\hat{L}(F) = \lim_{n \to \infty} L(F_n) = \check{L}(F)$$

and hence $F(\tau) \in \bar{\mathfrak{F}} = \mathfrak{F}$.

In particular, *if the domain* $\mathfrak{F}$ *of a complete operation is a regular field of functions, then it is closed with respect to uniform convergence.*

In order to convince ourselves of this, it is sufficient to let $G(\tau) = 1$.

If a linear operation $L(F)$ defined on a semifield of functions is monotone, then we define its *completion* to be its smallest extension to a complete operation. For the construction of this extension we can use:

Lemma 2.7.1. *A linear operation* $L(F)$ *on a semifield of functions* $\mathfrak{F}$ *is continued in a unique manner to a linear operation* $L_1(F)$ *on the field of functions*

$\mathfrak{F}_1$ *which is generated by the semifield* $\mathfrak{F}$. *If in addition, the operation* $L(F)$ *is monotone, then the extension* $L_1(F)$ *is positive.*

Indeed, the family $\mathfrak{F}_1$ of functions $F_1(\tau)$ of the form

$$F_1(\tau) = F(\tau) - G(\tau); \qquad F(\tau) \in \mathfrak{F}, \; G(\tau) \in \mathfrak{F},$$

is obviously a vector system, and it is also a field of functions since sup $\{F_1, 0\} = (\sup\{F, G\} - G) \in \mathfrak{F}_1$, and hence for $G_1(\tau) \in \mathfrak{F}$,

$$\sup\{F_1, G_1\} = (\sup\{F_1 - G_1, 0\} + G_1) \in \mathfrak{F}_1, \; \inf\{F_1, G_1\} = \\ = -\sup(-F_1, -G_1) \in \mathfrak{F}_1.$$

The formula

$$L_1(F_1) = L(F) - L(G)$$

defines a unique linear operation on $\mathfrak{F}_1$. In fact, if $F_1(\tau) = F'(\tau) - G'(\tau)$ is another representation of the function $F_1(\tau)$ in $\mathfrak{F}_1$, then $F(\tau) + G'(\tau) = F'(\tau) + G(\tau)$, and since $L(F)$ is a linear operation

$$L(F) + L(G') = L(F') + L(G).$$

Hence

$$L(F') - L(G') = L(F) - L(G).$$

On the other hand

$$L_1(F_1 + G_1) = L_1(F_1) + L_1(G_1), \; L_1(\alpha F_1) = \alpha L_1(F_1), \tag{19}$$

where $\alpha \geqslant 0$, and since

$$L_1(-F_1) = L(G) - L(F) = -L_1(F_1),$$

(19) is true for any real α. Finally, if the operation $L(F)$ is monotone, and if $F_1(\tau) \geqslant 0$, then $F(\tau) \geqslant G(\tau)$, $L(F) \geqslant L(G)$, and thus

$$L_1(F_1) = L(F) - L(G) \geqslant 0.$$

Every linear extension $L'(F)$ of an operation $L(F)$ defined on a field of functions $\mathfrak{F}' \supset \mathfrak{F}$ is also an extension of an operation $L_1(F)$ in $\mathfrak{F}_1$, and therefore in the case of a monotone operation $L(F)$, the operation $\bar{L}_1(F)$ is the smallest complete extension of the operation $L(F)$, i.e. $\bar{L}_1(F)$ is the completion of the operation $L(F)$ on the semifield $\mathfrak{F}$.

A monotone linear operation $L(F)$ on a semifield of functions is said to be *ordinary* if $\bar{L}(F)$ is an extension of the operation $L(F)$, i.e. if

$$[-\mathfrak{F}] \supset \mathfrak{F}, \; \bar{L}(-F) = -L(F), \qquad F(\tau) \in \mathfrak{F}. \tag{20}$$

In particular, $L(F)$ is a ordinary operation when $\mathfrak{F}$ is a field of functions.

$L(F)$ is a ordinary operation if and only if $\bar{L}(F)$ is an extension of the operation $L(F)$. In this case

$$[\mathfrak{F}] = [-\mathfrak{F}] = [\bar{\mathfrak{F}}]. \tag{21}$$

In fact, (20) is equivalent to the condition $\overline{\mathfrak{F}}\supset\mathfrak{F}$, and this means that $\bar{L}(F)$ is an extension of $L(F)$. It follows from $\overline{\mathfrak{F}}\supset\mathfrak{F}$ that $\overline{\mathfrak{F}}\supset-\mathfrak{F}$, and hence $[\overline{\mathfrak{F}}]\supset[\mathfrak{F}]$ and $[\overline{\mathfrak{F}}]\supset[-\mathfrak{F}]$, which together with the inclusions $[\overline{\mathfrak{F}}]\subset[\mathfrak{F}]$ and $[\overline{\mathfrak{F}}]\subset[-\mathfrak{F}]$ (which follow from the inclusions $\overline{\mathfrak{F}}\subset[\mathfrak{F}]$ and $\overline{\mathfrak{F}}\subset[-\mathfrak{F}]$) proves (21).

THEOREM 2.7.3. *The operation $\bar{L}(F)$ is the completion of the operation $L(F)$ if and only if the latter is ordinary.*

In fact, the completion is an extension of the operation $L(F)$, and hence the operation $L(F)$ is ordinary if its completion equals $\bar{L}(F)$. Conversely, if the operation $L(F)$ is ordinary, then the proof of the completeness and minimality of $\bar{L}(F)$ in the proof of Theorem 2.7.2 remains valid, so that in this case $\bar{L}(F)$ is the completion of the operation $L(F)$.

However, in the case of an ordinary operation $L(F)$ defined on a semifield, the operation $\bar{L}(F)$ is the completion of the operation $L(F)$. It is also suitable in that case to retain the name upper and lower operations for $\hat{L}(F)$ and $\check{L}(F)$.

§ 8. Linear Operations on σ-closed Semifields of Functions

The semifields $[\mathfrak{F}]$ *and* $[-\mathfrak{F}]$ *are σ-closed.* Indeed, let $\mathfrak{F}$ be a semifield of functions, and let $\{F_n(\tau)\}$ be a sequence of functions with the following properties:

$$\left.\begin{aligned} &F_n(\tau)\in[\mathfrak{F}],\ F_n(\tau)\leqslant F_{n+1}(\tau)\leqslant F'(\tau)\in[\mathfrak{F}],\\ &\lim_{n\to\infty} F_n(\tau)\doteq F(\tau). \end{aligned}\right\} \tag{1}$$

Then, for suitable $G_1'(\tau)$ and $G'(\tau)$ in $\mathfrak{F}$, we have

$$G_1'(\tau)\leqslant F_1(\tau)\leqslant F_n(\tau)\leqslant F'(\tau)\leqslant G'(\tau),$$
$$G_1'(\tau)\leqslant F(\tau)\leqslant G'(\tau),$$

i.e. $F(\tau)\in[\mathfrak{F}]$, and the semifield $[\mathfrak{F}]$ is σ-closed. In addition to $[\mathfrak{F}]$, the semifield $[-\mathfrak{F}]$ is also σ-closed.

Since the operation $\hat{L}(F)$ is monotone, we have

$$\hat{L}(F_n)\leqslant\hat{L}(F),$$

for a sequence $\{F_n(\tau)\}$ with the properties (1), and hence

$$\lim\hat{L}(F_n)\leqslant\hat{L}(F). \tag{2}$$

Similarly, for the operation $\check{L}(F)$, we obtain

$$\lim\check{L}(F_n)\leqslant\check{L}(F), \tag{2'}$$

if we replace $[\mathfrak{F}]$ by $[-\mathfrak{F}]$ in (1).

We now consider a monotone linear operation $L(F)$, defined on a σ-closed semifield of functions $\mathfrak{F}$, which satisfies the condition

$$\lim L(F_n) = L(F), \tag{3}$$

where $\{F_n(\tau)\}$ is a non-decreasing sequence in $\mathfrak{F}$ which converges to a function $F(\tau) \in \mathfrak{F}$. For such operations, we have:

THEOREM 2.8.1. *If a monotone linear operation $L(F)$, with values in a complete vector order Z, is defined on a σ-closed semifield of functions $\mathfrak{F}$ and satisfies the preceding condition, and if the vector order Z satisfies one of the conditions* I *or* II *of Sec* 1.9.2, *then $\bar{L}(F)$ is a positive Lebesgue operation.*

The proof of Theorem 2.8.1 is based on the following lemma.

LEMMA 2.8.1. *If an operation $L(F)$ and a vector order Z satisfy the hypotheses of Theorem* 2.8.1, *then under the condition* (1),

$$\lim \widehat{L}(F_n) = \widehat{L}(F). \tag{4}$$

By virtue of (2), it is sufficient to prove that

$$\lim \widehat{L}(F_n) \geqslant \widehat{L}(F). \tag{4'}$$

The proof of (4′) will be based on the following preliminary remark. *Let $F_k(\tau) \in [\mathfrak{F}]$, and let*

$$F_k(\tau) \leqslant G_k(\tau) \in \mathfrak{F}, \quad G'_k(\tau) = \sup\{G_1, G_2, \ldots, G_k\}. \tag{5}$$

Then

$$L(G'_n) \leqslant \widehat{L}(F_n) + \sum_{k=1}^{n} (L(G_k) - \widehat{L}(F_k)). \tag{6}$$

For $n = 1$, (6) is obviously satisfied. We show that it holds for $n = m+1$ if it holds for $n = m$. Indeed, $G'_{m+1}(\tau) = \sup\{G'_m, G_{m+1}\}$, and by letting $G''_m(\tau) = \inf\{G'_m, G_{m+1}\}$, and using the obvious relation $G'_m + G_{m+1} = G'_{m+1} + G''_m$, we have: $L(G'_m) + L(G_{m+1}) = L(G'_{m+1}) + L(G''_m)$.

Since $G''_m(\tau) \geqslant F_m(\tau)$, we have $L(G''_m) \geqslant \widehat{L}(F_m)$, and thus

$$L(G'_{m+1}) + \widehat{L}(F_m) \leqslant L(G'_m) + L(G_{m+1})$$

or after an obvious transformation

$$L(G'_{m+1}) \leqslant \widehat{L}(F_{m+1}) + (L(G'_m) - \widehat{L}(F_m)) + L(G_{m+1}) - \widehat{L}(F_{m+1}).$$

We now apply (6) to the middle term on the right-hand side, with $n = m$, and we obtain

$$L(G'_{m+1}) \leqslant \widehat{L}(F_{m+1}) + \sum_{k=1}^{m+1} (L(G_k) - \widehat{L}(F_k)),$$

i.e. (6) holds for $n = m+1$.

We now turn to the proof of (4′). We apply (6) to the functions $F_k(\tau)$ which satisfy the conditions (1). Here we will consider that the vector order

Z satisfies Condition II of Sec. 1.9.2 (in particular, this means that it is complete). The set $N_k \subset Z$ of values of the difference $L(G_k) - \widehat{L}(F_k)$, for all $G_k(\tau)$ which satisfy (5), is directed below; moreover, $\inf N_k = 0$. By virtue of Condition II of Sec. 1.9.2, $G_k(\tau)$ exist such that the series

$$\sum_{k=1}^{\infty} (L(G_k) - \widehat{L}(F_k))$$

converges. By replacing here, if necessary, the $G_k(\tau)$ by the functions $\inf (G_k, F') \in \mathfrak{F}$, where $F(\tau) \leqslant F'(\tau) \in \mathfrak{F}$, we can assume that $G_k(\tau) \leqslant F'(\tau) \in \mathfrak{F}$. Then $G'_n(\tau) \leqslant F'(\tau) \in \mathfrak{F}$, so that the limit $G'(\tau) = \lim G'_n(\tau)$ of the (non-decreasing) sequence $\{G'_n(\tau)\}$ belongs to $\mathfrak{F}$. By virtue of the assumptions about the operation $L(F)$,

$$\lim L(G'_n) = L(G'),$$

and $F_k(\tau) \leqslant G_k(\tau) \leqslant G'_k(\tau) \leqslant G'(\tau)$. Hence $F(\tau) \leqslant G'(\tau)$, $\widehat{L}(F) \leqslant L(G')$, and by passing to the limit in (6) for $n \to \infty$, we have

$$\widehat{L}(F) \leqslant L(G') \leqslant \lim \widehat{L}(F_n) + \sum_{k=1}^{\infty} (L(G_k) - \widehat{L}(F_k)). \tag{7}$$

The infimum of the sum in (7) equals zero (Lemma 1.9.1), and hence (4′) holds. Inequality (4) also holds in the case when Z has the property (II) of Sec. 1.9.2.

If Z satisfies the condition (I) of Sec. 1.9.2, then (4) is equivalent to

$$p(\widehat{L}(F)) = p\left(\lim_{n\to\infty} \widehat{L}(F_n)\right)$$

or

$$p(\widehat{L}(F)) = \lim_{n\to\infty} p(\widehat{L}(F_n)), \tag{8}$$

where p is any positive completely linear functional in Z. For each such p, the operation

$$l(F) = p(L(F))$$

is a monotone linear operation with real values which satisfies

$$\lim l(F_n) = l(F) \tag{3'}$$

on the σ-closed semifield $\mathfrak{F}$ if the sequence $F_n(\tau)$ in $\mathfrak{F}$ does not decrease and if it converges to $F(\tau) \in \mathfrak{F}$. Indeed,

$$\lim p(L(F_n)) = p(\lim L(F_n)) = p(L(F)).$$

For the operation $l(F)$, we have

$$\widehat{l}(F) = p(\widehat{L}(F)) \tag{9}$$

for $F(\tau) \in [\mathfrak{F}]$.

In fact, since the functional p is completely linear,

$$\inf l(F') = \inf p(L(F')) = p(\inf L(F')),$$

if

$$F(\tau) \in [\mathfrak{F}] \quad \text{and} \quad F(\tau) \leqslant F'(\tau) \in \mathfrak{F}.$$

The equations (8) can now be written in the form

$$\hat{l}(F) = \lim \hat{l}(F_n) \tag{8'}$$

and they are satisfied, by what has just been proved, for every functional $\hat{l}(F)$ given by (9), since $l(F)$ satisfies (3′) and the vector order of real numbers satisfies the condition (II) of Sec. 1.9.2. If the vector order Z has the property (I), then (4) follows since (8) holds for any functional p.

The proof of Theorem 2.8.1. Let a sequence $\{F_n(\tau)\}$ in $\mathfrak{F}$ satisfy the condition (1) with $[\mathfrak{F}]$ replaced by $\overline{\mathfrak{F}}$, and let $\overline{\mathfrak{F}} \subset [\mathfrak{F}] \cap [-\mathfrak{F}]$. Then we can apply both (4) and (2′) to $\{F_n(\tau)\}$, and since

$$\hat{L}(F_n) = \check{L}(F_n) = \bar{L}(F_n)$$

they can now be written in the form

$$\lim \bar{L}(F_n) = \hat{L}(F), \qquad \lim \bar{L}(F_n) \leqslant \check{L}(F). \tag{10}$$

Moreover, $F(\tau) \in [\mathfrak{F}] \cap [-\mathfrak{F}]$, since the families $[\mathfrak{F}]$ and $[-\mathfrak{F}]$ are σ-closed. It follows from (10) that $\hat{L}(F) \leqslant \check{L}(F)$, and hence

$$\hat{L}(F) = \check{L}(F) = \bar{L}(F), \qquad F(\tau) \in \overline{\mathfrak{F}}.$$

Thus the field $\overline{\mathfrak{F}}$ is σ-closed, i.e. it is a Baire field of functions, and the first equation in (10) gives

$$\lim \bar{L}(F_n) = \bar{L}(F),$$

i.e. the positive linear operation $\bar{L}(F)$ is continuous in $\overline{\mathfrak{F}}$.

A consequence of Theorem 2.8.1 is:

THEOREM 2.8.2. *The completion of a positive Lebesgue operation $L(F)$, with values in a complete vector order Z, which satisfies one of the conditions* (I) *or* (II) *of* Sec. 1.9.2, *is a positive Lebesgue operation.*

§ 9. The Extension of a Linear Operation to a Positive Lebesgue Operation

We now prove the assertion (made in Sec. 2.7.1) on the possibility of extending a positive linear operation $L(F)$, defined on a field or $*$-field of functions $\mathfrak{F}$, to a positive Lebesgue operation. We recall that it is assumed here that the necessary condition

$$\lim_{n \to \infty} L(G_n) = 0, \tag{1}$$

is satisfied when a non-decreasing sequence $\{G_n(\tau)\}$ in $\mathfrak{F}$ converges to zero. We will divide the proof into two parts:

1) We extend $L(F)$ to some monotone, linear operation $L_\sigma(F)$ which is defined on a σ-closed semifield of functions $\mathfrak{F}_\sigma$ and is continuous in the sense that

$$\lim L_\sigma(F_n) = L_\sigma(F), \tag{2}$$

when the non-decreasing sequence $\{F_n(\tau)\}$ in $\mathfrak{F}_\sigma$ converges to $F(\tau) \in \mathfrak{F}_\sigma$;

2) We will apply Theorem 2.8.1 to the operation $L_\sigma(F)$.

In order to define the operation $L_\sigma(F)$, the following lemma is necessary.

LEMMA 2.9.1. *Let a positive linear operation which satisfies the condition* (1) *be defined on a field $\mathfrak{F}$. Let $\{F_n(\tau)\}$ and $\{F'_n(\tau)\}$ be two non-decreasing sequences in $\mathfrak{F}$ such that*

$$F(\tau) = \lim F_n(\tau) \geqslant F'(\tau) = \lim F'_n(\tau)$$

and

$$L(F_n) \leqslant \eta \in Z. \tag{3}$$

Then

$$\lim L(F'_n) \leqslant \lim L(F_n). \tag{4}$$

Indeed, by letting

$$F_{mn}(\tau) = \sup\{F'_m - F_n, 0\},$$

and since

$$0 \leqslant F_{mn}(\tau) \leqslant F(\tau) - F_n(\tau),$$

we have:

$$F_{mn}(\tau) \in \mathfrak{F}, \quad F_{mn}(\tau) \geqslant F_{mn+1}(\tau), \quad \lim_{n \to \infty} F_{mn}(\tau) = 0.$$

Hence, by (1),

$$\lim_{n \to \infty} L(F_{mn}) = 0. \tag{5}$$

The inequality (4) is now obtained from (5), and the inequality

$$L(F'_m) - L(F_n) = L(F'_m - F_n) \leqslant L(F_{mn})$$

is obtained by successively passing to the limit for $n \to \infty$ and $m \to \infty$. The existence of the limits follows from (3).

COROLLARY. *If* $F'(\tau) = F(\tau)$ *in Lemma* 2.9.1, *then**

$$\lim_{m\to\infty} L(F'_m) = \lim_{n\to\infty} L(F_n). \tag{6}$$

We will let $\mathfrak{F}_\sigma$ denote the family of all finite limits $F(\tau) = \lim F_n(\tau)$ of non-decreasing sequences $\{F_n(\tau)\}$ in $\mathfrak{F}$ which satisfy (3), and on the family $\mathfrak{F}_\sigma$ (of functions $F(\tau)$) we will define the operation $L_\sigma(F)$ by the limit relation

$$L_\sigma(F) = \lim L(F_n). \tag{7}$$

This definition defines the operation $L_\sigma(F)$ uniquely since, by virtue of the corollary to Lemma 2.9.1, the value of the limit in (7) is independent of the choice of $\{F_n(\tau)\}$. The operation $L_\sigma(F)$ is an extension of the operation $L(F)$, since $\mathfrak{F}_\sigma \supset \mathfrak{F}$ and since $L_\sigma(F) = L(F)$ for $F(\tau) \in \mathfrak{F}$.

LEMMA 2.9.2. $\mathfrak{F}_\sigma$ *is a* σ*-closed semifield of functions, and the operation* $L_\sigma(F)$ *is linear, monotone, and continuous in the sense that* (2) *holds when the non-decreasing sequence* $\{F_n(\tau)\}$ *in* $\mathfrak{F}_\sigma$ *converges to* $F(\tau) \in \mathfrak{F}_\sigma$.

For the proof, we let $\mathfrak{I}$ denote the collection of those functions $F(\tau) = \lim F_n(\tau)$ for which the functions $F_n(\tau)$ are non-negative. The set $\mathfrak{I}$ is obviously a majorant system which is contained in $\mathfrak{F}_\sigma$, and $\mathfrak{F}_\sigma$ coincides with the σ-extension $\mathfrak{F}_\sigma^{(\mathfrak{I})}$ of the family $\mathfrak{F}$ with respect to $\mathfrak{I}$. Indeed, the function $F(\tau) - F_1(\tau) = \lim_{n\to\infty} (F_n(\tau) - F_1(\tau))$ belongs to $\mathfrak{I}$, and

---

* Equation (6) can also be obtained from the following proposition. Let $F(\tau) = \sum_{m=1}^{\infty} G_m(\tau) = \sum_{n=1}^{\infty} G'_n(\tau)$, $G_m(\tau) \geqslant 0$, $G'_n(\tau) \geqslant 0$, and let the functions $G_{mn}(\tau)$ be defined recurrently: $G'_{11}(\tau) = G_1(\tau)$, $G''_{11}(\tau) = G'_1(\tau)$,

$$G'_{mn}(\tau) = G_m(\tau) - \sum_{k=1}^{n-1} G_{mk}(\tau),$$

$$G''_{mn}(\tau) = G'_n(\tau) - \sum_{k=1}^{m-1} G_{kn}(\tau),$$

where $G_{mk}(\tau) = \inf\{G'_{mk}(\tau), G''_{mk}(\tau)\}$. Then

$$G_m(\tau) = \sum_{k=1}^{\infty} G_{mk}(\tau), \qquad G'_n(\tau) = \sum_{k=1}^{\infty} G_{kn}(\tau).$$

Whence, by virtue of (1) it follows that

$$\sum_{m=1}^{\infty} L(G_m) = \sum_m \sum_n L(G_{mn}) = \sum_{n=1}^{\infty} L(G'_n),$$

since whenever $G_m(\tau)$ and $G'_n(\tau)$ belong to a field of functions $\mathfrak{F}$, the functions $G_{mn}(\tau)$ also belong to $\mathfrak{F}$.

This was our initial way to prove (6) which, for greater brevity, was replaced by Lemma 2.9.1 which is due to F. Riesz.

$$F_n(\tau) = (F_n(\tau) - F_1(\tau)) + F_1(\tau) \leqslant (F(\tau) - F_1(\tau)) + F_1^+(\tau) \in \mathfrak{J},$$

so that $F(\tau) \in \mathfrak{F}_\sigma^{(\mathfrak{J})}$, i.e.

$$\mathfrak{F}_\sigma \subset \mathfrak{F}_\sigma^{(\mathfrak{J})}. \tag{8}$$

Conversely, if $F(\tau) \in \mathfrak{F}_\sigma^{(\mathfrak{J})}$ and if

$$F(\tau) = \lim F_n(\tau), \quad F_n(\tau) \in \mathfrak{F}, \quad F_n(\tau) \leqslant F_{n+1}(\tau) \leqslant G'(\tau) \in \mathfrak{J},$$

then by virtue of Lemma 2.9.1,

$$L(F_n) = L_\sigma(F_n) \leqslant L_\sigma(G') \in Z,$$

and hence $F(\tau) \in \mathfrak{F}_\sigma$, i.e. $\mathfrak{F}_\sigma^{(\mathfrak{J})} \subset \mathfrak{F}_\sigma$. Along with (8), this gives $\mathfrak{F}_\sigma = \mathfrak{F}_\sigma^{(\mathfrak{J})}$, and as we know (Theorem 2.1.2), $\mathfrak{F}_\sigma^{(\mathfrak{J})}$ is a σ-closed semifield of functions.

Since limit passage preserves vector operations, the operation $L_\sigma(F)$ defined on the semifield $\mathfrak{F}_\sigma$ is linear. It is also monotone: the inequality (4) means that $L_\sigma(F) \geqslant L_\sigma(F')$, when $F(\tau) \in \mathfrak{F}_\sigma$, $F'(\tau) \in \mathfrak{F}_\sigma$, and $F(\tau) \geqslant F'(\tau)$. It remains to prove that the operation $L_\sigma(F)$ is continuous.

Let $\{F_n(\tau)\}$ be a non-decreasing sequence in $\mathfrak{F}_\sigma$ such that the limit $\lim F_n(\tau) = F(\tau)$ belongs to $\mathfrak{F}_\sigma$, and for fixed n, let $\{F_{nm}(\tau)\}$ be a non-decreasing sequence in $\mathfrak{F}$ which converges to $F_n(\tau)$. By letting

$$\Phi_m(\tau) = \sup\{F_{1m}, F_{2m}, \ldots, F_{mm}\} \in \mathfrak{F}$$

and by noting that $F_{nm}(\tau) \leqslant F_{nm+1}(\tau)$, we have

$$\Phi_m(\tau) \leqslant \Phi_{m+1}(\tau)$$

and

$$F_{nm}(\tau) \leqslant \Phi_m(\tau) \leqslant \sup\{F_1, F_2, \ldots, F_m\} = F_m(\tau) \text{ for } m \geqslant n.$$

By taking the limit as $m \to \infty$, we have

$$F_n(\tau) = \lim_{m\to\infty} F_{nm}(\tau) \leqslant \lim_{m\to\infty} \Phi_m(\tau) \leqslant \lim_{m\to\infty} F_m(\tau) = F(\tau),$$

and hence

$$\lim \Phi_m(\tau) = F(\tau).$$

Since $\Phi_m(\tau) \leqslant F_m(\tau) \leqslant F(\tau)$, the inequality

$$L(\Phi_m) = L_\sigma(\Phi_m) \leqslant L_\sigma(F_m) \leqslant L_\sigma(F)$$

in the limit gives

$$L_\sigma(F) = \lim_{m\to\infty} L(\Phi_m) \leqslant \lim_{m\to\infty} L_\sigma(F_m) \leqslant L_\sigma(F),$$

and hence

$$\lim_{m\to\infty} L_\sigma(F_m) = L_\sigma(F),$$

i.e. the operation $L_\sigma(F)$ is continuous.

THEOREM 2.9.1. *A positive linear operation* $L(F)$, *defined on a field* $\mathfrak{F}$, *with values in a complete vector order* Z, *satisfying one of the conditions* (I) *or* (II) *of* Sec. 1.9.2 *can be extended to a positive Lebesgue operation if and only if it satisfies* (1).

In fact, by virtue of Lemma 2.9.2 and Theorem 2.8.1, $\bar{L}_\sigma(F)$ is a positive Lebesgue operation on the Baire field of functions $\bar{\mathfrak{F}}_\sigma$, and therefore it is sufficient to show that $\bar{L}_\sigma(F)$ is an extension of the operation $L(F)$. Let $F(\tau)\in\mathfrak{F}$. Then

$$F(\tau)\in\mathfrak{F}_\sigma, \quad -F(\tau)\in\mathfrak{F}\subset\mathfrak{F}_\sigma, \quad F(\tau)\in[\mathfrak{F}_\sigma]\cap[-\mathfrak{F}_\sigma]$$

and hence

$$\hat{L}_\sigma(F)=L_\sigma(F)=L(F); \quad \hat{L}_\sigma(-F)=L(-F)=-L(F),$$

i.e.

$$\hat{L}_\sigma(F)=\check{L}_\sigma(F)=L(F).$$

Thus, $F(\tau)\in\bar{\mathfrak{F}}_\sigma$, i.e. $\bar{\mathfrak{F}}_\sigma\supset\mathfrak{F}$, and $\bar{L}_\sigma(F)=L(F)$ for $F(\tau)\in\mathfrak{F}$.

REMARK. If the conditions of Theorem 2.9.1 are satisfied, then the set $\mathfrak{S}(L)$ of all positive Lebesgue operations which are extensions of the operation $L(F)$ is not empty. It is converted into an ordered set if we let $L_2 \geqslant L_1$ for two elements $L_1(F)$ and $L_2(F)$ in $\mathfrak{S}(L)$ when $L_2(F)$ is an extension of the operation $L_1(F)$.

The ordered set $\mathfrak{S}(L)$ *has a smallest element* $L'(F)$, *i.e. an element* $L'(F) \in \mathfrak{S}(L)$ *such that any operation in* $\mathfrak{S}(L)$ *is an extension of* $L'(F)$.

Indeed, let $\mathfrak{F}'$ be the Baire closure of the field of functions $\mathfrak{F}$ with respect to the majorant system $\mathfrak{I}=\mathfrak{F}^+$. Since $\mathfrak{F}'$ is the smallest Baire field which contains $\mathfrak{F}$, we have $\bar{\mathfrak{F}}_\sigma\supset\mathfrak{F}'$, and the operation $\bar{L}_\sigma(F)$ induces a positive Lebesgue operation $L'(F)$ on $\mathfrak{F}'$. $L'(F)$ is not only minimal, but it is also the smallest element of the set $\mathfrak{S}(L)$, since we show that for each function $F(\tau)$ in $\mathfrak{F}'$ the values $L''(F)$ of all operations L'' in $\mathfrak{S}(L)$ are equal, i.e. they are independent of $L''\in\mathfrak{S}(L)$. We will temporarily denote the collection of functions $F(\tau)$ in $\mathfrak{F}'$ which have this property by $\mathfrak{F}_1$. The family $\mathfrak{F}_1$ is a Baire system of functions with respect to $\mathfrak{I}=\mathfrak{F}^+$. In fact, if

$$F_n(\tau)\in\mathfrak{F}_1, \quad |F_n(\tau)|\leqslant F'(\tau)\in\mathfrak{F}^+, \quad \lim F_n(\tau)=F(\tau),$$

then $\lim L''(F_n)=L''(F)$. Whenever the $L''(F_n)$ are uniquely defined, $L''(F)$ is also uniquely defined. Since $\mathfrak{F}_1\supset\mathfrak{F}$, and since the closure $\mathfrak{F}'$ is minimal, $\mathfrak{F}_1\supset\mathfrak{F}'$, and hence $\mathfrak{F}'=\mathfrak{F}_1$.*

Thus, the extension of the operation $L(F)$, by a positive Lebesgue operation on F', is *unique*; namely, it is equal to $L'(F)$. By virtue of Theorem 2.8.2: *the*

* In essence, this argument is equivalent to the application of Lemma 2.4.1 in the simplest case of a one-term relation.

completion $\bar{L}'(F)$ *of an operation* $L'(F)$ *is the smallest extension of the operation* $L(F)$ *to a complete positive Lebesgue operation.*

THEOREM 2.9.2. *The assertion of Theorem* 2.9.1 *still holds if we replace the field* $\mathfrak{F}$ *by a* ∗*-field* $\mathfrak{F}$, *and if we replace the vector order Z by a vector* ∗*-order Z. In particular, when the* ∗*-field* $\mathfrak{F}$ *is regular, each extension of the operation* $L(F)$ *to a positive Lebesgue operation is an integral. In that case, the operation* $L(F)$ *on* $\mathfrak{F}$ *can be represented in the form*

$$L(F) = \int_T F(\tau)\, \mu(M_\tau), \tag{9}$$

where $\mu(M)$ *is a measure on some regular Borel field of subsets of the fundamental set T.*

Indeed, we will apply Theorem 2.9.1 to the operation $L_r(F)$ which is induced by the operation $L(F)$ on the field $\mathfrak{F}_r$ of all real functions in $\mathfrak{F}$. Let $L_r'(F)$ be an extension of the operation $L_r(F)$ to a positive Lebesgue operation which is defined on the Baire field of functions $\mathfrak{F}_r' \supset \mathfrak{F}_r$, and let $\mathfrak{F}'$ be the Baire ∗-field of functions $F(\tau)$ of the form $F(\tau) = F_1(\tau) + iF_2(\tau)$, where $F_1(\tau)$ and $F_2(\tau)$ belong to the field F_r'. Then

$$L'(F) = L_r'(F_1) + iL_r'(F_2)$$

defines an operation $L'(F)$ on $\mathfrak{F}'$, and $L'(F)$ is a positive Lebesgue operation which is an extension of the initial operation $L(F)$.

In the case when $\mathfrak{F}$ is a regular ∗-field, the Baire ∗-field $\mathfrak{F}'$ is also regular, and by virtue of Theorem 2.6.1, $L'(F)$ is an integral on $\mathfrak{F}'$ with respect to some measure $\mu(M)$ which is defined on a regular Borel field of sets $\mathfrak{B}$.

Theorem 2.9.2 has an interesting application in the case when the set T is a compact topological space and the functions of the ∗-field $\mathfrak{F}$ are continuous on T.

THEOREM 2.9.3. *A positive linear operation* $L(F)$, *with values in a complete vector* ∗*-order Z satisfying one of the conditions* (I) *or* (II) *of* Sec. 1.9.2, *which is defined on a regular* ∗*-field* $\mathfrak{F}$ *of continuous functions in a compact topological space T, can be extended to an integral with respect to a measure, and hence it admits the representation* (9).

Indeed, a non-increasing sequence $\{G_n(\tau)\}$ in $\mathfrak{F}$, which converges to zero on T, converges uniformly by virtue of the Dini* theorem. Since a positive

* The Dini theorem asserts that a monotone sequence $\{G_n(\tau)\}$ of continuous functions, on a compact topological space T, which converges to a continuous function $G(\tau)$, converges uniformly to $G(\tau)$. Without loss of generality, it can be assumed that $\{G_n(\tau)\}$ does not increase and that $G(\tau) = 0$. Then the sets $M_n = T(G_n < \varepsilon)$, where $\varepsilon > 0$, are open, $M_n \subset M_{n+1}$, and $T = \bigcup_n M_n$. Since the space T is compact, it is possible to choose from its covering sets, a finite covering, i.e. for some $n = k$, we have $M_k = T$. Therefore $0 \leqslant G_n(\tau) < \varepsilon$ for all $\tau \in T$ provided that $n \geqslant k$, i.e. $\{G_n(\tau)\}$ converges uniformly to zero.

linear operation $L(F)$ defined on a regular $*$-field $\mathfrak{F}$ is continuous with respect to uniform convergence, we have $\lim_{n\to\infty} L(G_n) = 0$. Thus, Theorem 2.9.3 follows immediately from Theorem 2.9.2.

§ 10. The Integral with Respect to a σ-finite Measure

10.1. Closed measures. A measure $\mu(M)$, defined on a Borel field $\mathfrak{B} \subset \mathfrak{P}(T)$, with values in a complete vector order Z is by definition *closed* if, for disjoint sets N_k in $\mathfrak{B}$, whenever the series $\sum_{k=1}^{\infty} \mu(N_k)$ converges, the union $M = \bigcup_k N_k$ belongs to $\mathfrak{B}$. Thus, for a closed measure $\mu(M)$ on $\mathfrak{B}$, and for a sequence $\{N_k\}$ of disjoint sets N_k in $\mathfrak{B}$ with $\bigcup_{k=1}^{\infty} N_k = M$, we have

$$\mu(M) = \sum_{k=1}^{\infty} \mu(N_k), \tag{1}$$

whenever the series in (1) converges.

A function $\mu'(M)$ on the system of sets $\mathfrak{M}'$ is said to be an *extension* or *continuation of the set function* $\mu(M)$ on the system $\mathfrak{M}$ if $\mathfrak{M}' \supset \mathfrak{M}$, and if $\mu'(M) = \mu(M)$ for sets M in $\mathfrak{M}$.

If the measure $\mu(M)$ is not closed on $\mathfrak{B}$, then in the following manner it is possible to obtain an extension $\mu_a(M)$ of $\mu(M)$ which will be a closed measure.

We define the function $\mu_a(M)$ by

$$\mu_a(M) = \sum_{k=1}^{\infty} \mu(N_k), \qquad M = \bigcup_{k=1}^{\infty} N_k \in \mathfrak{B}_a$$

on the collection $\mathfrak{B}_a$ of sets M which can be represented in the form of a countable (or finite) union of disjoint sets N_k in $\mathfrak{B}$ for which the series (1) converges. This definition is correct since the value $\mu_a(M)$ is independent of the choice of the representation of the set M: if $M = \bigcup_k N'_k$ is another representation for $M (N'_k \in \mathfrak{B}, N'_k \cap N'_m = \varnothing$ for $k \neq m)$, then since the measure $\mu(M)$ is σ-additive,

$$\sum_{m=1}^{\infty} \mu(N'_m) = \sum_{m=1}^{\infty} \sum_{k=1}^{\infty} \mu(N_k \cap N'_m) = \sum_{k=1}^{\infty} \sum_{m=1}^{\infty} \mu(N_k \cap N'_m) = \sum_{k=1}^{\infty} \mu(N_k).$$

The set function $\mu_a(M)$ is a closed measure on $\mathfrak{B}_a$.

Indeed, let

$$\left.\begin{aligned} &N_{km} \in \mathfrak{B},\ N_{km} \cap N_{km'} = \varnothing && \text{for } m \neq m', \\ &M_k = \bigcup_{m=1}^{\infty} N_{km} \in \mathfrak{B}_a,\quad M_k \cap M_n = \varnothing && \text{for } k \neq n, \end{aligned}\right\} \tag{2}$$

and let the series $\sum_{k=1}^{\infty} \mu_a(M_k)$ converge. Then $M = \bigcup_k M_k$ belongs to $\mathfrak{B}_a$, and

$$\mu_a(M) = \sum_{k=1}^{\infty} \mu_a(M_k). \tag{3}$$

In fact, $M = \bigcup_{k,m=1}^{\infty} N_{km}$, $N_{km} \in \mathfrak{B}$, and the series

$$\sum_{k,m} \mu(N_{km}) = \sum_{k=1}^{\infty} \sum_{m=1}^{\infty} \mu(N_{km}) = \sum_{k=1}^{\infty} \mu_a(M_k)$$

converges, so that $M \in \mathfrak{B}_a$, and (3) holds.

In particular, the series $\sum_{k=1}^{\infty} \mu_a(M_k)$ converges if the sets M_k in (2) satisfy the condition

$$M_k \subset M' \in \mathfrak{B}_a, \tag{4}$$

since in this case

$$\sum_{k=1}^{\infty} \mu_a(M_k) = \sum_{k,\, m=1}^{\infty} \mu(N_{km}) \leqslant \mu_a(M').$$

Therefore, it follows from (2) and (4) that $M = \bigcup_k M_k \in \mathfrak{B}_a$, i.e. the system $\mathfrak{B}_a$ is σ-closed.

$\mathfrak{B}_a$ is also a field of sets. Indeed, let

$$M = \bigcup_{k=1}^{\infty} N_k, \quad N_k \in \mathfrak{B}, \qquad M' = \bigcup_m N'_m, \quad N'_m \in \mathfrak{B}$$

be two representations of the sets of $\mathfrak{B}_a$. Then $M \cup M' \in \mathfrak{B}_a$. If $M' \supset M$, then

$$M' \setminus M = \bigcup_m (N'_m \setminus M) = \bigcup_m N''_m, \tag{5}$$

where

$$N''_m = N'_m \setminus \bigcup_k N_k = \bigcap_k (N'_m \setminus N_k) \in \mathfrak{B} \quad \text{and} \quad N''_m \subset M',$$

shows that $M' \setminus M \in \mathfrak{B}_a$ since the system $\mathfrak{B}_a$ is σ-closed. Thus, the family $\mathfrak{B}_a$ is a Borel field of sets, and the function $\mu_a(M)$ defined on $\mathfrak{B}_a$ is σ-additive, and hence it is a measure, which has been shown to be closed. The measure $\mu_a(M)$ is obviously an extension of the measure $\mu(M)$, and it is the smallest extension to a closed measure. We will say that the measure $\mu_a(M)$ is the *closure of the measure* $\mu(M)$. A measure $\mu(M)$ is closed if and only if it coincides with its closure $\mu_a(M)$. A measure $\mu(M)$ defined on a *regular* Borel field is closed.

10.2. σ-finite measures. A field of sets $\mathfrak{B} \subset \mathfrak{P}(T)$ is said to be *countably regular* if

$$T = \bigcup_{n=1}^{\infty} T_n, \quad T_n \in \mathfrak{B}. \tag{6}$$

Since each set T_n can be replaced by a union $\bigcup_{k=1}^{n} T_k$, which also belongs to $\mathfrak{B}$, we can consider that $\{T_n\}$ is already an increasing sequence.

Let $\mathfrak{B}$ be a countable regular Borel field of sets, and let $\mathfrak{B}'$ be the smallest regular field which contains $\mathfrak{B}$, i.e. the smallest Borel field which contains T and all of the sets of $\mathfrak{B}$. We show that $\mathfrak{B}'$ consists exactly of sets M of the form

$$M = \bigcup_{k=1}^{\infty} N_k, \qquad N_k \in \mathfrak{B}. \tag{7}$$

Indeed, by virtue of (6), T belongs to the sets of the form (7), and whenever M belongs to them, the set $T \setminus M$ also belongs to them, being the countable union of sets $T_m \setminus M \in \mathfrak{B}$ (cf. (5) with M' and N'_m replaced by T and T_m). Finally, a countable union of sets of the form (7) can again be represented in that form. Clearly, the sets N_k in (7) can be assumed to be disjoint.

If $M \in \mathfrak{B}'$, *and if* $N \in \mathfrak{B}$, *then* $M \cap N \in \mathfrak{B}$.

In fact,

$$M \cap N = \bigcup_{k=1}^{\infty} N_k \cap N, \qquad N_k \cap N \subset N \in \mathfrak{B}.$$

A measure $\mu(M)$ is said to be *σ-finite* if it is defined on a countably regular Borel field $\mathfrak{B}$, and it is said to be *finite* if, in addition, the set of its values is bounded in Z.

Closed finite measures coincide with the measures which have been defined on regular Borel fields (§2.6).

To every σ-finite measure $\mu(M)$ on $\mathfrak{B}$ there corresponds, for a fixed representation (6) in terms of an increasing sequence of sets T_n, a sequence of finite measures on $\mathfrak{B}'$,

$$\mu_n(M) = \mu(T_n \cap M), \qquad M \in \mathfrak{B}'. \tag{8}$$

It is clear that $\mu_n(T) = \mu(T_n)$, and

$$\lim_{n \to \infty} \mu_n(M) = \mu(M)$$

for $M \in \mathfrak{B}$. It is sufficient to give the definition of an integral $I(F)$ with respect to a σ-finite measure $\mu(M)$ for non-negative functions. In fact, according to

$$I(F) = I(F_1) - I(F_2) + iI(F_3) - iI(F_4)$$

it can be extended in a unique manner (cf. Lemma 2.7.1) to a function $F(\tau)$ of the form

$$F(\tau) = F_1(\tau) - F_2(\tau) + iF_3(\tau) - iI(F_4), \tag{9}$$

where the functions $F_k(\tau)$, $k = 1, 2, 3, 4$ are non-negative and are integrable in the sense of the definition given for non-negative functions.

For a fixed sequence (6), we will define the integral for non-negative functions in the following manner: Let $I_n(F)$ be the integral with respect to the measure $\mu_n(M)$, let the non-negative function $F(\tau)$ belong to all of the $\mathfrak{G}_{\mu_n}(\mathfrak{B}')$, and let

$$I_n(F) \leqslant \eta \in Z. \tag{10}$$

By definition, we will let

$$I(F) = \lim I_n(F). \tag{11}$$

In view of the inequality $I_n(F) \leqslant I_{n+1}(F)$, and (10), the limit in (11) exists.

The domain of the integral $I(F)$ with respect to a σ-finite measure is again denoted by $\mathfrak{G}_\mu(\mathfrak{B})$. The collection $\mathfrak{J} = \mathfrak{G}_\mu^+(\mathfrak{B})$ of non-negative functions in $\mathfrak{G}_\mu(\mathfrak{B})$ is obviously a majorant system, and every function $F(\tau)$ in $\mathfrak{G}_\mu(\mathfrak{B})$ can be represented in the form (9), where $F_k(\tau) \in \mathfrak{J}$. Therefore

$$|F(\tau)| \leqslant F_1(\tau) + F_2(\tau) + F_3(\tau) + F_4(\tau) = G'(\tau) \in \mathfrak{J}$$

and

$$F(\tau) \in (\mathfrak{G}_\mu(\mathfrak{B})/\mathfrak{J}) \subset (\tilde{\mathfrak{G}}(\mathfrak{B}')/\mathfrak{J}),$$

i.e.

$$\mathfrak{G}_\mu(\mathfrak{B}) \subset (\tilde{\mathfrak{G}}(\mathfrak{B}')/\mathfrak{J}). \tag{12}$$

Conversely, if

$$F(\tau) = F'(\tau) + iF''(\tau) \in \tilde{\mathfrak{G}}(\mathfrak{B}'), \qquad |F(\tau)| \leqslant G'(\tau) \in \mathfrak{J}$$

and if

$$F_1(\tau) = \sup\{F', 0\}, \qquad F_2(\tau) = \sup\{-F', 0\},$$
$$F_3(\tau) = \sup\{F'', 0\}, \qquad F_4(\tau) = \sup\{-F'', 0\},$$

then $F_k(\tau) \leqslant G'(\tau) \in \mathfrak{J}$, and the functions $F_k(\tau)$ in $\tilde{\mathfrak{G}}(\mathfrak{B}')$ belong to $\mathfrak{G}_\mu^+(\mathfrak{B})$, and hence $F(\tau) \in \mathfrak{G}_\mu(\mathfrak{B})$. Thus $(\tilde{\mathfrak{G}}(\mathfrak{B}')/\mathfrak{J}) \subset \mathfrak{G}_\mu(\mathfrak{B})$, and together with (12), this gives

$$\mathfrak{G}_\mu(\mathfrak{B}) = (\tilde{\mathfrak{G}}(\mathfrak{B}')/\mathfrak{J}), \quad \mathfrak{J} = \mathfrak{G}_\mu^+(\mathfrak{B}). \tag{13}$$

It follows from (13) that, whenever $\tilde{\mathfrak{G}}(\mathfrak{B}')$ is a Baire $*$-field of functions, the family $\mathfrak{G}_\mu(\mathfrak{B})$ is also a Baire $*$-field of functions (Theorem 2.3.1).

Since $\mathfrak{G}_\mu(\mathfrak{B})$ contains the $*$-field of functions $\mathfrak{E}_0(\mathfrak{B})$, which consists of all functions of the form $\sum_{k=1}^{n} \alpha_k \chi(E_k)$, where $E_k \in \mathfrak{B}$ $(k = 1, 2, \ldots, n)$, it also contains the Baire closure of $\mathfrak{E}_0(\mathfrak{B})$ with respect to the majorant system $\mathfrak{J} = \mathfrak{E}_0^+(\mathfrak{B})$, which we denote by $\mathfrak{G}(\mathfrak{B})$ (in the case of a regular field $\mathfrak{B}$, it coincides with $\mathfrak{G}(\mathfrak{B})$ in the old sense). In each Baire $*$-field of functions $\mathfrak{F}$ which satisfies

$$\mathfrak{G}(\mathfrak{B}) \subset \mathfrak{F} \subset \mathfrak{G}_\mu(\mathfrak{B}),$$

the integral $I(F)$ induces a positive linear operation on $\mathfrak{G}_\mu(\mathfrak{B})$ which we will also call the *integral $I(F)$ on $\mathfrak{F}$ with respect to the σ-finite measure $\mu(M)$.*

REMARK. If $\vartheta_n(\tau)$ is the characteristic function of the set T_n, then for $F(\tau)\in\mathfrak{G}_{\mu_n}(\mathfrak{B}')$,

$$I_n(F)=I(\vartheta_n F),$$

and hence for any function $F(\tau)$ in $\mathfrak{G}_\mu(\mathfrak{B})$,

$$I(F)=\lim_{n\to\infty} I(\vartheta_n F). \tag{14}$$

THEOREM 2.10.1. *If Z satisfies one of the conditions* I *or* II *of* Sec. 1.9.2, *then the integral $I(F)$ is a positive Lebesgue operation.*

In fact, let a non-increasing sequence $G_n(\tau)$ in $\mathfrak{G}_\mu(\mathfrak{B})$ converge to zero. Since

$$I(G_n)=I_k(G_n)+(I(G_n)-I_k(G_n)),$$

$$I(G_n)-I_{k'}(G_n)=I((1-\vartheta_k)G_n)\leqslant I((1-\vartheta_k)G_1)=I(G_1)-I_k(G_1),$$

we have

$$0\leqslant I(G_n)\leqslant I_k(G_n)+(I(G_1)-I_k(G_1)), \tag{15}$$

in which the first term on the right-hand side, according to Theorem 2.6.4, converges to zero for $n\to\infty$ ($G_n(\tau)\in\mathfrak{G}_{\mu_k}(\mathfrak{B}')$, and the measure $\mu_k(M)$ is finite). Therefore, in the limit (15) gives

$$0\leqslant\lim I(G_n)=\inf I(G_n)\leqslant I(G_1)-I_k(G_1)$$

and, by virtue of (11), $\lim I(G_n)=0$, which proves the continuity of the positive linear operation $I(F)$ defined on the Baire $*$-field of functions $\mathfrak{G}_\mu(\mathfrak{B})$.

COROLLARY. *The definition of the integral $I(F)$ is independent of the fixed representation in terms of monotone increasing T_n which was used in* (6).

Indeed, (14) remains true if $\vartheta_n(\tau)$ is the characteristic function of the set T_n of an arbitrary representation of this type.

§ 11. Extensions of Additive Set Functions

11.1. Additive set functions on separating families. A system $\mathfrak{N}$ of subsets of a set T is said to be a *separating family of sets* of T if:

1) the intersection of two sets of $\mathfrak{N}$ also belongs to $\mathfrak{N}$;

2) the difference of two sets of $\mathfrak{N}$ can be represented in the form of the union of a finite number of disjoint sets of $\mathfrak{N}$;

3) the union of all sets of $\mathfrak{N}$ is equal to T.

A separating family of sets of T is said to be *proper* if it contains T.

A field of sets is a separating family. Another example of a separating family is the collection of all intervals of the real line, or of n-dimensional space. (A one-dimensional interval can be open, half-open, closed, finite, infinite, or degenerate; an n-dimensional interval is the product of n one-dimensional intervals.)

We will associate a family $\mathfrak{E}(\mathfrak{N})$ of *elementary* ($\mathfrak{N}$)-*functions* with each separating family $\mathfrak{N}$, where $\mathfrak{E}(\mathfrak{N})$ is the collection of functions $\varphi(\tau)$ which can be represented in the form

$$\varphi(\tau) = \sum_{k=1}^{m} \alpha_k \chi_k(\tau), \tag{1}$$

where the $\chi_k(\tau)$ are the characteristic functions of disjoint sets N_k of $\mathfrak{N}$, and the coefficients α_k are any complex numbers.

We will say that (1) is a *representation of the function* $\varphi(\tau)$ *on the set* $N' = \bigcup_{k=1}^{m} N_k$ *in terms of the* $\mathfrak{N}$-*partition* $\{N_k\}$. If we add any number of terms $\alpha_k \chi_k(\tau)$ with coefficients $\alpha_k = 0$ to this representation, then we obtain a representation of $\varphi(\tau)$ on the set $N'' \supset N'$ which is called an *extension of the representation* (1).

LEMMA 2.11.1. *The representation* (1) *of a function* $\varphi(\tau)$ *on* N' *can be extended to a representation of the function* $\varphi(\tau)$ *on* $M' \cup N'$ *for any set* M' *of the form*

$$M' = \bigcup_{n=1}^{l} M_n, \quad M_n \in \mathfrak{N}, \quad M_k \cap M_n = \varnothing \quad \textit{for} \quad k \neq n. \tag{2}$$

Indeed, for any $M \in \mathfrak{N}$, the set

$$M \setminus M \cap N' = M \setminus \bigcup_{k=1}^{m} (M \cap N_k) = \bigcap_{k=1}^{m} (M \setminus M \cap N_k)$$

can be represented in the form of a sum of disjoint sets of $\mathfrak{N}$, since by virtue of 1) and 2), the sets $M \setminus M \cap N_k$ have this property. Thus we obtain an extension of the representation to the set $M \cup N' = N' \cup (M \setminus M \cap N')$. By successively continuing this process of extension with the sets M_1, M_2, $\ldots$, M_l in place of M, we obtain the desired extension of the representation to $M' \cup N'$.

THEOREM 2.11.1. *The family* $\mathfrak{E}(\mathfrak{N})$ *is a *-field of functions.*

In fact, we extend the representations of the functions $\varphi(\tau)$ and $\psi(\tau)$ from $\mathfrak{E}(N)$ in terms of the sets N' and M' to $M' \cup N'$ (Lemma 2.11.1). Then the intersection $\{M_n \cap N_k\}$ of corresponding ($\mathfrak{N}$)-partitions $\{N_k\}$ and $\{M_n\}$ of the set $M' \cup N'$ is suitable for the representation of both of the functions $\varphi(\tau)$

and $\psi(\tau)$. Whence it follows that the functions $\varphi(\tau)\pm\psi(\tau)$ belong to $\mathfrak{E}(\mathfrak{N})$, and if $\varphi(\tau)$ and $\psi(\tau)$ are real, then both sup $\{\varphi, \psi\}$ and inf $\{\varphi, \psi\}$ belong to $\mathfrak{E}(\mathfrak{N})$. Thus the family $\mathfrak{E}(\mathfrak{N})$ is a vector $*$-system of functions, and the real functions in $\mathfrak{E}(\mathfrak{N})$ form a lattice of functions.

THEOREM 2.11.2. *The collection* $\mathfrak{N}'$ *of sets* M' *of the form* (2) *is a field of sets.*

Indeed, the characteristic function of the set M' in $\mathfrak{N}'$ belongs to $\mathfrak{E}(\mathfrak{N})$ and, conversely, to any characteristic function $\chi(\tau)\in\mathfrak{E}(\mathfrak{N})$ there corresponds a set $M'=T\{\chi(\tau)=1\}$ in $\mathfrak{N}'$. In addition to the characteristic functions $\chi'(\tau)$ and $\chi''(\tau)$, the family $\mathfrak{E}(\mathfrak{N})$ also contains the characteristic functions sup (χ', χ'') and inf $\{\chi', \chi''\}$, and if $\chi''(\tau)\geqslant\chi'(\tau)$, then it also contains the characteristic function $\chi''(\tau)-\chi'(\tau)$. Thus, in addition to M' and M'', the collection $\mathfrak{N}'$ also contains the sets $M'\cup M''$ and $M'\cap M''$. If $M''\supseteq M'$, then the set $M''\setminus M'$ also belongs to $\mathfrak{N}'$, and $\mathfrak{N}'$ is a field of sets.

By letting $\mathfrak{N}'=\mathfrak{K}$, we obviously have $\mathfrak{E}(\mathfrak{N})=\mathfrak{E}(\mathfrak{K})$; moreover, the transition from $\mathfrak{N}$ to the field $\mathfrak{K}$ is manifested only by the appearance of new representations of $\varphi(\tau)$ of the form (1).

The definition of an additive and σ-additive set function $\mu(M)$ given in Sec. 2.5.2 is preserved when the field $\mathfrak{K}$ is replaced by a separating family of sets $\mathfrak{N}$.

If an additive function $\mu(M)$ is defined on a separating family of sets $\mathfrak{N}$, then the *formula*

$$I(\varphi)=\sum_{k=1}^{m}\alpha_k\mu(N_k), \tag{3}$$

where $\varphi(\tau)$ *is represented by* (1), *uniquely defines a linear operation* $I(\varphi)$ *on the* $*$*-field of functions* $\mathfrak{E}(\mathfrak{N})$.

Indeed, if

$$\varphi(\tau)=\sum_{n=1}^{r}\alpha'_n\chi'_n(\tau)$$

is another representation of $\varphi(\tau)$ on the set M in terms of the $(\mathfrak{N})$-partition $\{M_n\}$, then*

$$\sum_{k=1}^{m}\alpha_k\mu(N_k)=\sum_{k=1}^{m}\alpha_k\sum_{n=1}^{r}\mu(N_k\cap M_n)=$$

$$=\sum_{n=1}^{r}\alpha'_n\sum_{k=1}^{m}\mu(N_k\cap M_n)=\sum_{n=1}^{r}\alpha'_n\mu(M_n),$$

so that the quantity $I(\varphi)$ in (3) is independent of the representation of the function.

The uniquely defined operation $I(\varphi)$ is linear since, for two functions

---

* We remark that if $N_k\cap M_n$ is non-empty, then $\alpha_k=\alpha'_n$; if $N_k\cap M_n$ is empty, then $\mu(N_k\cap M_n)=0$.

$\varphi(\tau)$ and $\psi(\tau)$ in $\mathfrak{E}(\mathfrak{N})$, representations can be constructed (as for the proof of Theorem 2.11.1) on the same set with a common $(\mathfrak{N})$-partition.

THEOREM 2.11.3. *An additive function $\mu(M)$ on a separating family of sets $\mathfrak{N}$ can be extended to an additive set function $\mu'(M)$ on the field $\mathfrak{K} = \mathfrak{N}'$ in a unique manner. If the function $\mu(M)$ is σ-additive, then its extension $\mu'(M)$ is σ-additive.*

In fact, we let

$$\mu'(M) = I(\chi),$$

for $M \in \mathfrak{K}$, where $\chi(\tau)$ is the characteristic function of the set M. The uniquely defined set function $\mu'(M)$ on $\mathfrak{K}$ is additive in view of the linearity of the operation $I(\varphi)$, and it is an extension of the function $\mu(M)$.

The extension of the function μ to an additive function defined on the field $\mathfrak{K} = \mathfrak{N}'$ is unique. Indeed, if μ'' is another extension, then from (2) we obtain

$$\mu''(M') = \sum_{n=1}^{l} \mu''(M_n) = \sum_{n=1}^{l} \mu(M_n) = \sum_{n=1}^{l} \mu'(M_n) = \mu'(M'),$$

so that $\mu''(M') = \mu'(M)$.

Now let the function $\mu(M)$ also be σ-additive, let $M \in \mathfrak{N}$, let

$$M = \bigcup_{n=1}^{\infty} M_n, \quad M_n \in \mathfrak{K}, \quad M_n \cap M_k = \varnothing \quad \text{for} \quad k \neq n \tag{4}$$

and let $M_n = \bigcup_{k=1}^{r_n} M_{nk}$, where the sets M_{nk} are disjoint and belong to $\mathfrak{N}$. Then, since the function $\mu(M)$ is σ-additive on $\mathfrak{N}$,

$$\mu'(M) = \mu(M) = \sum_{n,k} \mu(M_{nk}) = \sum_{n} \mu'(M_n),$$

and the σ-additivity of the function $\mu'(M)$ is proved in the special case when $M \in \mathfrak{N}$. The general case reduces to this particular case. Indeed, for $M \in \mathfrak{K}$, let (4) hold and let $M = \bigcup_{k=1}^{m} N_k$, where the N_k are disjoint sets in $\mathfrak{N}$. Then

$$N_k = \bigcup_{n=1}^{\infty} N_k \cap M_n, \quad N_k \cap M_n \in \mathfrak{K},$$

so that by what has been proved

$$\mu'(N_k) = \sum_{n=1}^{\infty} \mu'(N_k \cap M_n)$$

and

$$\mu'(M) = \sum_{k=1}^{m} \sum_{n=1}^{\infty} \mu'(N_k \cap M_n) = \sum_{n=1}^{\infty} \sum_{k=1}^{m} \mu'(N_k \cap M_n) = \sum_{n=1}^{\infty} \mu'(M_n).$$

REMARK. The operation $I(\varphi)$ on $\mathfrak{E}(\mathfrak{N})$, with respect to $\mu(M)$, coincides with $I(\varphi)$ on $\mathfrak{E}(\mathfrak{K}) = \mathfrak{E}(\mathfrak{N})$ with respect to the extension $\mu'(M)$ of the function $\mu(M)$. If the separating family $\mathfrak{N}$ is regular, then the field $\mathfrak{K} = \mathfrak{N}'$ is also regular, and the operation $I(\varphi)$ on $\mathfrak{E}(\mathfrak{K})$, with respect to the function $\mu'(M)$, is identical on $\mathfrak{K}$ to the one already considered in Sec. 2.6.1.

11.2. The completion of a non-negative additive set function. In the case when an additive function $\mu(M)$ on a field $\mathfrak{K}$ (or a separating family $\mathfrak{N}$) is non-negative, the linear operation $I(\varphi)$ defined by (3) is positive. Under this supposition we will consider the operation $I(\varphi)$ on the field $\mathfrak{E}_r(\mathfrak{K})$ of real functions in $\mathfrak{E}(\mathfrak{K})$, and we will denote it by $I_r(\varphi)$. The completion $\bar{I}_r(F)$ of the operation $I_r(\varphi)$ to $\overline{\mathfrak{E}}_r(\mathfrak{K})$ has been defined on a field of functions $\mathfrak{F} = \overline{\mathfrak{E}}_r(\mathfrak{K}) \supset \mathfrak{E}_r(\mathfrak{F})$, and $\bar{I}_r(F)$ is by definition considered to be the (real) integral with respect to a non-negative additive function $\mu(M)$ on the field $\mathfrak{K}$ (or the separating family $\mathfrak{N}$). In the special case when the field of sets $\mathfrak{K}$ is regular, $\bar{I}_r(F)$ is an extension of the integral $I_r(F)$ (Sec. 2.6.1), since the domain of the complete operation $\bar{I}_r(F)$ on a regular field of functions is closed with respect to uniform convergence (Sec. 2.7.2). The collection $\overline{\mathfrak{K}}$ of sets M whose characteristic functions belong to the field of functions $\mathfrak{F} = \overline{\mathfrak{E}}_r(\mathfrak{K})$ contains the field $\mathfrak{K}$, and is also a field of sets: in addition to the characteristic functions $\chi(\tau)$ and $\chi'(\tau)$, the functions sup (χ, χ') and inf (χ, χ') also belong to $\mathfrak{F}$, and if $\chi'(\tau) \geqslant \chi(\tau)$, then $\mathfrak{F}$ also contains the difference $\chi'(\tau) - \chi(\tau)$. On the field $\overline{\mathfrak{K}}$, the formula

$$\bar{\mu}(M) = \bar{I}_r(\chi_M), \qquad M \in \overline{\mathfrak{K}}, \tag{5}$$

defines a set function $\bar{\mu}(M)$ which is non-negative and additive since the operation $\bar{I}_r(F)$ is positive and linear. Since

$$\overline{\mathfrak{K}} \supset \mathfrak{K}, \quad \bar{I}_r(\chi_M) = I(\chi_M) = \mu(M) \quad \text{for} \quad M \in \mathfrak{K},$$

the function $\bar{\mu}(M)$ is an extension of the function $\mu(M)$. By definition $\bar{\mu}(M)$ is the *completion of the function* $\mu(M)$. A non-negative additive function $\mu(M)$ on a field of sets $\mathfrak{K}$ is said to be *complete* if it coincides with its completion.

The completion $\bar{\mu}(M)$ *is a complete function.*

Indeed, the operation $I_r^{(\bar{\mu})}$ is an extension of the operation $I_r^{(\mu)}(\varphi) = I_r(\varphi)$, and $\bar{I}_r(\varphi)$ is the smallest complete extension of the operation $\bar{I}_r^{(\bar{\mu})}(\varphi)$. Therefore the completion $\bar{I}_r^{(\bar{\mu})}(\varphi)$ coincides with $I_r^{(\mu)}(\varphi) = \bar{I}_r(\varphi)$, so that $\overline{\overline{\mathfrak{K}}} = \overline{\mathfrak{K}}$ and, for $M \in \overline{\mathfrak{K}}$,

$$\bar{\bar{\mu}}(M) = \bar{I}_r^{(\bar{\mu})}(\chi_M) = \bar{I}_r^{(\mu)}(\chi_M) = \bar{\mu}(M),$$

i.e. $\bar{\mu}(M)$ is equal to its completion $\bar{\bar{\mu}}(M)$.

Non-additive extensions of the function $\mu(M)$ on the field $\mathfrak{K}$ can be obtained by using directly the upper and lower operations $\hat{I}_r(F)$ and $\check{I}_r(F)$

(the upper and lower integrals) with respect to $I_r(\varphi)$, namely, by letting

$$\widehat{\mu}(M)=\widehat{I}_r(\chi_M), \qquad \breve{\mu}(M)=\breve{I}_r(\chi_M), \tag{6}$$

where M is a set with the characteristic function $\chi_M(\tau)$ which belongs to the family $[\mathfrak{E}_r(\mathfrak{R})]$. In particular, if the field $\mathfrak{R}$ is regular, then the function identically equal to one belongs to $\mathfrak{E}_r(\mathfrak{R})$, and M is an arbitrary subset of the set T.

In the general case where $[\mathfrak{E}_r(\mathfrak{R})]$ is a Baire field of functions (Sec. 2.8.1) which contains $\mathfrak{E}_r(\mathfrak{R})$, the sets M form a Borel field of sets $[\mathfrak{R}]$ which contains $\mathfrak{R}$. In addition to each set M, the field $[\mathfrak{R}]$ also contains all subsets of the set M.

The non-negative set functions $\widehat{\mu}(M)$ and $\breve{\mu}(M)$ defined on the field $[\mathfrak{R}]$, which are monotone, are extensions of the function $\mu(M)$ ($\widehat{I}_r(\chi_M)=\breve{I}_r(\chi_M)$ $= I_r(\chi_M)=\mu(M)$, if $M\in\mathfrak{R}$). They are said to be, respectively, the *outer and inner functions with respect to* $\mu(M)$. It follows from the inequalities (2.7.3) and (2.7.3′) for upper and lower operations that

$$\left.\begin{aligned}\widehat{\mu}(M_1\cup M_2)+\widehat{\mu}(M_1\cap M_2)\leqslant\widehat{\mu}(M_1)+\widehat{\mu}(M_2),\\ \breve{\mu}(M_1\cup M_2)+\breve{\mu}(M_1\cap M_2)\geqslant\breve{\mu}(M_1)+\breve{\mu}(M_2),\end{aligned}\right\} \tag{7}$$

for upper and lower functions if we let $F(\tau)=\chi_{M_1}(\tau)$, $G(\tau)=\chi_{M_2}(\tau)$, and $L(F)=I_r(\varphi)$. It follows from (5) and (6) that the field $\overline{\mathfrak{R}}$ consists precisely of those sets M in $[\mathfrak{R}]$ for which

$$\widehat{\mu}(M)=\breve{\mu}(M),$$

and that $\bar{\mu}(M)=\widehat{\mu}(M)=\breve{\mu}(M)$ on $\overline{\mathfrak{R}}$.

For the outer function $\widehat{\mu}(M)$, it is possible to obtain the formula

$$\widehat{\mu}(M)=\inf_{M'\supset M}\mu(M') \qquad (M'\in\mathfrak{R}), \tag{8}$$

which expresses directly $\widehat{\mu}(M)$ in terms of the function $\mu(M)$. Indeed

$$\widehat{\mu}(M)=\inf I_r(\varphi), \qquad \chi_M(\tau)\leqslant\varphi(\tau)\in\mathfrak{E}_r(\mathfrak{R}), \tag{8′}$$

and if, in the representation (1) of the function $\varphi(\tau)$, we let $\alpha_k'=1$ when $\alpha_k\geqslant 1$, and let $\alpha_k'=0$ when $\alpha_k<1$, then

$$\chi_M(\tau)\leqslant\sum_{k=1}^{m}\alpha_k'\chi_k(\tau)=\chi'(\tau)\in\mathfrak{E}_r(\mathfrak{R}),$$

where $\chi'(\tau)$ is the characteristic function of the set M' in $\mathfrak{R}$. Therefore, the lower bound $\inf I_r(\varphi)$ in (8′) is already attained on the characteristic functions of the sets M in $\mathfrak{R}$ such that $M'\supset M$, i.e. (8) holds.

Remark. It follows from (8) that the field $[\mathfrak{R}]$, which is the domain of the function $\widehat{\mu}(M)$, consists precisely of all subsets of the sets of $\mathfrak{R}$.

In the case under consideration, the formula

$$\breve{\mu}(M) = \mu(N) - \hat{\mu}(N \setminus M), \qquad M \subset N \in \mathfrak{R}. \tag{9}$$

corresponds to the definition (2.7.8) of the operation $\breve{L}(F)$. In fact

$$\breve{\mu}(M) = -\hat{I}_r(-\chi_M) = -\inf I_r(\varphi), \quad -\chi_M(\tau) \leqslant \varphi(\tau) \in \mathfrak{E}_r(\mathfrak{R}),$$

and hence

$$\breve{\mu}(M) = I(\chi_N) - I(\chi_N) - \inf I_r(\varphi) = \mu(N) - \inf I(\chi_N + \varphi). \tag{9'}$$

But the inequality $-\chi_M(\tau) \leqslant \varphi(\tau) \in \mathfrak{E}_r(\mathfrak{R})$ is equivalent to

$$\chi_{N \setminus M}(\tau) = \chi_N(\tau) - \chi_M(\tau) \leqslant \varphi'(\tau) \in \mathfrak{E}_r(\mathfrak{R}),$$
$$\varphi'(\tau) = \chi_N(\tau) + \varphi(\tau),$$

so that (9′) coincides with (9).

From (9) and (8) we obtain

$$\breve{\mu}(M) = \sup_{M'' \subset M} \mu(M'') \qquad (M'' \in \mathfrak{R}) \tag{10}$$

which is similar to (8). Indeed,

$$\hat{\mu}(N \setminus M) = \inf_{M' \supset N \setminus M} \mu(M') \qquad (M' \in \mathfrak{R}).$$

If M' and M'' are sets in $\mathfrak{R}$ such that $M'' = N \setminus M'$, then the inclusion $M' \supset N \setminus M$ is equivalent to the inclusion $M'' \subset M$. Therefore

$$\breve{\mu}(M) = \mu(N) - \inf_{M'' \subset M} \mu(N \setminus M'') = \sup_{M'' \subset M} \mu(M'').$$

11.3. The extension to a measure.

Lemma 2.11.2. *If a non-negative function $\mu(M)$ on a field of sets $\mathfrak{R}$ (or a separating family $\mathfrak{N}$) is countably additive, then the operation $I(\varphi)$, defined on $\mathfrak{E}(\mathfrak{R})$ by* (3), *satisfies the condition*

$$\lim I(\varphi_n) = 0, \tag{11}$$

if the sequence $\{\varphi_n(\tau)\}$ in $\mathfrak{E}(\mathfrak{R})$ is non-increasing and converges to zero.

Proof. By analogy to (2.6.15), for any positive ε,

$$0 \leqslant I(\varphi_n) \leqslant \gamma\mu(M_n) + \varepsilon I(\varphi_1), \tag{12}$$

where $\gamma = \sup_{\tau} \{\varphi_1(\tau)\}$ and $M_n = T\ (\varphi_n > \varepsilon\varphi_1)$. Indeed, we consider the representation (1) of the functions $\varphi_1(\tau)$ and $\varphi_n(\tau)$ in terms of the same $\chi_k(\tau)$, i.e. with a common $(\mathfrak{N})$-partition $\{N_k\}$ (see the proof of Theorem 2.11.1). Then the set M_n is the union of some N_k and, on the remaining N_k, we have $\varphi_n(\tau) \leqslant \varepsilon\varphi_1(\tau)$. Therefore

$$0 \leqslant I(\varphi_n) = \sum_{N_k \subset M_n} \alpha_k^{(n)} \mu(N_k) + \sum_{N_k \not\subset M_n} \alpha_k^{(n)} \mu(N_k) \leqslant$$

$$\leqslant \gamma \sum_{N_k \subset M_n} \mu(N_k) + \varepsilon \sum_{N_k \not\subset M_n} \alpha_k^{(1)} \mu(N_k) \leqslant \gamma_M(M_n) + \varepsilon I(\varphi_1).$$

Since the sequence $\{\varphi_n(\tau)\}$ is non-increasing and converges to zero,

$$M_n \supset M_{n+1}, \quad \bigcap_{n=1}^{\infty} M_n = \varnothing,$$

and $\lim \mu(M_n) = 0$ since $\mu(M)$ is σ-additive. The estimate (12) in the limit now gives

$$0 \leqslant \lim I(\varphi_n) = \inf \{I(\varphi_n)\} \leqslant \varepsilon I(\varphi_1),$$

whence (11) now follows since ε was arbitrary.

Here and in the following sections, we will assume that the complete vector order Z, to which the values of the countably additive function $\mu(M)$ and the operations generated by it belong, satisfies one of the conditions (I) or (II) of Sec. 1.9.2.

THEOREM 2.11.4. *A set function $\mu(M)$ on a field $\mathfrak{K}$ (or a separating family $\mathfrak{N}$) admits an extension to a measure if and only if the function $\mu(M)$ is non-negative and is σ-additive. Extension to a measure is unique on the smallest Borel field $\mathfrak{B}$ which contains $\mathfrak{K}$ (or $\mathfrak{N}$).*

The necessity of the condition is obvious, and in proof of its sufficiency it is possible to restrict ourselves to the case of a field $\mathfrak{K}$, since a non-negative σ-additive function on $\mathfrak{N}$ is extended in a unique manner to such a function on the field $\mathfrak{K} = \mathfrak{N}'$ (Theorem 2.11.3).

If the function $\mu(M)$ on $\mathfrak{K}$ is non-negative and σ-additive, then the positive linear operation $I_r(\varphi)$, defined on $\mathfrak{E}_r(\mathfrak{K})$ by (3), satisfies (11) by Lemma 2.11.2, and therefore, by Theorem 2.8.1, it admits an extension to a positive Lebesgue operation $I'_r(F)$ defined on some Baire field of functions $\mathfrak{F}' \supset \mathfrak{E}_r(\mathfrak{K})$. By virtue of Lemma 2.5.4, the collection $\mathfrak{B}'$ of sets whose characteristic functions belong to $\mathfrak{F}'$ is a Borel field, and

$$\mu'(M) = I'_r(\chi_M), \quad M \in \mathfrak{B}',$$

defines a measure $\mu'(M)$ on $\mathfrak{B}'$ which is an extension of the function $\mu(M) = I_r(\chi_M)$, $M \in \mathfrak{K}$.

The measure $\mu'(M)$ induces a measure on $\mathfrak{B}$ for which we will retain the notation $\mu'(M)$. It is necessary to show that any extension of the function $\mu(M)$ to a measure $\mu''(M)$ on a Borel field $\mathfrak{B}''$ induces one and the same measure on $\mathfrak{B}$, independent of the method of extension. If we let $\mathfrak{B}_1$ denote the collection of those sets N in $\mathfrak{B}$ for which the value of $\mu''(N)$ is independent of the extension $\mu''(M)$ of $\mu(M)$, then the above statement is equivalent to the equation $\mathfrak{B} = \mathfrak{B}_1$, which also must be proved.

We let $\mathfrak{F}$, $\mathfrak{F}'$, $\mathfrak{F}''$, and $\mathfrak{F}_1$, denote the collections of characteristic set functions from $\mathfrak{R}$, $\mathfrak{B}$, $\mathfrak{B}''$, and $\mathfrak{B}_1$, respectively, and we let $L(\chi)$, $L'(\chi)$, and $L''(\chi)$, denote the operations on $\mathfrak{F}$, $\mathfrak{F}'$, and $\mathfrak{F}''$ defined by the formulas

$$L(\chi_M) = \mu(M), \qquad L'(\chi_M) = \mu'(M), \qquad L''(\chi_M) = \mu''(M),$$

where M belongs to $\mathfrak{R}$, $\mathfrak{B}$, and $\mathfrak{B}''$, respectively. The argument presented in the remark to Theorem 2.9.1 remains valid for such operations L, L', and L'', and it shows* that $\mathfrak{F}' = \mathfrak{F}_1$, i.e. $\mathfrak{B} = \mathfrak{B}_1$.

11.4. Inner and outer measures. The functions $\hat{\mu}(M)$ and $\breve{\mu}(M)$ (see (6)) are said to be, respectively, the *outer measure* and the *inner measure* if $\mu(M)$ is a measure on a Borel field $\mathfrak{B} \subset \mathfrak{P}(T)$. In this case, (7) admits the generalization to a countable union of sets. However, for functions $\mu(M)$ the corresponding inequality

$$\breve{\mu}(M) \geqslant \sum_{k=1}^{\infty} \breve{\mu}(M_k), \tag{13}$$

holds even in the case when the function $\mu(M)$ is defined on the field $\mathfrak{R}$, where the set M belongs to $[\mathfrak{B}]$ (or $[\mathfrak{R}]$), $M = \bigcup_k M_k$, and the sets M_k are pairwise disjoint. Indeed, for disjoint M_1 and M_2, the second of the inequalities in (7) can be written in the form

$$\breve{\mu}(M_1 \cup M_2) \geqslant \breve{\mu}(M_1) + \breve{\mu}(M_2)$$

and can be extended in an obvious manner to any finite number of pairwise disjoint terms. Therefore

$$\breve{\mu}(M) \geqslant \breve{\mu}\left(\bigcup_{k=1}^{n} M_k\right) \geqslant \sum_{k=1}^{n} \breve{\mu}(M_k)$$

and we obtain (13) by letting $n \to \infty$.

For an outer measure $\hat{\mu}(M)$, instead of (13), we have

$$\hat{\mu}(M) \leqslant \sum_{k=1}^{\infty} \hat{\mu}(M_k), \tag{14}$$

where $M \in [\mathfrak{B}]$ and $M = \bigcup_k M_k$ (under the assumption that the series in (14) converges). For the proof of (14), an additional condition on the vector order Z will be used. In the case of the additional condition (II) of Sec. 1.9.2, in view of (8), there can be found sets N_k such that $M_k \subset N_k \in \mathfrak{B}$, $N_k \subset M' \in \mathfrak{B}$, $M' \supset M$, and such that the series $\sum_k \mu(N_k)$ converges in Z. Then the sum

---

* The fact that now $\mathfrak{I} = \mathfrak{F}^+$ is not a majorant system is unessential, since $\mathfrak{F}^+$ is contained in the majorant system $(\mathfrak{E}_r(\mathfrak{R}))^+$.

$N' = \bigcup_{k=1}^{\infty} N_k$ also belongs to $\mathfrak{B}$, and by letting $N_k' = N_k \setminus \bigcup_{m=1}^{k-1} N_m$, we can write

$$N' = \bigcup_{k=1}^{\infty} N_k', \quad N_k' \cap N_m' = \varnothing \quad \text{for} \quad k \neq m, \quad N_k' \in \mathfrak{B}.$$

Since $N' \supset M$ and $N_k' \subset N_k$, we have

$$\hat{\mu}(M) \leqslant \mu(N') = \sum_{k=1}^{\infty} \mu(N_k') \leqslant \sum_{k=1}^{\infty} \mu(N_k).$$

Here, by virtue of the equation $\inf \mu(N_k) = \mu(M_k)$, and the remark after Lemma 1.9.1, the lower bound of the right-hand side is equal to the right-hand side in (14). This proves (14) in the case under consideration. If Z satisfies the condition (1) of Sec. 1.9.2, then (14) is satisfied for functions $\hat{\nu}(M) = p(\hat{\mu}(M))$, where p is any positive completely linear functional in Z, and hence it is satisfied for $\hat{\mu}(M)$.

We will say that a measure $\mu(M)$ is *complete* if it coincides with its completion $\bar{\mu}(M)$.

THEOREM 2.11.5. *The completion $\bar{\mu}(M)$ of a measure $\mu(M)$ is a complete measure.*

In fact, it is aleady known that $\bar{\mu}(M)$ is a complete function on the field of sets $\bar{\mathfrak{B}}$, and it is only necessary to show that $\bar{\mu}(M)$ is a measure. Let

$$M = \bigcup_{k=1}^{\infty} M_k, \quad M_k \cap M_n = \varnothing \quad \text{for} \quad k \neq n,$$

$$M_k \in \bar{\mathfrak{B}}, \quad M_k \subset M' \in \bar{\mathfrak{B}}.$$

Then, since $\bar{\mathfrak{B}} \subset [\mathfrak{B}]$, the set M belongs to the Borel field $[\mathfrak{B}]$, $\hat{\mu}(M_k) = \check{\mu}(M_k) = \bar{\mu}(M_k)$, and (13) and (14) give

$$\hat{\mu}(M) \leqslant \sum_{k=1}^{\infty} \bar{\mu}(M_k) \leqslant \check{\mu}(M).$$

Thus, $\hat{\mu}(M) = \check{\mu}(M) = \bar{\mu}(M)$, and therefore

$$M \in \bar{\mathfrak{B}}, \quad \bar{\mu}(M) = \sum_{k=1}^{\infty} \bar{\mu}(M_k),$$

i.e. $\bar{\mathfrak{B}}$ is a Borel field and the function $\bar{\mu}(M)$ is σ-additive on $\bar{\mathfrak{B}}$.

REMARK. With each non-negative σ-additive function $\mu(M)$ on a field of sets $\mathfrak{K}$ (or a separating family $\mathfrak{N}$), it is possible to relate the three functions

$$\bar{\mu}'(M), \quad \hat{\mu}'(M), \quad \check{\mu}'(M), \tag{15}$$

where $\mu'(M)$ is a uniquely defined measure which is an extension of the function $\mu(M)$ to the smallest Borel field which contains $\mathfrak{K}$ (or $\mathfrak{N}$) (Theorem 2.11.4). The functions (15) are said to be, respectively, the *complete measure*, *outer measure*, and *inner measure* generated by the function $\mu(M)$.

CHAPTER 3

Spaces

§ 1. Normed Spaces

1.1. The norm. A *norm* in a vector system $\mathfrak{B}$ (complex or real) is a real function $w(f)$, defined on $\mathfrak{B}(f \in \mathfrak{B})$, which satisfies the following conditions:

1) $w(f+g) \leqslant w(f)+w(g)$ for any f and g in $\mathfrak{B}$ (triangle inequality);
2) $w(\alpha f) = |\alpha| w(f)$ for $f \in \mathfrak{B}$ and α complex or real;
3) $w(f) > 0$, if $f \neq 0$.

We note that $w(0) = 0$, since $w(0) = w(0 \cdot 0) = 0\, w(0) = 0$.

By means of a norm, a vector system $\mathfrak{B}$ can be converted into a vector space (Sec. 1.8.2), i.e. in $\mathfrak{B}$ there can be introduced a notion of a limit which satisfies the conditions (a), (b), (c) of Sec. 1.8.1, and (d) of Sec. 1.8.2. For a given norm $w(f)$, *a sequence* $\{f_n\}$ *in* $\mathfrak{B}$ *has by definition a limit* $\lim f_n = f \in \mathfrak{B}$ *if* $\lim w(f_n - f) = 0$.

Thus, as defined, the limit obviously satisfies the conditions (b) and (c) of Sec. 1.8.1 and is unique. Indeed, from the equations $\lim f_n = f$ and $\lim f_n = f'$, and the triangle inequality

$$0 \leqslant w(f' - f) \leqslant w(f' - f_n) + w(f_n - f)$$

it follows that $w(f' - f) = 0$, or by the property 3) of the norm, $f' = f$.

From properties of the norm, it is possible to verify that the conditions (a) and (d) are satisfied. In fact,

$$f_n + g_n - (f + g) = (f_n - f) + g_n - g),$$
$$\alpha_n f_n - \alpha f = \alpha_n (f_n - f) + (\alpha_n - \alpha) f,$$

and hence by the properties 1) and 2) of the norm

$$w(f_n + g_n - (f + g)) \leqslant w(f_n - f) + w(g_n - g),$$
$$w(\alpha_n f_n - \alpha f) \leqslant |\alpha_n| w(f_n - f) + |\alpha_n - \alpha| w(f),$$

and therefore if $\lim f_n = f$, $\lim g_n = g$, and $\lim \alpha_n = \alpha$,

$$\lim (f_n + g_n) = f + g, \quad \lim_{n \to \infty} \alpha_n f_n = \alpha f,$$

q.e.d.

Two norms, $w_1(f)$ and $w_2(f)$, in the vector system $\mathfrak{V}$ are said to be *equivalent* if they define the same notion of limit in $\mathfrak{V}$, i.e. if together with $\mathfrak{V}$, they define the same vector space.

We fix one of the equivalent norms, and turn to the concept of normed space.

A complex (real) vector system $\mathfrak{V}$ with a norm $w(f)$ forms by definition a complex (real) normed space.

We denote a normed space by the symbol X, and we denote the value of the norm of a vector $f \in X$ by $\|f\|$. Elements of the space X will also be called *points*. When the vector system X coincides with (0) (Sec. 1.2.1), the space X will be said to be *degenerate* and we will write $X = (0)$.

In what follows, a normed space will always be considered to be a *vector space in which the concept of limit has been defined with the help of a norm.*

If a sequence $\{f_n\} \subset X$ has the limit $f_n = f$, i.e. if

$$\lim \|f_n - f\| = 0,$$

then we say that the sequence $\{f_n\}$ *converges strongly* to the element f. If the sequence $\{f_n\}$ *converges strongly* to f, then we will write

$$f_n \Rightarrow f.$$

Algebraic operations in X are continuous with respect to strong convergence, i.e. if

$$f_n \Rightarrow f, \quad g_n \Rightarrow g \quad \textit{and} \quad \lim \alpha_n = \alpha,$$

then

$$f_n + g_n \Rightarrow f + g, \quad \alpha_n f_n \Rightarrow \alpha f. \tag{1}$$

With the help of the norm it is possible to define the *distance* $d(f, g)$ between two points f and g of the space X. Namely, we can take

$$d(f, g) = \|f - g\|.$$

By the properties of the norm

$$\|f - g\| = \|(f - h) + (h - g)\| \leqslant \|f - h\| + \|h - g\|,$$
$$\|\alpha (f - g)\| = |\alpha| \, \|f - g\|,$$
$$\|f - g\| = 0 \ \textit{is equivalent to} \ f - g = 0.$$

Turning to $d(f, g)$, we obtain

1′) $d(f, g) \leqslant d(f, h) + d(h, g)$,

2′) $d(g, f) = d(f, g)$,

3′) $d(f, g) \geqslant 0$, where $d(f, g) = 0$ *is equivalent to* $f = g$. In general, a set X with a non-negative function $d(f, g)$ defined on pairs of elements $f, g \in X$, which satisfies the conditions 1')—3'), is called a *metric space*; its elements are *points*, and $d(f, g)$ is the *distance between two points f and g*.

In particular, *a normed space is a metric space*, moreover in this particular case the distance $d(f, g)$ has the additional properties.

4') $d(f+h, g+h) = d(f, g)$
(invariance with respect to the shift $f \to f+h$);

5') $d(\alpha f, \alpha g) = |\alpha| d(f, g)$.

The set of points f of a normed space X which satisfy the condition

$$d(f, f_0) = \|f - f_0\| < \delta, \tag{2}$$

is called an *open sphere with center f_0 and radius* δ, or a *spherical neighborhood* of the point f_0. The set of points f which satisfies

$$d(f, f_0) = \|f - f_0\| = \delta,$$

is called a *sphere*.

A spherical neighborhood (2) *induces a topology in the space X*, and as in any metric space, it induces the corresponding concepts of *limit*, *convergence*, *closed* and *open sets*, *closure*, *denseness*, *compactness*, etc. In particular, *convergence in a metric space X* coincides with strong convergence in the (normed) space X; a set $N \subset X$ is *closed* if the limit of any strongly convergent sequence in N belongs to N; a set $N \subset X$ is *compact* in X if each sequence from N contains a strongly convergent subsequence. A set N is *dense* in X if its closure coincides with X.

A function $F(f)$, defined on a subset of a space X and taking numerical values, is called a *functional*. Strong convergence in the space X corresponds to the concept of continuity of a functional. Namely, a functional $E(f)$ with domain $\Omega \subset X$ is by definition *continuous* at a point $f \in \Omega$ if $f_n \in \Omega$ and $f_n \Rightarrow f$ imply that

$$\lim F(f_n) = F(f).$$

The norm $\|f\|$ is a functional which is continuous at each point of the space. Indeed, from the property 1) of the norm we have

$$|\,\|f\| - \|g\|\,| \leqslant \|f - g\|, \tag{3}$$

which proves the continuity of the norm.

1.2. Vector manifolds. The dimensionality of a space. Every vector subsystem of a normed space X will be called a *vector manifold in X*, and if the manifold is closed it will be called a *subspace of the space X*.

The linear hull of an arbitrary subset M of a vector system X is a vector manifold in the space X. In the presence of a topology in X, the closure of the linear hull of a set M is of importance, i.e.—the so-called *closed linear hull of the set M*, which is the smallest of the subspaces of X which contain M.

A set M is said to be *fundamental in X* if the closed linear hull of M coincides with the whole space X. The smallest of the cardinalities of the fundamental sets is by definition the *dimensionality of a non-degenerate normed space X*, and the dimensionality of a degenerate space will be considered to be zero.

A space X is said to be *separable* if there exists a countable set M which is dense in X. The dimensionality of a separable space X is at most countable since a dense set is at the same time a fundamental set of X. Conversely, *if the dimensionality of a space X is finite or countable, then X is separable.*

Indeed, we form the set $\mathscr{L}_r(M)$ of all possible linear combinations, with rational complex (real) coefficients, of elements of some fundamental set M of the space X. Whenever the linear hull $\mathscr{L}(M)$ is dense in X, the set $\mathscr{L}_r(M)$ is also dense in X. In the case under consideration, where X has finite or countable dimensionality, a fundamental set M can be chosen which is, respectively, finite or countable. Then the set $\mathscr{L}_r(M)$, which is dense in X, is countable, and hence the space X is separable.

Thus, *separable spaces are spaces of finite or countable dimensionality.*

1.3. Isomorphisms and pseudoisomorphisms of spaces. The one-to-one mapping

$$f \to f' = Af, \qquad f \in X, \qquad f' \in X'$$

of a normed space X onto a normed space X' is said to be an *isomorphism* (*pseudoisomorphism*) if the mapping A is an isomorphism (pseudoisomorphism) of the vector systems X and X' and if the norm is preserved*, i.e.

$$\|f'\| = \|Af\| = \|f\|. \tag{4}$$

Two normed spaces X and X' are by definition *isomorphic* (*pseudoisomorphic*) if an isomorphism (pseudoisomorphism) can be established between them.

Under an isomorphic mapping (isomorphism) of the spaces X and X', a vector manifold and fundamental set in X correspond, respectively, to a vector manifold and fundamental set (of the same cardinality) in X', and hence the dimensionalities of two isomorphic spaces are equal.

If $X' = X$, then an isomorphic (pseudoisomorphic) mapping of the space X onto X' is by definition an *automorphism* (*pseudoautomorphism*). As in the case of a vector system (Sec. 1.4.2), *automorphisms form a group with*

---

* A normed space X can be considered to be a Φ-system (§1.3) where in addition to the relations of the vector system there has also been defined a one-term relation $\theta_a(f)$: $\|f\| = a$, where a is any non-negative number. Then the concept of isomorphism (pseudo-isomorphism) between X and X' coincides with the concept of isomorphism (pseudo-isomorphism) of corresponding Φ-systems.

respect to multiplication (*of transformations*). An automorphism (pseudo-automorphism) of a normed space X is said to be *involutive* if it is an involution of the vector system X.

A complex normed space X with a given involutive pseudoautomorphism will be called *a space with pseudoinvolution* or *a normed *-space*. In such a space, a given pseudoinvolution will be written in the form $f \to f^*$ so that

$$\left.\begin{array}{ll}(f+g)^* = f^* + g^*, & (\alpha f)^* = \bar{\alpha} f^*, \\ (f^*)^* = f, & \|f^*\| = \|f\|.\end{array}\right\} \tag{5}$$

1.4. The quasinorm. A real function $w(f)$ which satisfies the conditions 1) and 2) of the definition of norm in Sec. 3.1.1 will be called a *quasinorm on a vector system* $\mathfrak{V}$ (complex or real).

It follows from these conditions that a *quasinorm* $w(f)$ *is always non-negative*, i.e. $w(f) \geqslant 0$ for any $f \in \mathfrak{V}$. Indeed, by the properties 1) and 2) of the quasinorm, we have

$$w(0) = w(f + (-f)) \leqslant w(f) + w(-f) = 2w(f),$$

and since $w(0) = 0$, we have $w(f) \geqslant 0$. Moreover, as in the case of the norm

$$|w(f) - w(g)| \leqslant w(f - g). \tag{3'}$$

A quasinorm $w(f)$ *defined on a vector system* $\mathfrak{V}$ *is a linear equivalence relation* θ (Sec. 1.4.1), *namely*

$$f \equiv f' \,(\theta), \qquad \text{if} \qquad w(f - f') = 0. \tag{6}$$

Indeed, elements $h \in \mathfrak{V}$ which satisfy the equation

$$w(h) = 0,$$

form a vector subsystem $\mathfrak{V}_0$ of the system $\mathfrak{V}$, since whenever h_1, h_2, and h belong to $\mathfrak{V}_0$, the elements $h_1 + h_2$ and αh also belong to $\mathfrak{V}_0$:

$$0 \leqslant w(h_1 + h_2) \leqslant w(h_1) + w(h_2) = 0, \; w(\alpha h) = |\alpha| \, w(h) = 0.$$

The relation $w(f - f') = 0$ means that $f' - f \in \mathfrak{V}_0$, and the equivalence relation (6) defines (Sec. 1.4.1) a vector system $\mathring{\mathfrak{V}} = \mathfrak{V}/\mathfrak{V}_0$ (factor-system) of types $\mathring{f}$ with respect to θ. For two representatives f and f' of the same type $\mathring{f}$

$$w(f') = w(f), \tag{7}$$

since by (3′)

$$|w(f) - w(f')| \leqslant w(f - f') = 0.$$

A vector system $\mathfrak{V}$ *with a quasinorm* $w(f)$ *generates a normed space* X; $\mathring{\mathfrak{V}}$ *is a vector system of the space* X *and the norm* $\|\mathring{f}\|$ *of an element* $\|\mathring{f}\| \in \mathring{\mathfrak{V}}$ *is given by the formula*

$$\|\mathring{f}\| = w(f), \tag{8}$$

where f is a representative of the type $\mathring{f}$ with respect to θ.

In fact, by (7) the definition of norm given in (8) is independent of the choice of the representative f of the type $\mathring{f}$, and it is sufficient to verify that $\|\mathring{f}\|$ satisfies all three of the conditions of Sec. 3.1.1. Let f and g be representatives of the types $\mathring{f}$ and $\mathring{g}$. Then

1) $\|\mathring{f}+\mathring{g}\| = w(f+g) \leqslant w(f)+w(g) = \|\mathring{f}\|+\|\mathring{g}\|$,

2) $\|\alpha\mathring{f}\| = w(\alpha f) = |\alpha|\, w(f) = |\alpha|\, \|\mathring{f}\|$,

3) $\|\mathring{f}\| = w(f) \geqslant 0$,

and if $\|\mathring{f}\| = 0$, then $w(f) = 0$, and hence $\mathring{f} = 0$.

In correspondence with the remark in Sec. 1.1.1, it is sometimes expedient to consider $\mathfrak{V}$ to be a vector system of the space X which has been generated by the system $\mathfrak{V}$ and the quasinorm $w(f)$, and where elements f and f' in $\mathfrak{V}$ are considered to be equal if the congruence (6) holds.

§ 2. Banach Space

2.1. Completion of a space. In a normed space X, if a sequence of elements $\{f_n\}$ is strongly convergent to a vector f, then by the properties of the norm (Sec. 3.1.1)

$$\|f_n - f_m\| \leqslant \|f_n - f\| + \|f - f_m\|$$

and hence

$$\lim_{n,\, m\to\infty} \|f_n - f_m\| = 0. \tag{1}$$

A sequence $\{f_n\}$ in X which satisfies the condition (1) is called a *fundamental sequence* or a *Cauchy sequence.*

A normed space is said to be a *complete space* or a *Banach space* if every Cauchy sequence in it converges to some element of the space. If either of two isomorphic spaces is complete, then the other is complete.

A subspace of a complete space is also complete.

In fact, if the elements f_n of a Cauchy sequence $\{f_n\}$ belong to a subspace X' of a complete space X, then since X' is closed, the limit element $f \in X$ of the sequence $\{f_n\}$ belongs to X'.

A normed space X' is by definition an *extension of a normed space* X, if X is a vector subsystem of the system X', and if the value of the norm of an element in X is the same as its value in the space X'. Here, two extensions X' and X'' of a space X are not considered distinct if there exists an isomorphism of the space X' onto X'' which leaves each element of X unchanged.

THEOREM 3.2.1. *Every normed space X has complete extensions, and among them there is a unique extension in which X is dense.*

PROOF. The set $\mathfrak{B}$ of all Cauchy sequences of the space X is obviously a vector subsystem of the system $\mathfrak{G}(N_0; X)$ (Sec. 1.2.2), where N_0 is the set of natural numbers. By the inequality

$$| \|f_n\| - \|f_m\| | \leqslant \|f_n - f_m\|$$

(see (3.1.3)), for elements $\xi = \{f_n\} \in \mathfrak{B}$ the limit

$$w(\xi) = \lim \|f_n\|$$

exists.

We will show that $w(\xi)$ is a quasinorm in $\mathfrak{B}$. Indeed, for two elements $\xi = \{f_n\}$ and $\eta = \{g_n\}$ in $\mathfrak{B}$, we have:

$$w(\xi + \eta) = \lim \|f_n + g_n\| \leqslant \lim (\|f_n\| + \|g_n\|) = w(\xi) + w(\eta),$$
$$w(\alpha\xi) = \lim \|\alpha f_n\| = |\alpha| \lim \|f_n\| = |\alpha|\, w(\xi).$$

According to Sec. 3.1.4, the vector system $\mathfrak{B}$ with quasinorm $w(\xi)$ generates a normed space Y with a vector system $\mathring{\mathfrak{B}} = \mathfrak{B}/\mathfrak{B}_0$, where $\mathfrak{B}_0$ is the vector system of elements $\xi \in \mathfrak{B}$ which satisfy the equation $w(\xi) = 0$. In other words, $\xi = \{f_n\} \in \mathfrak{B}$ belongs to $\mathfrak{B}_0$ if and only if $f_n \Rightarrow 0$. We have

$$\|\mathring{\xi}\| = w(\xi) = \lim_{n\to\infty} \|f_n\|$$

for the norm of an element $\mathring{\xi} \in Y$ with representative $\xi = \{f_n\}$.

We consider elements in a normed space Y which have representatives ξ of the form $\xi = \{f_n\}$, where $f_n = f \in X$, $n = 1, 2, \ldots$ and in this case we will write $\mathring{f}$ instead of $\mathring{\xi}$. The set $\mathring{X}$ of all elements $\mathring{f}$ forms a vector manifold in Y, and $f \to \mathring{f}$ is an isomorphism of the space X onto the manifold $\mathring{X}$.

The vector manifold $\mathring{X}$ is dense in Y. Indeed, let $\mathring{\xi}$ be any element in Y and let $\xi = \{f_n\}$ be its representative. Since $\{f_n\}$ is a Cauchy sequence, it follows from

$$\|\mathring{\xi} - \mathring{f}_n\| = \lim_{m\to\infty} \|f_m - f_n\|$$

that

$$\lim_{n\to\infty} \|\mathring{\xi} - \mathring{f}_n\| = 0, \tag{2}$$

i.e. the sequence $\{\mathring{f}_n\}$ in $\mathring{X}$ converges strongly to $\mathring{\xi}$ (in Y).

The space Y is complete. In fact, let $\{\mathring{\xi}_n\}$ be a Cauchy sequence in Y and let $\mathring{f}_n$ be any element in $\mathring{X}$ such that

$$\|\mathring{\xi}_n - \mathring{f}_n\| < \frac{1}{n}. \tag{3}$$

Then, from $\mathring{f}_m - \mathring{f}_n = (\mathring{f}_m - \mathring{\xi}_m) + (\mathring{\xi}_m - \mathring{\xi}_n) + (\mathring{\xi}_n - \mathring{f}_n)$ it follows that

$$\|\mathring{f}_m - \mathring{f}_n\| \leqslant \|\mathring{\xi}_m - \mathring{\xi}_n\| + \frac{1}{m} + \frac{1}{n}.$$

Since $\{\mathring{\xi}_n\}$ is a Cauchy sequence in Y, $\{\mathring{f}_n\}$ is a Cauchy sequence in Y and thus $\{f_n\}$ is a Cauchy sequence in X, i.e. a representative of some type $\mathring{\xi} \in Y$. We show that

$$\lim \|\mathring{\xi}_n - \mathring{\xi}\| = 0. \tag{4}$$

Indeed, by (3),

$$\|\mathring{\xi}_n - \mathring{\xi}\| \leqslant \|\mathring{\xi}_n - \mathring{f}_n\| + \|\mathring{f}_n - \mathring{\xi}\| \leqslant \|\mathring{f}_n - \mathring{\xi}\| + \frac{1}{n}$$

and (4) follows from (2).

In Y, we identify the vector manifold $\mathring{X}$ with the space $\mathring{X}$ which is isomorphic to it, by replacing each element $\mathring{f} \in \mathring{X}$ in Y by a vector $f \in X$, and we let X' denote the space which is obtained as a result of this identification. Whenever Y is a complete normed space, X' is also a complete normed space, and X is dense in X' since $\mathring{X}$ is dense in Y. Hence, *the complete normed space X' is an extension of the space X, and X is dense in X'.*

Let X'' be another extension of the space X with these properties. Then an isomorphism can be established between the spaces X' and X'' in the following way. A sequence $\{f_n\}$ in X which converges strongly in X' to an element ξ' is a Cauchy sequence in X which converges strongly in X'' to an element $\xi'' \in X''$. The mapping $\xi' \to \xi''$ of the space X' onto X'' is one-to-one and leaves elements from X unchanged. Therefore, in view of the preservation of vector operations under passage to the limit, and the relation

$$\|\xi'\| = \lim \|f_n\| = \|\xi''\|,$$

the mapping $\xi' \to \xi''$ is an isomorphism of the space X' onto X'' which leaves X unchanged. This proves the uniqueness of the complete extension in which X is dense.

The complete extension of the space X in which X is dense is called the *completion of the space X.*

REMARK. A normed space X generated by a vector system $\mathfrak{V}$ with a given quasinorm $w(f)$ can be extended until complete if X is not complete. If X is a complete space, then, in correspondence with the remark in Sec. 3.1.4, we will usually consider $\mathfrak{V}$ itself to be a vector system of this space.

2.2. Examples of complete normed spaces.

1) We define the norm $\|f\|$ in the vector system $\mathfrak{G}(T)$ of bounded functions $f = f(\tau)$ on the set T by the formula

$$\|f\| = \sup_{\tau \in T} |f(\tau)|.$$

It is easy to verify that the conditions 1)–3) of Sec. 3.1.1 are satisfied by this norm. The norm $\|f\|$ given above converts the vector system $G(T)$ into a normed space X_T. In this space, strong convergence coincides with uniform convergence of functions on the set T. Hence X_T is a complete space.

2) Let a *numerical* finite or σ-finite measure $\mu(M)$ be defined on a Borel field $\mathfrak{B} \subset \mathfrak{P}(T)$. In the vector system $\mathfrak{G}_\mu(\mathfrak{B})$ of functions integrable with respect to the measure $\mu(M)$ (§2.6 and §2.10), it is easily seen that

$$w(f) = \int_T |f(\tau)| \, \mu(\underset{\tau}{M}); \qquad f(\tau) \in \mathfrak{G}_\mu(\mathfrak{B}),$$

defines a quasinorm $w(f)$. The normed space generated by the vector system $\mathfrak{G}_\mu(\mathfrak{B})$ and the quasinorm $w(f)$ will be denoted by $X^{(\mu)}$. Elements of the space $X^{(\mu)}$ are types $\mathring{f}$ of functions $f = f(\tau) \in \mathfrak{G}_\mu(\mathfrak{B})$ with respect to the equivalence

$$f(\tau) \equiv g(\tau), \quad (\mu) \tag{5}$$

i.e. in correspondence with the remark in Sec. 3.1.4, they are elements of the vector system $\mathfrak{G}_\mu(\mathfrak{B})$ for which equality is defined by the congruence (5).

Now let $\mathfrak{B}'$ be a Borel field generated by T and by the sets of $\mathfrak{B}$ ($\mathfrak{B}' = \mathfrak{B}$ if the measure $\mu(M)$ is finite (see Sec. 2.10.2)), let $\tilde{\mathfrak{G}}(\mathfrak{B}')$ be the vector system of all $\mathfrak{B}'$-measurable functions on T, let $\mathfrak{G}_p^{(\mu)}(\mathfrak{B})$ $(p > 1)$ be the collection of functions $f(\tau) \in \tilde{\mathfrak{G}}(\mathfrak{B}')$ which satisfy the condition

$$|f(\tau)|^p \in \mathfrak{G}_\mu(\mathfrak{B}),$$

and let $\mathfrak{J}^{(p)}$ be the set of non-negative functions of the family $\mathfrak{G}_p^{(\mu)}(\mathfrak{B})$.

The set $\mathfrak{J}^{(p)}$ is a majorant of the system of functions, and

$$\mathfrak{G}_p^{(\mu)}(\mathfrak{B}) = (\tilde{\mathfrak{G}}(\mathfrak{B}')/\mathfrak{J}^{(p)}). \tag{6}$$

Indeed, if $f(\tau)$ and $g(\tau)$ belong to $\mathfrak{J}^{(p)}$, then $\alpha f(\tau) \in \mathfrak{J}^{(p)}$ for $\alpha \geqslant 0$, and since

$$[f(\tau) + g(\tau)]^p \leqslant 2^p (\sup\{f, g\})^p \leqslant 2^p [(f(\tau))^p + g(\tau))^p] \tag{7}$$

whenever $f(\tau)$ and $g(\tau)$ belong to $\mathfrak{J}^{(p)}$, the function $f(\tau) + g(\tau)$ also belongs to $\mathfrak{J}^{(p)}$. Formula (6) now follows from (2.10.13).

From the representation (6), it follows that $\mathfrak{G}_p^{(\mu)}(\mathfrak{B})$ *is a Baire $*$-field of functions, and in particular, it is a vector $*$-system of functions. In this system, the formula*

$$w_p(f) = \left[\int_T |f(\tau)|^p \, \mu(\underset{\tau}{M})\right]^{1/p} = [I(|f|^p)]^{1/p},$$

where $f(\tau) \in \mathfrak{G}_p^{(\mu)}(\mathfrak{B})$, *defines a quasinorm*. Indeed, the equation $w_p(\alpha f) = |\alpha| w_p(f)$ is obvious. We will prove the *Minkowski inequality*:

$$w_p(f + g) \leqslant w_p(f) + w_p(g). \tag{8}$$

In order to do this we note that the inequality*

---

* The validity of (9), with $a \geqslant 0$, $b \geqslant 0$, $p > 1$, $q > 1$ and $\frac{1}{p} + \frac{1}{q} = 1$, follows from the

$$ab \leqslant \frac{a^p}{p} + \frac{b^q}{q}, \tag{9}$$

where $a, b \geqslant 0$, $p > 1$, $q > 1$ and $\frac{1}{p} + \frac{1}{q} = 1$, shows (for $a = |f(\tau)|$, $b = |g(\tau)|$), that the product of the functions $f(\tau) \in \mathfrak{G}_p^{(\mu)}(\mathfrak{B})$ and $g(\tau) \in \mathfrak{G}_q^{(\mu)}(\mathfrak{B})$ belongs to $\mathfrak{G}^{(\mu)}(\mathfrak{B})$ and that

$$\left| \int_T f(\tau)\, g(\tau)\, \mu\,(\underset{\tau}{M}) \right| \leqslant \int_T |f(\tau)|\,|g(\tau)|\, \mu\,(\underset{\tau}{M}) \leqslant \frac{1}{p} + \frac{1}{q} = 1,$$

if

$$\int_T |f(\tau)|^p\, \mu\,(\underset{\tau}{M}) = 1, \qquad \int_T |g(\tau)|^q\, \mu\,(\underset{\tau}{M}) = 1.$$

Therefore, for any $f(\tau) \in \mathfrak{G}_p^{(\mu)}(\mathfrak{B})$ and $g(\tau) \in \mathfrak{G}_q^{(\mu)}(\mathfrak{B})$

$$\left| \int_T f(\tau)\, g(\tau)\, \mu\,(\underset{\tau}{M}) \right| \leqslant w_p(f)\, w_q(q). \tag{10}$$

For functions $f(\tau)$ and $g(\tau)$ in $\mathfrak{G}_p^{(\mu)}(\mathfrak{B})$ we now have

$$\left.\begin{aligned} |f(\tau) + g(\tau)| &\leqslant |f(\tau)| + |g(\tau)| = h(\tau), \\ w_p(f+g) &\leqslant w_p(h), \\ (w_p(h))^p &= \int_T |f(\tau)|\,(h(\tau))^{p-1}\, \mu\,(\underset{\tau}{M}) + \\ &\quad + \int_T |g(\tau)|\,(h(\tau))^{p-1}\, \mu\,(\underset{\tau}{M}), \end{aligned}\right\} \tag{11}$$

where the function $(h(\tau))^{p-1}$ belongs to $\mathfrak{G}_q^{(\mu)}(\mathfrak{B})$ since $q(p-1) = p$. By applying the inequality (10) to the integrals in (11), we obtain

$$\begin{aligned} (w_p(h))^p &\leqslant w_p(f)\, w_q(h^{p-1}) + w_p(g)\, w_q(h^{p-1}) = \\ &= w_q(h^{p-1})\,(w_p(f) + w_p(g)). \end{aligned} \tag{12}$$

Since $(h(\tau))^{(p-1)q} = (h(\tau))^p$, and hence

$$w_q(h^{p-1}) = (w_p(h))^{\frac{p}{q}} = (w_p(h))^{p-1},$$

dividing (12) by $w_q(h^{p-1})$ yields

$$w_p(h) \leqslant w_p(f) + w_p(g),$$

and by (11), the inequality (8) follows.

The vector system $\mathfrak{G}_p^{(\mu)}(\mathfrak{B})$ with the quasinorm $w_p(f)$ generates a *normed*

---

fact that the function $\Phi(\tau) = \frac{t^p}{p} - bt + \frac{b^q}{q}$, $t \in [0, +\infty)$, is non-negative in so far as its minimum on the interval $[0, +\infty)$ equals zero.

space $X_p^{(\mu)}$, whose elements can be considered to be functions $f(\tau)$ in $\mathfrak{G}_p^{(\mu)}(\mathfrak{B})$ for which equality in $\mathfrak{G}_p^{(\mu)}(\mathfrak{B})$ is defined by the congruence (5) of types $\mathring{f}$ of functions $f(\tau)$ with respect to the equivalence (5) in $\mathfrak{G}_p^{(\mu)}(\mathfrak{B})$. In addition, by letting $\mathfrak{G}_1^{(\mu)}(\mathfrak{B}) = \mathfrak{G}_\mu(\mathfrak{B})$ and $X_1^{(\mu)} = X^{(\mu)}$, we obtain the definition of the normed spaces $X_p^{(\mu)}$ for all (finite) $p \geqslant 1$. We extend this definition to $p = \infty$.

Let $\mathfrak{G}_\infty^{(\mu)}(\mathfrak{B})$ be the collection of all functions $f(\tau)$ which satisfy the congruence (5), where $g(\tau)$ ranges over the family of functions $\mathfrak{G}(\mathfrak{B})$. For $f(\tau) \in \mathfrak{G}_\infty^{(\mu)}$, we denote the upper bound of $|f(\tau)|$, to within a set of measure zero with respect to $\mu(M)$, by $\overset{(\mu)}{\sup}\, |f(\tau)|$: $|f(\tau)| \leqslant \overset{(\mu)}{\sup}\, |f(\tau)|$ almost everywhere with respect to $\mu(M)$, and if $\gamma < \overset{(\mu)}{\sup}\, |f(\tau)|$ and if M_γ is the set of all $\tau \in T$ for which $|f(\tau)| > \gamma$, then $\mu(M_\gamma) > 0$.

The vector system $\mathfrak{G}_\infty^{(\mu)}(\mathfrak{B})$ with the quasinorm

$$w_\infty(f) = \overset{(\mu)}{\sup}\, |f(\tau)|, \qquad f(\tau) \in \mathfrak{G}_\infty^{(\mu)}(\mathfrak{B}),$$

generates by definition a *space* $X_\infty^{(\mu)}$ *with norm*

$$\|f\| = w_\infty(f),$$

if the family $\mathfrak{G}_\infty^{(\mu)}(\mathfrak{B})$ is a vector system of the space $X_\infty^{(\mu)}$ in which equality of functions $f(\tau)$ and $g(\tau)$ has been defined by the congruence (5).

Similarly, in the space $X_p^{(\mu)}$, for any $p \geqslant 1$, we define a norm by the formula

$$\|f\| = w_p(f). \tag{13}$$

The space $X_p^{(\mu)}$ *is complete.*

In fact, let $\{f_n\}$ be a Cauchy sequence in $X_p^{(\mu)}$, $1 < p < \infty$ i.e.

$$\lim \|f_m - f_n\| = \lim w_p(f_m - f_n) = 0, \quad m \to \infty, \quad n \to \infty, \tag{14}$$

and let the series $\sum_{k=1}^{\infty} \varepsilon_k$, where $\varepsilon_k > 0$, converge. Then it is possible to find a number n_k such that

$$w_p(f_m - f_n) \leqslant \varepsilon_k \quad \text{for} \quad m \geqslant n_k, \quad n \geqslant n_k, \tag{14'}$$

and hence

$$\sum_{k=1}^{\infty} w_p(f_{n_{k+1}} - f_{n_k}) \leqslant \sum_{k=1}^{\infty} \varepsilon_k < \infty. \tag{15}$$

We will prove that *the sequence* $\{f_{n_k}\}$ *converges except perhaps on a set of* μ*-measure zero*. It is sufficient to show (since $\mu(M)$ is σ-finite) that convergence holds on sets of finite measure. Let $T' \subset T$ be a set of finite measure in $\mathfrak{B}$ and let

$$w_p(f;\, T') = \left(\int_{T'} |f(\tau)|^p \, \mu(M_\tau) \right)^{\frac{1}{p}}.$$

Then by (10) ($g(\tau) = 1$, $T = T'$)

$$w(f;\, T') = w_1(f;\, T') \leqslant w_p(f;\, T')(\mu(T'))^{\frac{1}{q}},$$

hence (see (15)),

$$\sum_{k=1}^{\infty} w\left(f_{n_{k+1}} - f_{n_k};\, T'\right) \leqslant \sum_{k=1}^{\infty} w_p\left(f_{n_{k+1}} - f_{n_k}\right)(\mu(T'))^{\frac{1}{q}} < \infty, \quad (15')$$

and $w(f;\, T')$ can be considered as an integral $I(|f|)$ with respect to the finite measure $\mu'(M) = \mu(M \cap T')$. From (15′) and Theorem 2.6.7, it now follows that the series

$$\sum_{k=1}^{\infty} \left| f_{n_{k+1}}(\tau) - f_{n_k}(\tau) \right|$$

converges almost everywhere on T', with respect to the measure $\mu'(M)$. Therefore the series $\sum_{k=1}^{\infty}(f_{n_{k+1}}(\tau) - f_{n_k}(\tau))$ also converges, i.e. except on a set of measure zero, $\lim f_{n_k}(\tau) = f(\tau)$ exists in T' and also in T, since $T = \bigcup_n T_n$, where each of the sets T_n of finite measure can play the role of T'.

By applying Lemma 2.6.2 to the inequality

$$w_p\left(f_m - f_{n_r};\, T'\right) \leqslant \varepsilon_k, \quad m \geqslant n_k, \quad r > k,$$

we obtain

$$w_p(f_m - f;\, T') \leqslant \varepsilon_k.$$

If $T_1 \subset T_2 \subset \ldots$, then as $n \to \infty$ in the inequality

$$w_p(f_m - f,\, T_n) \leqslant \varepsilon_k, \text{ where } m \geqslant n_k,$$

we obtain

$$w_p(f_m - f) = w_p(f_m - f;\, T) \leqslant \varepsilon_k, \quad m \geqslant n_k.$$

Therefore $f = (f - f_m) + f_m \in X_p^{(\mu)}$, and in view of (13)

$$\lim \|f_m - f\| = \lim w_p(f_m - f) = 0,$$

i.e. the space $X_p^{(\mu)}$ is complete.

In the case $p = 1$ we have $w_1(f;\, T') \leqslant w_1(f)$, and hence instead of (15′), we have the inequality

$$\sum_k w_1\left(f_{n_{k+1}} - f_{n_k};\, T'\right) \leqslant \sum_k w_1\left(f_{n_{k+1}} - f_{n_k}\right) < \infty,$$

and this holds in what follows. The completeness of the space $X_\infty^{(\mu)}$ is almost obvious, since it follows from (14) with $p=\infty$ that $\{f_n(\tau)\}$ converges uniformly on the set $T_0 \subset T$, where $\mu(T \setminus T_0)=0$.

If T is a set of Euclidean space, and if $\mu(M)$ is Lebesgue measure on T, then we will write $L_p(T)$ in place of $X_p^{(\mu)}$.

3) Let a *countable* set T be given in the form of a sequence $\{n\}$ of natural numbers, and let a *weight* $\rho_n = \rho(n) > 0$ be given at each point n. Then the equation

$$\mu(M) = \sum_{n \in M} \rho_n \tag{16}$$

defines a *point measure* $\mu(M)$ on the collection $\mathfrak{B}$ of all sets $M \subset T$ for which the series (16) converges. In the case of such a point measure $\mu(M)$, the space $X_p^{(\mu)} = X_p^{(\rho)}$ consists precisely of all countable sequences $f=\{f(n)\}=\{\alpha_n\}$ which satisfy the condition

$$\sum_{n=1}^{\infty} \rho_n |\alpha_n|^p < \infty,$$

and the norm in $X_p^{(\rho)}$ is defined by

$$\|f\| = \|\{\alpha_n\}\| = \left(\sum_{n=1}^{\infty} \rho_n |\alpha_n|^p\right)^{\frac{1}{p}}. \tag{17}$$

In particular, for $\rho_n = 1$.

$$\|f\| = \|\{\alpha_n\}\| = \sum_{n=1}^{\infty} |\alpha_n|^p. \tag{17'}$$

If T is the finite set $\{1, 2, \ldots, m\}$, then $\mathfrak{B}$ consists of all subsets of T, and (17) and (17′) hold if the summation is taken from $n=1$ to $n=m$.

4) We now define the space $X_p^{(\rho)}$ in the case of an uncountable set T of weight functions $\rho(\tau) > 0$. With this aim, we form the set $X_p^{(\rho)}$ of functions $f=f(\tau)$ in $\mathfrak{G}^{(1)}(T, K_c)$ (Sec. 1.2.2) which satisfy the condition

$$\sum_{\tau \in T} \rho(\tau) |f(\tau)|^p < \infty, \quad 1 \leqslant p < \infty.$$

Summation of the series makes sense here since the function $f(\tau)$ is different from zero on a more than countable set of values τ. Furthermore, we let

$$w_p(f) = \left(\sum_{\tau} \rho(\tau) |f(\tau)|^p\right)^{\frac{1}{p}} \tag{18}$$

for a function $f(\tau) \in X_p^{(\rho)}$. Then, in the case of uncountable T, the fact that $X_p^{(\rho)}$ is a complete normed space does not formally follow from the proof in Sec. 2, but it is possible to give a similar proof. In fact, whenever $f(\tau)$ belongs to $X_p^{(\rho)}$, $\alpha f(\tau)$ also belongs to $X_p^{(\rho)}$, moreover $w_p(\alpha f) = |\alpha| w_p(f)$ and the inequality (7) shows that whenever $f(\tau)$ and $g(\tau)$ belong to $X_p^{(\rho)}$, the function $f(\tau)+g(\tau)$ also belongs to $X_p^{(\rho)}$. From (9), we now have

$$\left|\sum_{\tau} \rho(\tau) f(\tau) g(\tau)\right| \leqslant w_p(f)\, w_q(g), \tag{10'}$$

if $f \in X_p^{(\rho)}$, $g \in X_q^{(\rho)}$, $p>1$, $q>1$ and $\frac{1}{p}+\frac{1}{q}=1$. This inequality can be proved in the same way as the inequality (10) was proved. In the same way as in the case of space $X_p^{(\mu)}$, we obtain (8) by replacing the integrals by sums. The function $w_p(f)$ now satisfies not only the conditions 1) and 2) for the norm, but also the condition 3), since from $w_p(f)=0$ and $\rho(\tau)>0$, it follows that $f(\tau)=0$ for all $\tau \in T$. Thus $w_p(f)$ is a norm and $X_p^{(\rho)}$ is a normed space with the norm $\|f\| = w_p(f)$.

The space $X_p^{(\rho)}$ is complete. Indeed, let $\{f_n(\tau)\}$ be a Cauchy sequence from $X_p^{(\rho)}$, i.e.

$$\lim \|f_m - f_n\| = 0 \quad m \to \infty, \quad n \to \infty. \tag{19}$$

Then $\lim f_n(\tau) = f(\tau)$ exists for all τ, since

$$|f_m(\tau) - f_m(\tau)| \leqslant \frac{w_p(f_m - f_n)}{(\rho(\tau))^{1/p}} = \frac{\|f_m - f_n\|}{(\rho(\tau))^{1/p}}.$$

We will show that $f(\tau) \in X_p^{(\rho)}$ and that $\lim \|f_n - f\| = 0$. In fact, (19) means that for any $\varepsilon > 0$ there exists a number $n_0(\varepsilon)$ such that

$$\sum_\tau \rho(\tau) |f_n(\tau) - f_m(\tau)|^p < \varepsilon^p, \tag{19'}$$

if $n \geqslant n_0(\varepsilon)$ and $m \geqslant n_0(\varepsilon)$. We leave a finite number of terms in (19) and then pass to the limit $m \to \infty$. Then

$$\sum{}' \rho(\tau) |f_n(\tau) - f(\tau)|^p \leqslant \varepsilon^p, \quad n \geqslant n_0(\varepsilon),$$

where $\sum'$ is an arbitrary *finite* sum. The reverse passage to the full sum gives

$$\sum_\tau \rho(\tau) |f_n(\tau) - f(\tau)|^p \leqslant \varepsilon^p.$$

Whence it follows that the function $f(\tau) - f_n(\tau)$, and also the function $f(\tau) = (f(\tau) - f_n(\tau)) + f_n(\tau)$, belongs to $X_p^{(\rho)}$, and that

$$\|f_n - f\| \leqslant \varepsilon.$$

Since ε is arbitrary, $\lim \|f_n - f\| = 0$. Thus the Cauchy sequence $\{f_n(\tau)\}$ converges strongly to the function $f(\tau)$.

In the case of the weight function $\rho(\tau) = 1$,

$$\|f\| = \left(\sum_\tau |f(\tau)|^p\right)^{\frac{1}{p}};$$

the space $X_p^{(\rho)}$ with this norm will be denoted by $X_p(T)$.

5) In the examples 2) and 4), the case $p = q = 2$ is of particular significance, i.e. the case when $X_p^{(\mu)} = X_q^{(\mu)} = X_2^{(\mu)}$ and $X_p^{(\rho)} = X_q^{(\rho)} = X_2^{(\rho)}$. For these spaces, $X_2^{(\mu)}$ and $X_2^{(\rho)}$, it is possible to form the integral and, correspondingly, the sum

$$W(f, g) = \int_T f(\tau) \bar{g}(\tau) \mu(M_\tau);$$

$$W(f, g) = \sum_\tau \rho(\tau) f(\tau) \bar{g}(\tau), \tag{20}$$

where $f(\tau)$ and $g(\tau)$ are arbitrary functions in $X_2^{(\mu)}$, or correspondingly in $X_2^{(\rho)}$.

In both cases, $W(f, g)$ is a linear form with respect to the argument f and a pseudolinear form with respect to g, and

$$\|f\|^2 = W(f, f). \tag{21}$$

The spaces $X_2^{(\mu)}$ and $X_2^{(\rho)}$ are very important examples of so-called unitary spaces whose general definition will be given in the following section by starting from properties of the function $W(f, g)$, and in particular, from (21).

§ 3. Unitary Space

3.1. Bilinear and quadratic forms. The unitary norm. A complex function $W(f, g')$, where the variables f and g' range independently over *complex* vector systems $\mathfrak{V}$ and $\mathfrak{V}'$, will be called a *bilinear form* if it is linear in the first argument f:

$$W(f_1 + f_2, g') = W(f_1, g') + W(f_2, g');$$
$$W(\alpha f, g') = \alpha W(f, g'),$$

and is pseudolinear in the second:

$$W(f, g_1' + g_2') = W(f, g_1') + W(f, g_2');$$
$$W(f, \beta g') = \bar{\beta} W(f, g').$$

The collection $B(\mathfrak{V}, \mathfrak{V}')$ of these bilinear forms is converted into a complex vector system if we let

$$(W_1 + W_2)(f, g') = W_1(f, g') + W_2(f, g');$$
$$(\alpha W)(f, g') = \alpha W(f, g').$$

In the special case when $\mathfrak{V}' = \mathfrak{V}$, we will write $W(f, g)$ in place of $W(f, g')$ and $B(\mathfrak{V})$ instead of $B(\mathfrak{V}, \mathfrak{V})$. A form $W(f, g)$ in $B(\mathfrak{V})$ is called a bilinear form in the vector system $\mathfrak{V}$.

To a bilinear form $W = W(f, g)$ in $B(\mathfrak{V})$ there corresponds a *quadratic form*

$$Q(f) = W(f, f),$$

where f ranges over the vector system $\mathfrak{V}$. For the quadratic form $Q(f)$,

$$Q(\alpha f) = |\alpha|^2 Q(f). \tag{1}$$

Indeed, $W(\alpha f, \alpha f) = \bar{\alpha}\alpha W(f, f)$.

We now show that a bilinear form W in $B(\mathfrak{V})$ can be uniquely recovered from the quadratic form $Q(f)$ which it generates. In fact, if $W = W(f, g)$

is a bilinear form and $Q(f)$ is the quadratic form which it generates, then

$$Q(f+g)=Q(f)+W(f, g)+W(g, f)+Q(g), \tag{2}$$

$$Q(f-g)=Q(f)-W(f, g)-W(g, f)+Q(g). \tag{2'}$$

By replacing g by ig, we obtain two additional identities, and by forming a linear combination of these four identities, we obtain

$$Q(f+g)-Q(f-g)+iQ(f+ig)-iQ(f-ig)=4W(f, g).$$

We let

$$h_1=\frac{f+g}{2}, \quad h_2=\frac{f-g}{2}, \quad h_3=\frac{f+ig}{2}, \quad h_4=\frac{f-ig}{2}, \tag{3}$$

and finally obtain

$$W(f, g)=Q(h_1)-Q(h_2)+iQ(h_3)-iQ(h_4), \tag{4}$$

which solves the problem of recovering $W(f, g)$ from $Q(f)$.

In particular, if follows from (4) that: *if* $Q(f)=W(f,f)=0$ *identically, then* $W(f, g)=0$, i.e. $W=0$.

By combining (2) and (2′), we obtain the so-called *parallelogram equation*

$$Q\left(\frac{f+g}{2}\right)+Q\left(\frac{f-g}{2}\right)=\frac{Q(f)+Q(g)}{2}, \tag{5}$$

which together with (1), and a certain form of continuity, is a characteristic property of the quadratic form $Q(f)$ (see below, Sec. 3.6.1).

We place the form $W\in B(\mathfrak{B})$ into correspondence with a form W^* which is defined by

$$W^*(f, g)=\overline{W(g, f)}. \tag{6}$$

The mapping $W\to W^*$ *is a pseudoinvolution of the system* $B(\mathfrak{B})$, *and it converts* $B(\mathfrak{B})$ *into a vector* $*$*-system*. Indeed, the form $\overline{W(g, f)}$ is pseudolinear in the first argument g and linear in the second argument f, i.e. $W^*\in B(\mathfrak{B})$. Furthermore, it is easy to verify that

$$(W_1+W_2)^*=W_1^*+W_2^*, \quad (\alpha W)^*=\bar{\alpha}W^*, \quad (W^*)^*=W,$$

i.e., that the mapping $W\to W^*$ is a pseudoinvolution in $B(\mathfrak{B})$.

We have

$$W(f, g)=\overline{W(g, f)} \tag{7}$$

for the Hermitian elements ($W^*=W$) of the $*$-system $B(\mathfrak{B})$. The bilinear form $W^*(f, g)$ is said to be *adjoint* to the form $W(f, g)\in B(\mathfrak{B})$. If $W(f, g)$ coincides with its adjoint $W^*(f, g)$, i.e. if it satisfies (7), then it is called *a self-adjoint bilinear form*.

Self-adjoint bilinear forms in $\mathfrak{B}$ thus coincide with the Hermitian elements of the vector $*$-system $B(\mathfrak{B})$, and hence *they form a real vector system*.

From the decomposition (1.5.3) it now follows that: *every bilinear form $W(f, g)$ can be uniquely represented in the form*

$$W(f, g) = W_1(f, g) + iW_2(f, g), \tag{8}$$

where W_1 and W_2 are self-adjoint forms.

For $g = f$, (7) gives

$$Q(f) = \overline{Q(f)},$$

so that the quadratic form $Q(f)$ which is generated by the self-adjoint form $W(f, g)$ has only real values. The converse is also true, namely:

THEOREM 3.3.1. *A bilinear form $W(f, g)$ is self-adjoint if and only if the quadratic form $Q(f)$ generated by it is real.*

In fact, by the representation (8),

$$Q(f) = W(f, f) = W_1(f, f) + iW_2(f, f) = Q_1(f) + iQ_2(f),$$

where $Q_1(f)$ and $Q_2(f)$ are real since $W_1(f, g)$ and $W_2(f, g)$ are self-adjoint. Therefore $W(f, g)$ is real if and only if $Q_2(f) = 0$ identically for all $f \in \mathfrak{B}$. But then $W_2 = 0$ and $W = W_1$, which is a self-adjoint form.

A quadratic form $Q(f)$ and the corresponding bilinear form $W(f, g)$ are said to be *positive* if

$$Q(f) \geqslant 0, \quad Q \not\equiv 0,$$

and *positive definite* if

$$Q(f) > 0 \quad \text{for} \quad f \neq 0.$$

THEOREM 3.3.2. *If $W(f, g)$ is a positive (positive definite) bilinear form in $\mathfrak{B}$, then the function*

$$w(f) = \sqrt{Q(f)} = \sqrt{W(f, f)} \tag{9}$$

is a quasinorm (norm).

In fact, for $Q(f) = W(f, f) \geqslant 0$, Equation (1) gives $w(\alpha f) = |\alpha| w(f)$ and it remains to show that

$$w(f + g) \leqslant w(f) + w(g). \tag{10}$$

We begin with the proof of the inequality

$$|W(f, g)| \leqslant w(f)\, w(g). \tag{11}$$

In the inequality

$$\alpha\bar{\alpha}Q(f) + \alpha\bar{\beta}W(f, g) + \bar{\alpha}\beta W(g, f) + \beta\bar{\beta}Q(g) = Q(\alpha f + \beta g) \geqslant 0,$$

which follows from (2), we will consider the left-hand side for fixed f and g to be a Hermitian form of the variables α and β. From the positivity of this form, it follows that

$$Q(f)Q(g)-W(f,\ g)W(g,\ f)=\begin{vmatrix} Q(f) & W(g,\ f) \\ W(f,\ g) & Q(g) \end{vmatrix}\geqslant 0,$$

which is equivalent to (11). Inequality (10) now follows from the inequality (11). Indeed,

$$Q(f+g)\leqslant Q(f)+2w(f)w(g)+Q(g)=[w(f)+w(g)]^2,$$

and by taking the square root we obtain (10). If $Q(f)$ is positive definite, then $w(f)=0$ implies that $Q(f)=0$ and $f=0$, i.e. in this case $w(f)$ is a norm.

A norm (or quasinorm) $w(f)$ in a complex vector system $\mathfrak{V}$ is said to be *unitary* if its square $(w(f))^2$ is a quadratic form in $\mathfrak{V}$. Equation (9) gives the general form of such a unitary norm or quasinorm. The bilinear form $W(f, g)$ which generates the square $(w(f))^2$ of a unitary norm or quasinorm $w(f)$ is positive, and hence is self-adjoint.

3.2. The scalar product. *A complex normed space is called a unitary space if it is complete and if its norm is unitary.*

Unitary spaces will be denoted by the letter H.

The bilinear form $O(f, g)$ which is generated by the square of the norm $\|f\|^2$ in a unitary space H will be called the *fundamental form* of this space. The value of $O(f, g)$ is the *scalar product of the vectors* f and g, and we usually write (f, g) instead of $O(f, g)$.

The bilinear form $O(f, g)$ *is positive definite*, since

$$O(f,\ f)=(f,\ f)=\|f\|^2,$$

and since the form $O(f, g)$ is self-adjoint, we have

$$(g,\ f)=\overline{(f,\ g)}. \tag{12}$$

Conversely, *a positive definite bilinear form* $O(f, g)=(f, g)$ *in a vector system* $\mathfrak{V}$ *is the fundamental form of a unitary space with the norm*

$$\|f\|=\sqrt{(f,\ f)},\qquad f\in\mathfrak{V}, \tag{13}$$

if and only if the normed space defined by the system $\mathfrak{V}$ *and the norm* (13) *is complete.* Indeed, by Theorem 3.3.2, Equation (13) defines a unitary norm in $\mathfrak{V}$.

For a fundamental form of a unitary space, (11) can be written in the form

$$|(f,\ g)|\leqslant\|f\|\ \|g\|. \tag{14}$$

This inequality was proved by Cauchy (for finite sequences) and by Bunyakovski (for functions). In what follows we will call it the *Cauchy-Bunyakovski inequality*.

From the Cauchy-Bunyakovski inequality we obtain

$$|(f', g') - (f, g)| \leqslant \|g'\| \, \|f' - f\| + \|f\| \, \|g' - g\|. \tag{15}$$

Indeed

$$|(f', g') - (f, g)| \leqslant |(f', g') - (f, g')| + |(f, g') - (f, g)|,$$

and by (14)

$$|(f', g') - (f, g')| = |(f' - f, g')| \leqslant \|f' - f\| \, \|g'\|$$

and similarly

$$|(f, g') - (f, g)| \leqslant \|f\| \, \|g' - g\|.$$

The inequality (15) proves, in particular, the continuity of the fundamental form (f, g) (the scaler product) in both variables f and g. However, this also follows from the continuity of the norm and from (4):

$$(f, g) = \|h_1\|^2 - \|h_2\|^2 + i\|h_3\|^2 - i\|h_4\|^2, \tag{16}$$

where the elements h_k are given by (3).

In the definition of isomorphism and pseudoisomorphism $f \to f'$ $(f \in H, f' \in H')$ of two unitary spaces H and H' with fundamental forms O and O', the condition $\|f'\| = \|f\|$ of the preservation of the norm can be replaced by the corresponding conditions

$$O'(f', g') = O(f, g) \quad \text{and} \quad O'(f', g') = \overline{O(f, g)}, \tag{17}$$

i.e., *under an isomorphism the scalar product is preserved, and for a pseudo-isomorphism it is transformed according to the second equation of* (17). In fact, each of the equations in (17) imply (3.1.7), and in view of (16), the condition (3.1.7) is equivalent to (17) (in the case of pseudoisomorphism, the elements h_3 and h_4 are mapped respectively into $h_4' = \dfrac{f' - ig'}{2}$ and $h_3' = \dfrac{f' + ig'}{2}$).

REMARK. If a positive definite bilinear form $O(f, g) = (f, g)$ is defined in the complex vector system $\mathfrak{V}$, but if the normed space defined by the system $\mathfrak{V}$ and the norm (13) is not complete, then its completion (Sec. 3.2.1) is unitary. For the proof it is sufficient to show that under completion the norm remains unitary. Let $\mathring{\xi}$ and $\mathring{\eta}$ be two elements of the extended space with representatives $\xi = \{f_n\}$ and $\eta = \{g_n\}$ $(f_n, g_n \in \mathfrak{V})$. The equation

$$W(\mathring{\xi}, \mathring{\eta}) = \lim_{n \to \infty} (f_n, g_n) \tag{18}$$

defines a bilinear form in the vector system of the extended space which coincides on $\mathfrak{V}$ with (f, g). Indeed, the existence of the limit in (18) follows from (15) if we let $f = f_m$, $g = g_m$, $f' = f_n$, and $g' = g_n$. The fact that the limit is independent of the choice of the representatives ξ and η of the elements $\mathring{\xi}$ and $\mathring{\eta}$ also follows from (15) if we let $f = f_n$, $g = g_n$, $f' = f_n'$, and $g' = g_n'$, where $\xi = \{f_n\}$, $\xi' = \{f_n'\}$ and $\eta = \{g_n\}$, $\eta' = \{g_n'\}$ are arbitrary representatives of the elements $\mathring{\xi}$ and $\mathring{\eta}$.

It now follows from (18) that the function $W(\xi, \eta)$ is linear in $\mathring{\xi}$ and pseudolinear in $\mathring{\eta}$. By letting $\mathring{\eta} = \mathring{\xi}$ in (18), we have

$$W(\mathring{\xi}, \mathring{\xi}) = \lim_{n\to\infty} (f_n, f_n) = \lim_{n\to\infty} \| f_n \|^2 = \| \mathring{\xi} \|^2.$$

Hence the norm $\|\mathring{\xi}\|$ is unitary, i.e. the completed space is unitary and its fundamental form is $(\mathring{\xi}, \mathring{\eta}) = W(\mathring{\xi}, \mathring{\eta})$.

If the bilinear form $W(f, g)$ defined on $\mathfrak{V}$ is only positive, then it also generates a unitary space. In this case the function

$$w(f) = \sqrt{W(f, f)},$$

is, according to Theorem 3.3.2, a (unitary) quasinorm in $\mathfrak{V}$ and it generates a normed space X of types $\mathring{f}$ with the norm (f is the representative of the type $\mathring{f}$)

$$\|\mathring{f}\| = w(f).$$

The norm $\|\mathring{f}\|$ *is unitary*. Indeed, we let (f and g are representatives of the types $\mathring{f}$ and $\mathring{g}$)

$$(\mathring{f}, \mathring{g}) = O(\mathring{f}, \mathring{g}) = W(f, g).$$

This definition of the form O is independent of the choice of the representatives f and g: if f' is another representative of the type $\mathring{f}$, then

$$|W(f', g) - W(f, g)| = |W(f' - f, g)| \leqslant w(f' - f)\, w(g)$$

and $W(f', g) = W(f, g)$, since $w(f' - f) = 0$. The form $(\mathring{f}, \mathring{g})$ is positive definite on the vector system $\mathring{\mathfrak{V}}$, and if the normed space X which is defined by $\mathring{\mathfrak{V}}$ and by the norm

$$\|\mathring{f}\| = \sqrt{(\mathring{f}, \mathring{f})} = w(f),$$

is not complete, then by extending it until it is complete, we obtain a unitary space generated by the system $\mathfrak{V}$ and the positive bilinear form in $\mathfrak{V}$.

3.3. Examples of unitary spaces. The complete normed spaces $X_2^{(\mu)}$ and $X_2^{(\rho)}$ (Sec. 3.2.2) are unitary: the scalar product in these spaces is defined by the formulas (3.2.20) ($(f, g) = W(f, g)$).

We remark that the space $X_2(T)$ is a special case of the space $X_2^{(\rho)}$ where the formula for the scalar product has the form

$$(f, g) = \sum_{\tau} f(\tau)\, \overline{g}(\tau). \tag{19}$$

In particular, if T is the set of natural numbers, or some subset $\{1, 2, \ldots, m\}$, then

$$(f, g) = \sum_{n=1}^{m} \alpha_n \bar{\beta}_n, \text{ where } f = \{\alpha_n\}, \quad g = \{\beta_n\}. \tag{19'}$$

The spaces $X_2^{(\mu)}$, $X_2^{(\rho)}$, and $X(T)$ will be denoted, respectively, by $H^{(\mu)}$, $H^{(\rho)}$, and $H(T)$. For countable T, $H(T)$ is a unitary space of sequences $\{\alpha_n\}$ of the form which first appeared in the works of Hilbert (for the most part for real α_n).

In the case when T is the set K_r of real numbers, the space $H(T)$ can be exemplified by a class of almost periodic functions. In $\mathfrak{G}(T)$, we consider the linear hull $\mathfrak{B}$ of functions $e^{i\lambda x}$, $\lambda \in K_r$ and $x \in T = K_r$. A function $f = f(x)$ in $\mathfrak{B}$ will be written in the form

$$f(x) = \sum_{\lambda} a(\lambda) e^{i\lambda x}, \tag{20}$$

where only a finite number of (complex) coefficients $a(\lambda)$ (λ ranges over all of K_r) are different from zero. We define a scalar product in the system $\mathfrak{B}$ ($f(x)$ and $g(x)$ in $\mathfrak{B}$)

$$(f, g) = \lim_{N \to +\infty} \frac{1}{2N} \int_{-N}^{N} f(x) \overline{g(x)}\, dx. \tag{21}$$

A simple calculation shows that

$$(e^{i\lambda x}, e^{i\mu x}) = \begin{cases} 0, & \lambda \neq \mu, \\ 1, & \lambda = \mu, \end{cases}$$

and hence

$$a(\lambda) = (f, e^{i\lambda x}) = \lim_{N \to +\infty} \frac{1}{2N} \int_{-N}^{N} f(x) e^{-i\lambda x}\, dx. \tag{22}$$

Therefore

$$(f, g) = \sum_{\lambda} \bar{\beta}(\lambda)(f, e^{i\lambda x}) = \sum_{\lambda} \alpha(\lambda) \bar{\beta}(\lambda), \tag{23}$$

if $g(x) = \sum_{\lambda} b(\lambda) e^{i\lambda x}$. In particular,

$$(f, f) = \sum_{\lambda} |a(\lambda)|^2, \tag{24}$$

so that $(f, f) \geqslant 0$, and $(f, f) = 0$ implies that $a(\lambda) = 0$ for all λ, i.e. that $f(x) = 0$. Thus (f, g) is a positive definite bilinear form in $\mathfrak{B}$, and by completion of the normed space H_0, which is defined by $\mathfrak{B}$ and by the unitary norm $\|f\| = (f, f)$, we obtain a unitary space H.

If $\xi = \{f_n\}$ and $\eta = \{g_n\}$, where $f_n, g_n \in \mathfrak{B}$ are representatives of the elements $\mathring{\xi}$ and $\mathring{\eta}$ in H, then

$$(\mathring{\xi}, \mathring{\eta}) = \lim_{n \to \infty} (f_n, g_n) = \sum_{\lambda} a(\lambda) \overline{b(\lambda)},$$

$$a(\lambda) = (\overset{\circ}{\xi}, e^{i\lambda x}) = \lim_{n\to\infty} (f_n, e^{i\lambda x});$$

$$b(\lambda) = (\overset{\circ}{\eta}, e^{i\lambda x}).$$

Conversely, to a function $a(\lambda) \in H(T)$ $(T = K_r)$ there corresponds an element $\overset{\circ}{\xi} \in H$ whose representative is the sequence of partial sums of the series $\sum_\lambda a(\lambda) e^{i\lambda x}$. Thus the space H is isomorphic to $H(T)$ $(T = K_r)$.

On the other hand, in $\mathfrak{G}(T)$ $(T = K_r)$ we consider a function $f(x)$ which satisfies the condition: in every finite interval $\Delta_N = (-N, N)$ both $f(x)$ and $|f(x)|^2$ are Lebesgue integrable, and

$$\overline{\lim_{N\to+\infty}} \frac{1}{2N} \int_{-N}^{N} |f(x)|^2 \, dx < \infty.$$

Functions which satisfy these conditions form a vector system $\mathfrak{V}_1$, and

$$w(f) = \left(\overline{\lim_{N\to+\infty}} \frac{1}{2N} \int_{-N}^{N} |f(x)|^2 \, dx \right)^{1/2}$$

is obviously a quasinorm in $\mathfrak{V}_1$. The vector system $\mathfrak{V}_1$, with quasinorm $w(f)$, defines a normed space X_1 with the vector system $\mathfrak{V}_1$, in which the equation $w(f - f') = 0$ defines the equality of elements f and f' from $\mathfrak{V}_1$, and the norm $\|f\| = w(f)$. It is possible to show that X_1 is a complete space.

Let X be the closed linear hull of the set of functions $e^{i\lambda x}$ in the space X_1. Then the complete normed space X is an extension of the space H_0, and H_0 is dense in X. By Theorem 3.2.1 there exists an isomorphism of the space X onto H which leaves H_0 unchanged, and hence X is a unitary space which is isomorphic to $H(T)$. In later sections (see Sec. 3.4.2 and Sec. 3.4.4) we will show that the functions $e^{i\lambda x}$ form an orthogonal basis in X and that the dimensionality of the space X is equal to the cardinality of the continuum.

Functions $f(x) \in X$ are said to be B_2-almost periodic, and they satisfy the equations (21)–(24). The series $\sum_\lambda a(\lambda) e^{i\lambda x}$ with coefficients (22) is called the *Fourier series of the function* $f(x) \in X$.

§ 4. Orthogonality

4.1. Orthogonal systems of vectors. In a unitary space H, the leading role does not belong to the norm, but to the fundamental form, i.e. to the scalar product. The fundamental form defines not only a distance, but also another metric concept—the angle. We restrict ourselves to the definition of right angle, i.e. of orthogonality.

A vector $g \in H$ is by definition *orthogonal* to a vector f if

$$(f, g) = 0.$$

By (3.3.12), the relation of orthogonality is symmetric, and we can talk about the *mutual orthogonality* of two vectors.

If the elements f and g are orthogonal, then

$$\|f+g\|^2 = \|f\|^2 + \|g\|^2, \tag{1}$$

as follows from (3.3.2) when applied to the fundamental form ($W(f, g) = (f, g)$).

Two sets are said to be *mutually orthogonal* if any element of one set is orthogonal to any element of the other set.

LEMMA 3.4.1. *If a set M is orthogonal to a set N, then it is orthogonal to the closed linear hull of N.*

Indeed, let $f_1, f_2, \ldots, f_m$ be any elements in N and let $g \in M$. Then $(f_k, g) = 0$, and for a linear combination $f' = \sum \alpha_k f_k$, we have

$$(f', g) = \sum_k \alpha_k (f_k, g) = 0,$$

i.e., the set M is orthogonal to the linear hull of the set N. Now let the sequence $\{f_n'\}$ in this hull be strongly convergent to an arbitrary element f of its closure. Then $(f_n', g) = 0$ for any n, and by the continuity of the scalar product,

$$(f, g) = \lim (f'_n, g) = 0,$$

i.e. M is orthogonal to the closed linear hull of the set N.

The collection of elements f in H which are orthogonal to a set $M \subset H$, will be called the *orthogonal complement* of M.

From Lemma 3.4.1 it follows that:

The orthogonal complement of any set is a subspace.

If the subspace H' is a closed linear hull of a set $N \subset H$, and if H'' is its orthogonal complement, then by Lemma 3.4.1, the subspace H'' is the orthogonal complement of H'. Two subspaces H' and H'' form by definition an orthogonal pair $[H', H'']$, if H'' is the orthogonal complement of H'. The intersection of the subspaces H' and H'' of an orthogonal pair is $\{0\}$, since for any vector h in this intersection, $\|h\|^2 = (h, h) = 0$. Therefore the vector systems H' and H'' are linearly independent, and if there exists a representation of a vector $f \in H$ of the form

$$f = f' + f''; \quad f' \in H', \quad f'' \in H'', \tag{2}$$

then it is *unique*. We will give the decomposition (2) with respect to an orthogonal pair $[H', H'']$ a geometric meaning, i.e. we will call f' the *orthogonal projection* of the vector f, and f'' its *normal* to the space H'.

It turns out that the decomposition (2) holds for any vector $f \in H$, i.e. that H is the direct sum of the vector systems H' and H''. It is possible to indicate a unique process for the construction of the orthogonal projection f'.

A non-empty set of vectors of a unitary space is by definition an *orthogonal system* if it does not contain the zero vector and if all of its elements are mutually orthogonal.* If every element h of the orthogonal system is normalized (i.e., $\|h\| = 1$) then it is called a normed orthogonal system, or an *orthonormal system*.

We now give a construction of the orthogonal projection of any vector f on the subspace H' in the case when H' is given as a closed linear hull of the orthonormal system N. In turn, we will consider N to be finite, countable, and uncountable.

1) If $N = \{h_1, h_2, \ldots, h_m)$ is a finite orthonormal system, then the linear hull of the set N coincides with its closure H', and for an element

$$f' = \sum_{k=1}^{m} \alpha_k h_k \tag{3}$$

in H', the coefficients α_k are *uniquely* defined, namely

$$\alpha_k = (f', h_k). \tag{3'}$$

In fact,

$$(h_r, h_k) = 0, \quad r \neq k; \qquad (h_k, h_k) = 1.$$

Hence, by forming the scalar products (f', h_k), we have

$$(f', h_k) = \sum_{r=1}^{m} \alpha_r (h_r, h_k) = \alpha_k.$$

Whence it follows that *every orthogonal system is linearly independent.*

Now let f be an arbitrary vector space and form the *Fourier coefficients*

$$\alpha_k = (f, h_k) \tag{4}$$

with respect to the orthonormal system N, and form the linear combination (3) with these coefficients.

The vector $g = f - f'$ *belongs to the orthogonal complement* H'' *of* H'. Indeed, by (4) and (3')

$$(g, h_k) = (f - f', h_k) = (f, h_k) - (f', h_k) = 0 \tag{5}$$

and g belongs to the orthogonal complement of N. Thus

$$f = f' + g = \sum_{k=1}^{m} \alpha_k h_k + g \tag{2'}$$

is the decomposition (2) with respect to the orthogonal pair $[H', H''] (g = f'')$,

* According to this definition, a set which consists of one non-zero vector is an orthogonal system.

and hence the vector f', defined by (3) and (4), is the orthogonal projection of the vector f on H'.

2) For the case of *countable* N we have:

LEMMA 3.4.2. *The series* $\sum_{k=1}^{\infty} g_k$, *with mutually orthogonal terms, converges strongly, or diverges with the series*

$$\sum_{k=1}^{\infty} \| g_k \|^2. \tag{6}$$

Indeed, by (1) the condition of convergence of $\sum_{k=1}^{\infty} g_k$,

$$\lim_{m,\, n \to \infty} \left\| \sum_{k=m}^{n} g_k \right\|^2 = \lim_{m,\, n \to \infty} \sum_{k=m}^{n} \| g_k \|^2 = 0$$

coincides with the condition of convergence of the series (6).

Whence it follows that the strong convergence of such an orthogonal series (and in the case of convergence, the sum of the series) is independent of the order of the sequence of terms in the series.

In the case of a countable *orthonormal* system $N = \{h_k,\}$, we form the decomposition (2′) with respect to its first m terms, and by (1) we have

$$\| f \|^2 = \| f' \|^2 + \| g \|^2 = \sum_{k=1}^{m} |\alpha_k|^2 + \| g \|^2$$

and hence

$$\sum_{k=1}^{m} |(f,\, h_k)|^2 = \sum_{k=1}^{m} |\alpha_k|^2 \leqslant \| f \|^2.$$

By letting $m \to \infty$, we obtain the so-called *Bessel inequality*:

$$\sum_{k=1}^{\infty} |(f,\, h_k)|^2 \leqslant \| f \|^2, \tag{7}$$

which, by Lemma 3.4.2 and the equation

$$\sum_{k=1}^{\infty} \| (f,\, h_k)\, h_k \|^2 = \sum_{k=1}^{\infty} |(f,\, h_k)|^2,$$

guarantees the (strong) convergence of the *Fourier series*

$$\sum_{k=1}^{\infty} (f,\, h_k)\, h_k. \tag{8}$$

The element $f' = \sum_{k=1}^{\infty} (f,\, h_k) h_k$ *is the orthogonal projection of the vector* f *on the subspace* H'.

In fact, the vector f' belongs to the closed linear hull of the system N, i.e. to the subspace H', and it is only necessary to show that in the decomposition

$$f = f' + g = \sum_{k=1}^{\infty} (f, h_k) h_k + g \tag{9}$$

the vector g is orthogonal to the system $N = \{h_k\}$, i.e. $g \in H''$. Let

$$f_n = \sum_{k=1}^{n} (f, h_k) h_k = \sum_{k=1}^{n} \alpha_k h_k.$$

Then $f_n \Rightarrow f'$, and since $(f_n, h_k) = \alpha_k$ for $n \geqslant k$, in the limit,

$$(f', h_k) = \lim_{n \to \infty} (f_n, h_k) = \alpha_k = (f, h_k).$$

Hence, $(g, h_k) = (f, h_k) - (f', h_k) = 0$.

(3) In the case of an uncountable orthonormal system $N = \{h_\tau\}$, among the coefficients

$$\alpha_\tau = (f, h_\tau) \tag{4'}$$

of every (fixed) vector f in H, only a finite or countable number differ from zero. Indeed, the number of elements h_τ for which $|\alpha| \geqslant \frac{1}{n}$, by virtue of (7), does not exceed $n^2 \|f\|^2$, i.e. it is finite. We associate with each element $f \in H$ not more than a countable orthonormal system N_f of those elements h_τ in N for which $\alpha_\tau = (f, h_\tau) \neq 0$. With respect to this countable or finite system N_f, we have

$$f' = \sum_{\tau} (f, h_\tau) h_\tau, \tag{8'}$$

$$f = f' + g = \sum_{\tau} (f, h_\tau) h_\tau + g, \tag{9'}$$

moreover the summations in (8′) and (9′) can be extended to values of the index τ not only of the system N_f, but to all of the system N, since $(f, h_\tau) = 0$ for elements h_τ which do not belong to N_f. The vector g is orthogonal not only to the system N_f, but also to the whole system N, since if the vector h_τ in N did not belong to N_f, then $(f, h_\tau) = (f', h_\tau) = 0$.

Thus, when the subspace H' is a closed linear hull of the orthonormal system $N = \{h_\tau\}$, Equations (8′) and (9′), respectively, define the orthonormal projection of the vector f on H, and its decomposition (2) with respect to the orthogonal pair $[H', H'']$.

4.2. Orthogonal bases. An orthonormal system N is called an *orthogonal basis* of a space H if it is a fundamental set of H. In our previous notation, the property that an orthonormal system $N = \{h_n\}$ or $N = \{h_\tau\}$ be an orthogonal basis of H can be written in the form $H' = H$, which is equivalent to the condition

$$H'' = (0). \tag{10}$$

Indeed, the orthogonal complement of $H' = H$ is (0). If $H' \neq H$, and if f is an element of H which does not belong to H', then in the decompositions (9) or (9′), the vector g is not zero and belongs to H'', i.e. $H'' \neq (0)$.

An orthonormal system N which satisfies (10), i.e. which does not admit an extension by non-zero vectors orthogonal to N, will be said to be *complete.*

THEOREM 3.4.1. *An orthonormal system $N = \{h_n\}$ (or $N = \{h_\tau\}$) is an orthogonal basis of a space H if and only if one of the following conditions is satisfied*:

1) *for any $f \in H$,*

$$f = \sum_k (f, h_k) h_k \quad \left(or \quad f = \sum_\tau (f, h_\tau) h_\tau\right) \tag{11}$$

2) *N is a complete orthonormal system*;

3) *for every $f \in H$,*

$$\|f\|^2 = \sum_k |(f, f_k)|^2 \quad \left(or \quad \|f\|^2 = \sum_\tau |(f, h_\tau)|^2\right). \tag{12}$$

In fact, 1) is equivalent to the equation $H = H'$, and 2) is equivalent to the equation $H'' = (0)$. Since the series $\sum_k (f, h_k)h_k$ converges strongly, and since

$$\left\| \sum_k (f, h_k) h_k \right\| = \sum_k |(f, h_k)|^2,$$

for $f' = \sum_k (f, h_k)h_k$, we have

$$\|f'\| = \sum_k |(f, h_k)|^2. \tag{12′}$$

In the case of an uncountable normalized system $N = \{h_\tau\}$, by applying (12′) to the system N_f (see the end of Sec. 3.4.1), we obtain

$$\|f'\|^2 = \sum_\tau |(f, h_\tau)|^2. \tag{12″}$$

The decompositions (9) and (9′) now give

$$\|f\|^2 = \sum_k |(f, h_k)|^2 + \|g\|^2, \tag{13}$$

$$\|f\|^2 = \sum_\tau |(f, h_\tau)|^2 + \|g\|^2, \tag{13′}$$

and hence (12) is equivalent to (11). Thus 3) is equivalent to 1) and the theorem is proved.

An orthonormal system N is said to be closed if the condition 3) of Theorem 3.4.1 is satisfied. From this theorem it follows that *in a unitary space the concepts of closed orthogonal system, of complete orthonormal system, and of orthogonal basis are equivalent.*

Theorem 3.4.1 stipulated the existence of an orthogonal basis in the space

H. The following theorem shows that this condition is satisfied for any (non-degenerate) unitary space H.

THEOREM 3.4.2. *Every orthonormal system N_1 in H can be continued to an orthogonal basis of the space H.*

In fact, by Theorem 3.4.1 it is sufficient to show that the system N_1 can be extended to a complete orthonormal system, and that this can be done in the same way as in the case of a basis of a vector system (Sec. 1.2.3). In the collection $\mathfrak{N}$ of orthonormal systems in H, we establish a proper order relation which converts $\mathfrak{N}$ into an ordered set. Namely, we let $N' \geqslant N$ if the orthonormal system N is a subset of the orthonormal system N'. According to Lemma 1.1.1, a chain from one of the elements N_1 can be continued to an upper maximal chain $\mathfrak{Z}$ in $\mathfrak{N}$.

The chain $\mathfrak{Z}$ possesses a greatest element which is also an orthogonal basis of H (which contains the orthonormal system N_1). Indeed, the union $\bar{N}$ of sets N of $\mathfrak{Z}$ is an orthonormal system in H, i.e. $\bar{N} \in \mathfrak{N}$. Since $N \leqslant \bar{N}$ for every $N \in \mathfrak{Z}$, and since $\mathfrak{Z}$ is maximal, the system $\bar{N}$ belongs to the chain $\mathfrak{Z}$, and hence is its greatest element. This means that it is the maximal element of the set $\mathfrak{N}$. Therefore it is impossible to supplement the system $\bar{N}$ by normalized vectors which are orthogonal to it, i.e. $\bar{N}$ is an orthogonal basis of the space H.

REMARK. In a (non-degenerate) space H, it is possible to take any normalized vector as N_1, and Theorem 3.4.2 thus asserts the existence of an orthogonal basis of H.

In the case when H is finite or countable, an orthogonal basis can be constructed in a more efficient manner—*by the process of orthogonalization* of some fundamental set in H.

We assume first that H has finite dimensionality and that the elements $f_1, f_2, \ldots, f_n$ of the fundamental set are linearly independent, i.e. that they form a basis of the vector system H. We now let $g_1 = f_1$ and let

$$h_1 = \frac{g_1}{\| g_1 \|}, \quad h_2 = \frac{g_2}{\| g_2 \|}, \quad \ldots, \quad h_n = \frac{g_n}{\| g_n \|}, \tag{14}$$

where

$$g_m = f_m - \sum_{k=1}^{m-1} (f_m, h_k)\, h_k.$$

For the recurrent definition of the elements $h_1, h_2, \ldots, h_n$ by (14), the vectors g_m are different from zero and the linear hull of the vectors $h_1, h_2, \ldots, h_m$ coincides with the linear hull $\mathfrak{V}_m$ of the elements $f_1, f_2, \ldots, f_m$ ($m = 1, 2, \ldots, n$). Indeed, for $m = 1$ this is obvious. We will show that our assertion is true for m if it is true for $m-1$. We assume that the latter is true, and note that the vector $\sum_{k=1}^{m-1} (f_m, h_k)h_k$ belongs to $\mathfrak{V}_{m-1}$ and that the vector g_m in (14), by virtue of the linear independence of the vectors f_1, $f_2, \ldots, f_m$, is different from zero and belongs to $\mathfrak{V}_m$. Therefore $h_m \in \mathfrak{V}_m$, and since by (14), the vector f_m is contained in the linear hull of the elements $h_1, h_2, \ldots, h_m$, this hull coincides with $\mathfrak{V}_m$. Moreover, it follows from (14) that g_m is perpendicular to the vector f_m in $\mathfrak{V}_{m-1}$. Therefore $h_1, h_2, \ldots, h_n$ is an orthonormal system with the linear hull $\mathfrak{V}_n = H$, i.e. it is an orthogonal basis of H.

In the case when H is of countable dimensionality, we can assume that the countable fundamental set $\{f_k\}$ is linearly independent by removing, if necessary, elements f_k which are linearly dependent on preceding elements. Then the equations (14) hold for all natural m, and now determine a countable orthonormal system $\{h_m\}$. Since according to what has been shown, the linear hull of the elements $\{h_1, h_2, \ldots, h_m\}$ coincides for all m with the linear hull of the elements $\{f_1, f_2, \ldots, f_m\}$, whenever $\{f_m\}$ is a fundamental set in H, $\{h_m\}$ is also a fundamental set in H, i.e. $\{h_m\}$ is an orthogonal basis of the space H.

4.3. A theorem or orthogonal projection.*

THEOREM 3.4.3. *Every element $f \in H$ has an orthogonal projection and a perpendicular on any subspace H' of the space H.*

In fact, according to Theorem 3.4.2, H' has an orthogonal basis N, and the construction of the orthogonal projection and perpendicular when H' has an orthogonal basis was given in Sec. 3.4.1.

In the decomposition (2) of an element $f \in H$ with respect to an orthogonal pair $[H', H'']$, the vector f'' is the orthogonal projection of the element f on the subspace H'', and hence f' is perpendicular to H''. Therefore H' is the orthogonal complement of H'', i.e. whenever $[H', H'']$ is an orthogonal pair, $[H'', H']$ is also an orthogonal pair.

It follows from (2) that H is the direct sum of the vector systems H' and H''. In view of the orthogonality of H' and H'', we will say that H is the *orthogonal sum of the subspaces H'* and H'', and we will write

$$H = H' \oplus H''.$$

An important consequence of Theorem 3.4.3 is

THEOREM 3.4.4. *A set M is a fundamental set of the space H if and only if its orthogonal complement is* (0)

In fact, a vector g which is orthogonal to the set M is also orthogonal to the closed linear hull $\tilde{M}$ of the set M, and if $\tilde{M} = H$, then $g = 0$. If $\tilde{M} \neq H$, and if f is an element of H which does not belong to $\tilde{M}$, then the

* See also Sec. 3.6.2.

perpendicular f' of f on the subspace $H' = \tilde{M}$ is a non-zero element which is orthogonal to $\tilde{M}$, and thus also to M.

4.4. The classification of unitary spaces. The distance $\|f-g\|$ between normalized, orthogonal elements f and g in H is a constant, and equals $\sqrt{2}$, since

$$\| f-g \|^2 = \| f \|^2 + \| g \|^2 = 2.$$

Therefore the spherical neighborhoods

$$\| f - h_\tau \| < \frac{1}{\sqrt{2}} \tag{15}$$

of elements h_τ of an arbitrary orthonormal system

$$N = \{h_\tau\}$$

are pairwise disjoint. We now assume that the unitary space H is infinite dimensional, and that the system N is its orthogonal basis, Hence:

1) *the set N does not contain a (strongly) convergent sequence, and therefore the unit sphere $S = \{\|f\| = 1\}$, on which N lies, is not compact.*

2) *the cardinality of any set M which is dense in H is not less than the cardinality of the orthonormal system N.*

Indeed, in each of the neighborhoods (15) of the element h_τ, there exists a vector $f_\tau \in M$, and all of the f_τ will be distinct since the neighborhoods (15) are pairwise disjoint.

If the space H is separable, then there exists in H a *countable* dense set M, and hence by 2) every orthogonal basis is countable if the dimensionality of H is infinite. Thus, in the case of a separable space H, the cardinality of an orthogonal basis coincides with the dimensionality of H. This is also true in the general case.

THEOREM 3.4.5. *The dimensionality of a (non-degenerate) space H is equal to the cardinality of its orthogonal basis.*

It is sufficient to outline the proof under the assumption that H is infinite dimensional.

Let N be a fixed orthogonal basis of the space H and let M be one of its fundamental sets. We form the collection M' of those linear combinations of the set M with *rational* (complex) coefficients. Whenever the linear hull of the set M is dense in H, the set M' is also dense in H, and since M is infinite, the cardinalities of the sets M and M' are equal. By the property 2), the cardinality of M', and thus of any fundamental set of M is not less than the cardinality of the orthogonal basis N. But the basis N itself is a fundamental set, and hence the cardinality of the basis N is minimal, i.e. it is by definition equal to the dimensionality of the space H.

COROLLARY. *All orthogonal bases of a unitary space have the same cardinality.*

In the decomposition (11)

$$f = \sum_{\tau} \alpha_\tau h_\tau, \qquad \alpha_\tau = \alpha(\tau;\ f) = (f,\ h_\tau) \tag{11'}$$

of an arbitrary vector $f \in H$ with respect to an orthogonal basis $\{f_\tau\}$ of the space H, the coefficients $\alpha_\tau = \alpha(\tau)$, considered as functions of the variable τ which ranges over an index set T, are different from zero only on a finite or countable subset of T. They satisfy the condition (see (12))

$$\sum_{\tau} |\alpha(\tau)|^2 < \infty,$$

i.e. $\alpha(\tau)$ belongs to the space $H(T)$ with norm

$$\|\alpha(\tau)\| = \Big(\sum_{\tau} |\alpha(\tau)|^2\Big)^{1/2}.$$

Conversely, for each function $\alpha(\tau) = \alpha_\tau$ in $H(T)$, the series (11′) converges strongly to an element $f \in H$, moreover $\alpha(\tau) = (f, h_\tau)$ and (see (12))

$$\|f\|^2 = \sum_{\tau} |\alpha(\tau)|^2 = \|\alpha(\tau)\|^2. \tag{16}$$

Since $\alpha(\tau) = \alpha(\tau; f)$ is a linear form with respect to f, the one-to-one correspondence $f \to \alpha(\tau)$ established between vectors $f \in H$ and functions $\alpha(\tau) \in H(T)$ is an isomorphism of the vector systems H and $H(T)$, and in view of (16), it is an isomorphism of the spaces H and $H(T)$.

Thus, we have proved

THEOREM 3.4.6. *A unitary space H is isomorphic to the space $H(T)$ if the cardinality of the set T is equal to the dimensionality of H.*

We remark that for the scalar product, we have

$$(f,\ g) = \sum_{\tau} \alpha(\tau) \overline{\beta(\tau)},$$

if $f \to \alpha(\tau)$, $g \to \beta(\tau)$ under the isomorphism.

From Theorem 3.4.6 we have

THEOREM 3.4.7. *Two unitary spaces are isomorphic if and only if their dimensionalities are identical.*

In other words, the dimensionality of a unitary space is its complete system of invariants with respect to an isomorphism.

§ 5. Orthogonal Sums of Spaces

In the decomposition (3.4.8′), the separate terms are the projections of the vector f on the one-dimensional, mutually orthogonal subspaces (h_τ)

which have been generated by the vectors h_τ of the orthogonal system $\{h_\tau\}$. The replacement of the one-dimensional subspaces (h_τ) by an arbitrary system of mutually orthogonal subspaces of the space H leads to a natural generalization of the results of Sec. 3.4.1.

A set $\{H_\tau\}$ of mutually orthogonal subspaces H_τ of the space H, none of which are (0), will be called an *orthogonal system of subspaces*.

For an orthogonal system $\{H_\tau\}$, we form the closed linear hull H' of the union of all the H_τ (i.e. the closure of the sum of the vector systems H_τ), and for an arbitrary vector $f \in H$, we form the sum*

$$f' = \sum_\tau f_\tau, \tag{1}$$

where f_τ is the orthogonal projection of the vector f on the subspace H_τ.

The sum f' is the orthogonal projection of the vector f on the subspace H'. In fact, by taking the union of the elements of the orthogonal bases N_τ of the subspaces H_τ, we obtain an orthogonal basis $N = \bigcup_\tau N_\tau$ of the space H'. The elements of the set N form a complete orthonormal system in H', since a non-zero vector g which is orthogonal to N is also orthogonal to all of the N_τ, and thus to all of the H_τ, and hence (Lemma 3.4.1) to the subspace H'. Therefore it cannot belong to H'.

In (1) we now represent the projection f_τ by its decomposition (of the form (3.4.8′)) with respect to the orthonormal system N_τ. The right-hand side of (1) is then changed into the decomposition (3.4.8′) with respect to $N = \bigcup_\tau N_\tau$, and hence is equal to the orthogonal projection of the vector f on H. In place of (3.4.9′) we now have

$$f = f' + g = \sum_\tau f_\tau + g,$$

and the vector f belongs to H' if and only if $g = 0$, i.e. if

$$f = \sum_\tau f_\tau. \tag{2}$$

Moreover, in this case we have (equivalent to (2))

$$\|f\|^2 = \sum_\tau \|f_\tau\|^2, \tag{3}$$

which can be obtained from the corresponding equation (3.4.12) by grouping the terms on the right-hand side which involve the same N_τ.

The collection of all vectors f which are representable in the form of a convergent series (2), where f_τ belongs to the element H_τ of the orthogonal system, will be called the *orthogonal sum of the subspaces* H_τ, and we will

* Since elements f_τ which are different from zero are orthogonal projections of the vector f on the mutually orthogonal, one-dimensional spaces (f_τ), there are not more than a countable number of them.

write $\sum_\tau \oplus H_\tau$. As has been shown above, this orthogonal sum coincides with the subspace H', i.e.

$$H' = \sum_\tau \oplus H_\tau. \tag{4}$$

When the index set T is finite and equal to n, or is countable, we will write

$$H' = \sum_{k=1}^{n} \oplus H_k, \qquad H' = \sum_{k=1}^{\infty} \oplus H_k.$$

In the case of finite T, the condition for convergence of the series (2) (Lemma 3.4.2)

$$\sum_\tau \|f_\tau\|^2 < \infty \tag{5}$$

is known to be fulfilled, and the vector system of orthogonal sums $\sum_{k=1}^{n} \oplus H_k$ is the direct sum of the vector systems H_k.

An orthogonal system of subspaces $\{H_\tau\}$ is said to be *complete* if there exists no subspace different from (0) which is orthogonal to all of the H_τ. For the complete system $\{H_\tau\}$, the orthogonal complement of the union of the sets H_τ is (0), so that $H' = H$, and (4) is transformed into a representation

$$H = \sum_\tau \oplus H_\tau$$

of the space H as an orthogonal sum of subspaces. Equations (2) and (3) now hold for any vector $f \in H$, moreover each of these equations holds for any $f \in H$ if and only if $\{H_\tau\}$ is a complete system.

Thus, to each vector $f \in H$ there corresponds an element $[f_\tau]$ of the vector system $\mathfrak{G}^{(1)}(T;\, H_\tau)$ (Sec. 1.2.2), for which the condition (5) is satisfied. Conversely, each element $[f_\tau] \in \mathfrak{G}^{(1)}(T;\, H_\tau)$ which satisfies (5) corresponds to some vector $f \in H$; indeed, it follows from (5) that the series (2) converges to an element $f \in H$. The mapping $f \to [f_\tau]$ is obviously one-to-one and linear. By (3), we have

$$(f,\, g) = \sum_\tau (f_\tau,\, g_\tau), \tag{6}$$

for the scalar product if $f \to [f_\tau]$ and $g \to [g_\tau]$.

THEOREM 3.5.1. *Let $\{H_\tau\}$ be a complete orthogonal system, let $H_\tau = H'_\tau \oplus H''_\tau$, and let*

$$H' = \sum_\tau \oplus H'_\tau, \qquad H'' = \sum_\tau \oplus H''_\tau.$$

Then

$$H = H' \oplus H''.$$

In fact, H'_τ and H''_σ are orthogonal for any τ and σ in T, and at the same time the subspaces H' and H'' are orthogonal. Let g be an element of H which is orthogonal to H', and let

$$g=\sum_{\tau} g_{\tau}$$

be its decomposition with respect to the complete system $\{H_{\tau}\}$. By applying (6) to g, and to those elements $f \in H'$ for which $f_{\tau}=0$ for $\tau \neq \sigma$, we obtain

$$(f_{\sigma}, g_{\sigma})=(f, g)=0,$$

where f_{σ} is any vector in H'_{σ}. Therefore $g_{\sigma} \in H''_{\sigma}$, and hence $g \in H''$, so that H'' coincides with the orthogonal complement of H'.

REMARK. In the system $\mathfrak{S}(H)$ of all subspaces of the space H, we can introduce a proper order relation in a natural way: $H_2 \geqslant H_1$ if H_1 is contained in H_2. This order relation converts $\mathfrak{S}(H)$ into a complete, bounded lattice. Indeed, H and (0) are the greatest and least elements of $\mathfrak{S}(H)$, and if $\{H_{\tau}\}$ is an arbitrary set of subspaces, then as is easily seen, (1) inf $\{H_{\tau}\}$ exists and coincides with the intersection of the subspaces H_{τ}, (2) sup $\{H_{\tau}\}$ exists and equals the closed linear hull of the union of the sets H_{τ}.

It follows from (4) that: *if the subspaces H_{τ} are mutually orthogonal, then*

$$\sup_{\tau}\{H_{\tau}\}=\sum_{\tau} \oplus H_{\tau}. \tag{7}$$

The concept of orthogonal sum can be generalized in a natural manner to the system $\{\mathfrak{V}_{\tau}\}$ of mutually orthogonal vector manifolds $\mathfrak{V}_{\tau}$ in H. In this case, by the orthogonal sum

$$\mathfrak{V}=\sum_{\tau} \oplus \mathfrak{V}_{\tau}$$

we will mean the vector manifold $\mathfrak{V}$ of elements which are represented by convergent series of the form (2), where $f_{\tau} \in \mathfrak{V}_{\tau}$. For fixed $f \in \mathfrak{V}$, not more than a countable number of the elements f_{τ} in (2) are different from zero, and again the condition of convergence (5) is satisfied. For the closures $\overline{\mathfrak{V}}_{\tau}$ and $\overline{\mathfrak{V}}$, we have

$$\overline{\mathfrak{V}}=\sum_{\tau} \oplus \overline{\mathfrak{V}}_{\tau}. \tag{8}$$

Indeed, by letting $H_{\tau}=\overline{\mathfrak{V}}_{\tau}$ we reduce (8) to (4), since $\overline{\mathfrak{V}}=\overline{H}'=H'$.

When the unitary spaces H_{τ} are not subspaces of the given space H, but are given independently, we will define their orthogonal sum by reference to the mapping $f \rightarrow [f_{\tau}]$ (see Sec. 1.3.2). We will consider the orthogonal sum

$$H=\sum_{\tau} \oplus H_{\tau} \tag{9}$$

to define a vector system $\hat{H}$ of elements

$$\hat{f}=[f_{\tau}]=f(\tau)$$

in $\mathfrak{G}^{(1)}(T, H_\tau)$ which satisfy (5), and with the norm.*

$$\|\hat{f}\| = \left(\sum_\tau \|f_\tau\|^2\right)^{1/2}.$$

The norm $\|\hat{f}\|$ is unitary, since its square is a quadratic form generated by the bilinear form (scalar product)

$$(\hat{f}, \hat{g}) = \sum_\tau (f_\tau, g_\tau); \quad \hat{f} = [f_\tau], \quad \hat{g} = [g_\tau]$$

in the vector system $\hat{H}$.

$\hat{H}$ is a unitary space.

For the proof, only the completeness of the space $\hat{H}$ is necessary. We will establish this by isomorphically mapping $\hat{H}$ onto some space $H(T)$. In each H_τ we choose an orthogonal basis N_τ, and for elements of the set $N = \bigcup_\tau N_\tau$, we introduce the notation h_σ, where σ ranges over some set T'. Each element h_σ belongs to one and only one H_τ, and to each element $\hat{f} = [f_\tau] \in \hat{H}$, there corresponds a function

$$\alpha(\sigma) = (f_\tau, h_\sigma), \quad h_\sigma \in H_\tau$$

on the set T'. The mapping $\hat{f} \to \alpha(\sigma)$ preserves vector operations and is one-to-one ($\alpha(\sigma) = 0$ for all $\sigma \in T'$ implies that $f_\tau = 0$ for all $\tau \in T$). Moreover,

$$\sum_{\sigma \in T'} |\alpha(\sigma)|^2 = \sum_{\tau \in T} \sum_{h_\sigma \in H_\tau} |(f_\tau, h_\sigma)|^2 = \sum_{\tau \in T} \|f_\tau\|^2 = \|\hat{f}\|^2,$$

so that $\hat{f} \to \alpha(\sigma)$ is an isomorphism of the space $\hat{H}$ onto the space $H(T')$ of functions $\alpha(\sigma)$ with norm

$$\|\alpha(\sigma)\| = \left(\sum_{\sigma \in T'} |\alpha(\sigma)|^2\right)^{1/2}.$$

Under this mapping, a function $\alpha(\sigma) \in H(T')$ corresponds to an element $\hat{f} = \{f_\tau\} \in \hat{H}$, where

$$f_\tau = \sum_{h_\sigma \in H_\tau} \alpha(\sigma) h_\sigma.$$

The space $\hat{H}$ is isomorphic to the unitary space $H(T')$ (see Sec. 3.3.3 and Sec. 3.2.2) and is itself unitary.

REMARK 1. Let $\hat{H}_\tau$ be the subspace of all elements $\hat{f} \in \hat{H}$ for which $f(\tau') = 0$ for $\tau' \neq \tau$. Then $\hat{H}_\tau$ is isomorphic to H_τ, and $\hat{H} = \sum_\tau \oplus \hat{H}_\tau$ in the sense of the original definition.

REMARK 2. If, for the subspaces H_τ in the decomposition (4), we form the corresponding space $\hat{H}$ which is determined by (9), then it will be iso-

* It is easy to verify that the conditions 1)-3) in the definition of the norm (Sec. 3.1.1) are satisfied.

morphic to the space H. Thus, to each decomposition of the space H into an orthogonal sum (4), there corresponds an isomorphism $f \to [f_\tau]$ of the corresponding space $\hat{H}$.

§ 6. Supplementary Material

6.1. Quadratic forms. A quadratic form $Q(f)$ in a complex vector system $\mathfrak{V}$ satisfies:

1) the parallelogram equation

$$Q\left(\frac{f+g}{2}\right)+Q\left(\frac{f-g}{2}\right)=\frac{1}{2}(Q(f)+Q(g)), \tag{1}$$

2) the homogenity condition

$$Q(\alpha f)=|\alpha|^2 Q(f), \tag{2}$$

3) the continuity condition

$$\lim_{\varepsilon\to 0} Q(f+\varepsilon h)=Q(f), \tag{3}$$

where h is any element in $\mathfrak{V}$.

Properties 1) and 2) are already known (see (3.3.5) and (3.3.1)), and (3) follows from (see (3.3.2))

$$Q(f+\varepsilon h)=Q(f)+\bar{\varepsilon}W(f, h)+\varepsilon W(g, f)+|\varepsilon|^2 Q(h),$$

where $W(f, g)$ is the bilinear form which generates $Q(f)$.

THEOREM 3.6.1. *A function $Q(f)$ on a complex vector system $\mathfrak{V}$ is a quadratic form if and only if it satisfies the conditions* 1), 2) *and* 3).

In fact, let $Q(f)$ satisfy 1), 2), and 3). By using (3.3.4), we show that the function

$$W_1(f, g)=Q\left(\frac{f+g}{2}\right)-Q\left(\frac{f-g}{2}\right)=Q(h_1)-Q(h_2) \tag{4}$$

satisfies the equations:

$$W_1(g, f)=W_1(f, g), \qquad W_1(if, g)=-W_1(f, ig) \tag{5}$$

and (α is real)

$$\begin{gathered} W_1(f'+f'', g)=W_1(f', g)+W_1(f'', g), \\ W_1(\alpha f, g)=\alpha W_1(f, g). \end{gathered} \tag{6}$$

Indeed, by (4) and (2), with $\alpha=-1$, we obtain the first equation in (5):

$$W_1(g, f)=Q(h_1)-Q(-h_2)=Q(h_1)-Q(h_2)=W_1(f, g),$$

and with $\alpha=i$, we obtain the second:

$$W_1(if, g) = Q\left(\frac{if+g}{2}\right) - Q\left(\frac{if-g}{2}\right) =$$
$$= Q\left(\frac{f-ig}{2}\right) - Q\left(\frac{f+ig}{2}\right) = -W_1(f, ig).$$

For the right-hand side of the first of the equations in (6), we have

$$W_1(f', g) + W_1(f'', g) =$$
$$= (Q(h_1') - Q(h_2')) + (Q(h_1'') - Q(h_2'')) =$$
$$= (Q(h_1') + Q(h_1'')) - (Q(h_2') + Q(h_2'')), \tag{7}$$

where, according to (4)

$$h_1' = \frac{f'+g}{2}, \quad h_1'' = \frac{f''+g}{2}, \quad h_2' = \frac{f'-g}{2}, \quad h_2'' = \frac{f''-g}{2}.$$

We now let

$$h_1' = \frac{f_1+g_1}{2}, \quad h_1'' = \frac{f_1-g_1}{2}, \quad h_2' = \frac{f_2+g_2}{2}, \quad h_2'' = \frac{f_2-g_2}{2},$$

i.e.

$$\left.\begin{aligned} f_1 &= h_1' + h_1'' = \frac{f'+f''}{2} + g, \\ f_2 &= h_2' + h_2'' = \frac{f'+f''}{2} - g, \\ g_1 &= h_1' - h_1'' = \frac{f'-f''}{2} = h_2' - h_2'' = g_2. \end{aligned}\right\} \tag{8}$$

By transforming the right-hand side of (7) with the help of (1), we obtain

$$W_1(f', g) + W_1(f'', g) =$$
$$= \frac{1}{2}(Q(f_1) + Q(g_1)) - \frac{1}{2}(Q(f_2) + Q(g_2)) =$$
$$= \frac{1}{2}(Q(f_1) - Q(f_2)),$$

and by (8), and the definition (4),

$$W_1(f', g) + W_1(f'', g) = \frac{1}{2} W_1(f'+f'', 2g). \tag{9}$$

By letting $f' = f$ and $f'' = 0$, we have

$$W_1(f, 2g) = 2W_1(f, g), \tag{10}$$

since $W_1(0, g) = Q\left(\frac{g}{2}\right) - Q\left(\frac{-g}{2}\right)$, and hence

$$W_1(0, g) = 0. \tag{11}$$

By comparing (9) with (10), for $f=f'+f''$, we obtain the first equation in (6), from which the second follows in turn for positive rational α. In view of

$$W_1(-f,\ g)=Q(-h_2)-Q(-h_1)=Q(h_2)-Q(h_1)=-W_1(f,\ g)$$

and (11), the second equation in (6) holds for all rational α.

Now let $\{\alpha_n\}$ be a sequence of rational numbers which converges to a real α, and let $\alpha=\alpha_n+\varepsilon_n$. Then $\lim\varepsilon_n=0$, and

$$W_1(\alpha f,\ g)=\alpha_n W_1(f,\ g)+W_1(\varepsilon_n f,\ g), \tag{12}$$

and by (3)

$$\lim W_1(\varepsilon_n f,\ g)=\lim\left[Q\left(\frac{\varepsilon_n f+g}{2}\right)-Q\left(\frac{\varepsilon_n f-g}{2}\right)\right]=$$
$$=Q\left(\frac{g}{2}\right)-Q\left(-\frac{g}{2}\right)=0.$$

Therefore, for $n\to\infty$, (12) gives the second equation in (6) for any real α.

We now let (see (3.3.4) and (4))

$$W(f,\ g)=W_1(f,\ g)+iW_1(f,\ ig) \tag{13}$$

and we will show that $W(f, g)$ *is a bilinear form in* $\mathfrak{B}$ *and that* $Q(f)$ *is the quadratic form generated by it*. Indeed, by (6) and the first equation in (5),

$$W(f'+f'',\ g)=W(f',\ g)+W(f'',\ g),$$
$$W(\alpha f,\ g)=\alpha W(f,\ g),$$
$$W(f,\ g'+g'')=W(f,\ g')+W(f,\ g''),$$
$$W(f,\ \alpha g)=\alpha W(f,\ g),$$

for real α, and in order to prove that $W(f, g)$ is bilinear, it is sufficient to verify that

$$W(if,\ g)=iW(f,\ g),\qquad W(f,\ ig)=-iW(f,\ g). \tag{14}$$

By virtue the second equation in (5), and (2) with $\alpha=i$, we have

$$W(if,\ g)=W_1(if,\ g)+iW_1(if,\ ig)=$$
$$=-W_1(f,\ ig)+i(Q(ih_1)-Q(ih_2))=$$
$$=-W_1(f,\ ig)+iW_1(f,\ g)=iW(f,\ g),$$

This is the first equation in (14); the second now follows from (13) and the equation

$$W(f,\ ig)=W_1(f,\ ig)+iW_1(f_1-g)=W_1(f,\ ig)-iW_1(f,\ g).$$

Finally, by (5)

$$W_1(f,\ if)=W_1(if,\ f)=-W_1(f,\ if)=0,$$

and since $W_1(f, f)=Q(f)$, we have $Q(f)=W(f, f)$ for any bilinear form (13).

REMARK. Condition 2) of Theorem 3.6.1 can be weakened, replacing it by the condition

$$Q(-f)=Q(f), \quad Q(if)=Q(f), \tag{2'}$$

since in the proof of the theorem, (2) was applied only for $\alpha = -1$ and $\alpha = \pm i$.

If $w(f)$ is a norm or a quasinorm in the vector system $\mathfrak{V}$, then the function $Q(f) = (w(f))^2$ obviously satisfies the conditions 2) and 3) of Theorem 3.6.1, and hence we have:

THEOREM 3.6.2. *A norm or quasinorm $w(f)$ is unitary if and only if it satisfies the condition*

$$\left(w\left(\frac{f+g}{2}\right)\right)^2+\left(w\left(\frac{f-g}{2}\right)\right)^2=\frac{1}{2}((w(f))^2+(w(g))^2). \tag{15}$$

In particular, *a Banach space is a unitary space if and only if its norm $\|f\| = w(f)$ satisfies* (15).

6.2. Convex sets. A subset M of a complex or real vector system $\mathfrak{V}$ is called a *convex set* if whenever the elements f and g belong to M, the "segment" joining them, i.e., the set of points of the form

$$h=\alpha f+\beta g, \quad \alpha\geqslant 0, \quad \beta\geqslant 0, \quad \alpha+\beta=1,$$

also belongs to M. The points h with $\alpha>0$ and $\beta>0$ are called *interior points of the segment*, and f and g are its *end points.*

If $w(f)$ is a quasinorm in $\mathfrak{V}$, then the set M of elements f which satisfy the inequality

$$w(f)\leqslant\delta \quad (\delta>0), \tag{16}$$

is convex. Indeed,

$$w(\alpha f+\beta g)\leqslant\alpha w(f)+\beta w(g)\leqslant(\alpha+\beta)\delta=\delta$$

for $w(f)\leqslant\delta$, $w(g)\leqslant\delta$, $\alpha\geqslant 0$, $\beta\geqslant 0$, and $\alpha+\beta = 1$. This conclusion still holds if the condition 2) in the definition of quasinorm is weakened; namely, by requiring it to hold only for non-negative α. Functions $w(f)$ which satisfy the condition

$$w(f+g)\leqslant w(f)+w(g), \quad w(\alpha f)=\alpha w(f), \quad \alpha\geqslant 0, \tag{17}$$

are called *convex forms* in $\mathfrak{V}$, and are called *convex functionals* if $\mathfrak{V}$ is a Banach space or a vector manifold of a Banach space.

A convex set M which is defined by the inequality (16), where $w(f)$ is a non-negative convex form in $\mathfrak{V}$, has the following properties:

1) The zero vector belongs to M, and every segment which is contained

in M, one end of which is the point $f = 0$, can be continued in M so that $f = 0$ becomes an interior point of the extended segment.

In fact, it is sufficient to show that for any non-zero $f \in \mathfrak{V}$, and for a sufficiently small γ, the segment with end points $-\gamma f$ and γf is contained in M. Let $\gamma_1 = \dfrac{\delta}{w(f)}$ if $w(f) \neq 0$, and let $\gamma_1 = \delta$ if $w(f) = 0$. Furthermore, let $\gamma_2 = \dfrac{\delta}{w(-f)}$ if $w(-f) \neq 0$, and let $\gamma_2 = \delta$ if $w(-f) = 0$. We let $\gamma = \min(\gamma_1, \gamma_2)$. Then for $\alpha \leqslant \gamma$ and $\alpha \geqslant 0$,

$$w(\alpha f) = \alpha w(f) \leqslant \delta, \qquad w(-\alpha f) = \alpha w(-f) \leqslant \delta$$

and the segment with end points $-\gamma f$ and γf is contained in M.

2) If $\alpha_n f \in M$ and $\alpha_n \geqslant 0$, and if $\lim \alpha_n = \alpha$, then $\alpha f \in M$. Indeed, $w(\alpha_n, f) \leqslant \delta$ implies that

$$w(\alpha f) = \alpha w(f) = \lim_{n\to\infty} (\alpha_n w(f)) = \lim_{n\to\infty} w(\alpha_n f) \leqslant \delta.$$

Now let a convex set M which satisfies 1) and 2) be given in a Banach space X. Then the set $\mathfrak{V}_M$ of elements $f \in X$ for which numbers $t > 0$ exist such that $tf \in M$ is, by 1), a vector manifold and

$$w(f;\ M) = \inf \frac{1}{t}, \qquad tf \in M, \qquad t > 0, \tag{18}$$

defines a functional $w(f;\ M)$ (the Minkowski functional) on the vector manifold $\mathfrak{V}_M$.

THEOREM 3.6.3. *$w(f;\ M)$ is a convex functional in $\mathfrak{V}_M$, and the convex set M is defined by the inequality*

$$w(f) = w(f;\ M) \leqslant 1. \tag{19}$$

In fact, $tf \in M$ for any $t > 0$ if and only if $w(f) = 0$, so that in particular, $w(0) = 0$. Now $t'\alpha f \in M$ ($\alpha > 0$), and $tf \in M$ for $t = t'\alpha$, are equivalent, and hence

$$w(\alpha f) = \inf \frac{1}{t'} = \inf \frac{\alpha}{t} = \alpha w(f),$$

i.e. the second equation in (17) holds.

Now let $tf \in M$ and $ug \in M$, $t > 0$, $u > 0$. Then

$$\frac{tu}{t+u}(f+g) = \frac{u}{t+u}(tf) + \frac{t}{t+u}(ug) \in M$$

and, according to the definition (18),

$$w(f+g) \leqslant \frac{t+u}{tu} = \frac{1}{t} + \frac{1}{u}.$$

Now, by replacing the expression on the right-hand side by its lower bound, we obtain the first equation in (17).

For any $f \in \mathfrak{V}_M$, a sequence $\{t_n\}$ exists such that

$$t_n f \in M, \quad t_n > 0, \quad \lim \frac{1}{t_n} = w(f;\ M),$$

and in view of property 2), which the convex set M has by assumption, $\frac{f}{w(f)} \in M$, if $w(f) \neq 0$. Therefore, for $w(f) \leqslant 1$, the element f lies on the segment with end points 0 and $\frac{f}{w(f)}$, and hence it belongs to the set M. Conversely, if $f \in M$, then (19) holds by the definition (18) of the functional $w(f;\ M)$. If, for each $f \in X$ and for sufficiently small $\alpha > 0$, the convex set M contains the segment with end points 0 and αf, then $\mathfrak{V}_M = X$, and the functional $w(f;\ M)$ is defined everywhere in X. In particular, this holds if $f = 0$ is an interior point of the set M.

The point $f = 0$ is an interior point of a convex set M if and only if the functional $w(f;\ M)$ is defined for all $f \in X$ and is bounded, i.e.

$$w(f;\ M) \leqslant \gamma \|f\|, \quad 0 < \gamma < \infty. \tag{20}$$

Indeed, let the sphere $\|f\| \leqslant \delta$ be contained in M. Then $\frac{\delta f}{\|f\|} \in M\ (f \neq 0)$, and hence for all $f \in X$,

$$w(f;\ M) \leqslant \frac{1}{\delta} \|f\|.$$

Thus (20) holds with $\gamma = \frac{1}{\delta}$. Conversely, if (20) holds, then for all $f \in X$, we obtain $\frac{f}{(\gamma+\varepsilon)\|f\|} \in M$, $\varepsilon > 0$, and hence all points of the sphere $\|f\| < \frac{1}{\gamma+\varepsilon}$ belong to M.

In the normed space X, we will define the *distance* $d(f;\ M)$ of an element f from the set $M \subset X$ by

$$d(f;\ M) = \inf_{h \in M} \|f - h\|.$$

For convex sets of a unitary space, we have:

THEOREM 3.6.4. *Let M be a closed convex set of a unitary space H. Then the distance $d(f;\ M)$ of an arbitrary element $f \in H$ from the set M is taken on at a point of the set M, i.e. for some $h_0 \in M$,*

$$d = d(f;\ M) = \|h_0 - f\|. \tag{21}$$

In fact, if $f \in M$, then $d = 0$ and $h_0 = f$. Therefore, we can assume that f does not belong to M. In this case $d > 0$, since M is a closed set. First, let $f = 0$; then

$$d(0;\ M) = \inf_{h \in M} \|h\| = d > 0$$

and a sequence $\{h_n\}$ in M exists with the property

$$\lim \|h_n\| = d. \tag{22}$$

We will show that $\{h_n\}$ is a Cauchy sequence.

The application of Equation (1) to $Q(f) = \|f\|^2$, and to the elements $f = h_m$ and $g = h_n$, gives

$$\|h_m - h_n\|^2 = 2\|h_m\|^2 + 2\|h_n\|^2 - 4\left\|\frac{h_m + h_n}{2}\right\|^2.$$

Here $\left\|\frac{h_m+h_n}{2}\right\| \geqslant d$, since $\frac{h_m+h_n}{2} \in M$. Therefore

$$\|h_m - h_n\|^2 \leqslant 2\|h_m\|^2 + 2\|h_n\|^2 - 4d^2,$$

and in the limit

$$\lim_{n,\, m\to\infty} \|h_m - h_n\|^2 = 2d^2 + 2d^2 - 4d^2 = 0.$$

Since H is complete, $h_n \Rightarrow h_0$, and hence

$$d = \lim \|h_n\| = \|h_0\|.$$

Since M is a closed set, $h_0 \in M$, and (21) is proved for the special case when $f = 0$. The general case can be reduced to this case; namely, it is sufficient to replace the set M by the set M' of elements g of the form

$$g = h - f, \qquad h \in M,$$

which like M, is closed and convex. Let $g_0 \in M'$ and let $d(0;\, M') = \|g_0\|$. Then $g_0 = h_0 - f$, where $h_0 \in M$ and

$$d(f,\, M) = d(0,\, M') = \|g_0\| = \|h_0 - f\|.$$

Remark. If, in Theorem 3.6.4, we consider M to be a subspace, then $h_0 = f'$ and $f - h_0 = f''$ are, respectively, the orthogonal projection and the perpendicular of the vector f on H'. Indeed, the vector $f'' = f - h_0$ is orthogonal to H'. Let g' be any (non-zero) element from H' and let

$$h' = f' + \varepsilon g' = h_0 + \varepsilon g' \in H'.$$

Then for any complex ε,

$$\|f'' - \varepsilon g'\|^2 = \|f - h'\|^2 \geqslant (d(f;\, H'))^2 = \|f''\|^2$$

or

$$-\bar{\varepsilon}(f'',\, g') - \varepsilon(g',\, f'') + \varepsilon\bar{\varepsilon}(g',\, g') \geqslant 0,$$

and since ε was arbitrary, we have $(f'', g') = 0$.

Thus we have obtained another proof of Theorem 3.4.3 for the existence of an orthogonal projection on an arbitrary subspace H'. We remark that the starting point of this proof is the property that the perpendicular gives the least distance from a subspace.

6.3. The Beppo-Levi inequality. When the convex set M is a subspace, Theorem 3.6.4 can be proved with the help of the following inequality.

Let $\mathfrak{B}$ be a vector manifold in H, and let g_1 and g_2 be two of its elements. Furthermore, let f be an element in H which is removed from $\mathfrak{B}$ by the distance $d = d(f; \mathfrak{B})$. Then

$$\|g_1 - g_2\| \leqslant \sqrt{\|f - g_1\|^2 - d^2} + \sqrt{\|f - g_2\|^2 - d^2}. \qquad (23)$$

Indeed, for any complex α and β such that $\alpha+\beta \neq 0$, the element $\dfrac{\alpha g_1 + \beta g_2}{\alpha+\beta}$ belongs to $\mathfrak{B}$, and therefore

$$\left\| \frac{\alpha (f - g_1) + \beta (f - g_2)}{\alpha+\beta} \right\| = \left\| f - \frac{\alpha g_1 + \beta g_2}{\alpha+\beta} \right\| \geqslant d.$$

Whence it follows that the Hermitian form

$$\alpha\bar{\alpha} [\|f - g_1\|^2 - d^2] + \alpha\bar{\beta} [(f - g_1, f - g_2) - d^2] + \\ + \bar{\alpha}\beta [(f - g_2, f - g_1) - d^2] + \beta\bar{\beta} [\|f - g_2\|^2 - d^2]$$

is positive. Therefore

$$|(f-g_1, f-g_2)-d^2|^2 \leqslant [\|f-g_1\|^2-d^2] [\|f-g_2\|^2-d^2]. \qquad (24)$$

Furthermore

$$\|g_1 - g_2\|^2 = (f - g_1 - f + g_2, f - g_1 - f + g_2) = \\ = \|f-g_1\|^2 + \|f-g_2\|^2 - (f-g_1, f-g_2) - (f-g_2, f-g_1) = \\ = [\|f - g_1\|^2 - d^2] + [\|f - g_2\|^2 - d^2] - \\ - [(f - g_1, f - g_2) - d^2] - [(f - g_2, f - g_1) - d^2],$$

whence

$$\|g_1 - g_2\|^2 \leqslant [\|f - g_1\|^2 - d^2] + [\|f - g_2\|^2 - d^2] + \\ + 2 [(f - g_1, f - g_2) - d^2]. \qquad (25)$$

By comparing (25) with (24), we obtain (23).

The inequality (23) is called the *Beppo-Levi inequality*. It can also be used for the proof of the existence of an orthogonal projection and perpendicular. Indeed, if $\{h_n\}$ is a sequence in M such that $\lim \|f-h_n\| = d(f; M) = d$. then

$$\|h_n - h_m\| \leqslant \sqrt{\|f-h_n\|^2-d^2} + \sqrt{\|f - h_m\|^2-d^2} \xrightarrow[n,\, m \to \infty]{} 0.$$

Thus, the sequence $\{h_n\}$ itself converges, and the element $h_0 = \lim h_n$ gives the distance from f to the set M.

CHAPTER 4

Continuous Linear and Pseudolinear Transformations

§ 1. Bounded Transformations

1.1 Bounded quasinorms. A quasinorm $w(f)$, defined on a normed space X or on a vector manifold $\mathfrak{V}$ of X, is said to be *bounded* if

$$\sup_{\|f\|=1} w(f) = \gamma(w) < \infty. \tag{1}$$

If the vector f in (1) is not assumed to have been normalized, then (1) can be written in the form

$$w(f) \leqslant \gamma(w) \|f\|, \tag{1'}$$

where $\gamma(w)$ is the smallest number γ for which the inequality

$$w(f) \leqslant \gamma \|f\| \tag{2}$$

holds. Indeed, for $f \neq 0$ it is sufficient to let $f = \|f\|g$ and to apply (1) to the normalized vector g:

$$w(f) = \|f\| \, w(g) \leqslant \gamma(w) \|f\|;$$

by the very definition of the number $\gamma(w)$, it has the indicated minimal property.

LEMMA 4.1.1. *A quasinorm $w(f)$ defined on a vector manifold $\mathfrak{V}$ of a normed space is bounded if and only if it is continuous for $f = 0$.*

Let the quasinorm $w(f)$ be bounded and let $f_n \Rightarrow 0$ as $n \to \infty$ $(f_n \in \mathfrak{V})$. Then by (1′) or (2), $\lim w(f_n) = 0$, i.e. $w(f)$ is continuous for $f = 0$. If the quasinorm $w(f)$ is not bounded, then a sequence $\{f_n\} \in \mathfrak{V}$ can be found such that $\|f_n\| = 1$ and

$$\lim_{n \to \infty} w(f_n) = \infty.$$

We let $g_n = \dfrac{f_n}{w_n}$, where $w_n = w(f_n)$. Then

$$\|g_n\| = \frac{1}{w_n}, \qquad w(g_n) = 1,$$

and $w(g_n)$ does not converge to zero although $g_n \Rightarrow 0$.

LEMMA 4.1.2. *A quasinorm $w(f)$ is continuous everywhere if and only if it is bounded.*

Indeed, according to Lemma 4.1.1, the necessity of the condition already follows from the continuity of $w(f)$ for $f = 0$, and the sufficiency follows from the inequality

$$|w(f) - w(g)| \leqslant w(f-g) \leqslant \gamma \|f - g\|. \tag{3}$$

REMARK. The region of application of Lemmas 4.1.1 and 4.1.2 can be extended somewhat. In Lemma 4.1.1 it is sufficient to require that the function $w(f)$ satisfy the condition $w(\alpha f) = \alpha w(f)$ for $\alpha \geqslant 0$, or even the more general condition

$$w(\alpha f) = \alpha^\rho w(f), \quad \alpha \geqslant 0, \quad \rho > 0, \tag{4}$$

by letting $w_n = |w(f_n)|^{\frac{1}{\rho}}$ in the proof. A function $w(f)$ in a vector system $\mathfrak{V}$ which satisfies the (positive) homogeneity condition (4) will be called a *form.* The number ρ, which will be assumed to be an integer, is called the *weight of the form $w(f)$.*

The concept of form generalizes to a function $w(f_1, f_2, \ldots, f_m)$ of m arguments, where the variables $f_k (k = 1, 2, \ldots, m)$ range independently over complex (or real) vector systems $\mathfrak{V}_k$. The function $w(f_1, f_2, \ldots, f_m)$ is called a form if it satisfies (4) in each argument for fixed values of the remaining arguments. In this connection, each argument f_k of the form has its own weight $\rho = \rho_k$. When the vector systems $\mathfrak{V}_k$ are normed spaces X_k, or vector manifolds of such spaces, the form $w(f_1, f_2, \ldots, f_m)$ is said to be bounded if

$$\sup |w(f_1, f_2, \ldots, f_m)| = \gamma(w) < \infty$$

for $f_k \in \mathfrak{V}_k$ and $\|f_k\| = 1$. This is equivalent to the condition

$$|w(f_1, f_2, \ldots, f_m)| \leqslant \gamma \|f_1\|^{\rho_1} \|f_2\|^{\rho_2} \ldots \|f_m\|^{\rho_m} \tag{2'}$$

with finite γ, moreover $\gamma(w)$ is the smallest of the numbers γ for which (2′) holds for all $f_k \in \mathfrak{V}_k$. The form $w(f_1, f_2, \ldots, f_m)$ is bounded if and only if it is continuous for $f_1 = 0, f_2 = 0, \ldots, f_m = 0$.

Lemma 4.1.2 remains true if the function $w(f)$ is a convex functional (Sec. 3.6.2). In this case it is necessary to replace the inequality (3) by the following:

$$|w(f) - w(g)| \leqslant \sup \{w(f-g), w(g-f)\}.$$

1.2. The extension of bounded transformations. Let X and X' be complex or real Banach spaces.

A linear (pseudolinear) transformation A, whose domain Ω_A is a vector manifold Ω in X, and whose range $\Omega_A^{-1} = A\Omega$ is contained in X', is called a *linear (pseudolinear) transformation from X to X'*. In the case when $X' = X$, the transformation A will be called a *linear (pseudolinear) operator* in the space X.

Since the domain of a linear (pseudolinear) transformation may not coincide with the entire space X, it is natural to introduce the concept of *extension* or *continuation* of a linear (pseudolinear) transformation. A linear (pseudolinear) transformation B is an *extension* of a linear (pseudolinear) transformation A if $\Omega_B \supset \Omega_A$, and if $Bf = Af$ for all $f \in \Omega_A$. Then we will write $B \supset A$. If B is an extension of A, then we say that B induces A on $\Omega_A \subset \Omega_B$.

A linear (pseudolinear) transformation A from X to X' is said to be *bounded* if

$$\sup_{\|f\|=1} \|Af\| = \gamma_A < \infty, \qquad f \in \Omega_A. \tag{5}$$

The function

$$w(f) = \|Af\| \tag{6}$$

is a quasinorm in Ω_A since

$$w(f+g) = \|Af + Ag\| \leqslant \|Af\| + \|Ag\| = w(f) + w(g),$$
$$w(\alpha f) = \|\alpha A f\| = |\alpha|\,\|Af\| = |\alpha|\,w(f).$$

Thus, the boundedness of the transformation A implies the boundedness of the quasinorm (6), and conversely, the condition (5) is equivalent to the condition

$$\|Af\| \leqslant \gamma \|f\|, \tag{7}$$

where the smallest value of γ is equal to γ_A.

A transformation A is *continuous at the point* $f \in \Omega_A$ if

$$Af_n \Rightarrow Af,$$

when $f_n \in \Omega_A$ and $f_n \Rightarrow f$.

Since $\|Af_n - Af\| = w(f_n - f)$, the continuity of the transformation A is equivalent to the continuity of the corresponding quasinorm (6). From Lemmas 4.1.1 and 4.1.2 it now follows that:

THEOREM 4.1.1. *A linear (pseudolinear) transformation A which is continuous at one point is continuous everywhere, and this is true if and only if it is bounded.*

Indeed, the continuity of the transformation A for $f = f_0$ implies its continuity for $f = 0$: if $f_n \Rightarrow 0$, then $f_n + f_0 \Rightarrow f_0$, and $A(f_n + f_0) \Rightarrow Af_0$ implies that $Af_n \Rightarrow 0$.

THEOREM 4.1.2. *A continuous linear (pseudolinear) transformation A, with range in a Banach space X', has a unique continuous extension $\bar{A}$ with domain $\Omega_{\bar{A}} = \bar{\Omega}_A$, moreover $\gamma_{\bar{A}} = \gamma_A$.*

In fact, let $f \in \bar{\Omega}_A$, $f_n \in \Omega_A$ and $f_n \Rightarrow f$. Since A is a bounded transformation, by (7)

$$\|Af_n - Af_m\| = \|A(f_n - f_m)\| \leqslant \gamma \|f_n - f_m\|$$

and like $\{f_n\}$, $\{Af_n\}$ is a Cauchy sequence. By virtue of the completeness of the space X', the sequence $\{Af_n\}$ converges to some element $f' \in X'$, and the formula $\bar{A}f = f'$ defines a transformation $\bar{A}$ on $\bar{\Omega}_A$, since the definition of $\bar{A}f$ is independent of the choice of the sequence $\{f_n\}$ which converges strongly to f. Indeed, if $\{g_n\}$ is another such sequence, then $f_n - g_n \Rightarrow 0$, and by the continuity of A,

$$Af_n - Ag_n = A(f_n - g_n) \Rightarrow 0 \cdot$$

For an element $f \in \Omega_A$ we can let $f_n = f$, so that $\bar{A}f = Af$ for $f \in \Omega_A$; thus $\bar{A}$ is an extension of A. Strong convergence preserves the properties of linearity, and $\bar{A}$ is a linear (pseudolinear) transformation. It is continuous, since

$$\|\bar{A}f\| \leqslant \gamma_A \|f\|, \tag{8}$$

for $f \in \bar{\Omega}_A = \Omega_{\bar{A}}$, as follows from the inequality

$$\|Af_n\| \leqslant \gamma_A \|f_n\|,$$

where $\{f_n\}$ is a sequence in Ω_A with limit f. At the same time (8) shows that $\gamma_{\bar{A}} \leqslant \gamma_A$, and according to the very definition of γ_A, the inequality $\gamma_{\bar{A}} \geqslant \gamma_A$ holds, so that $\gamma_{\bar{A}} = \gamma_A$. Finally, it is obvious that every continuous extension of the transformation A coincides with $\bar{A}$ on $\bar{\Omega}_A$.

1.3. The normed spaces $\mathfrak{A}(X, X')$ and $\bar{\mathfrak{A}}(X, X')$. If the domain Ω_A of a transformation A is dense in a space X, i.e if $\bar{\Omega}_A = X$, then A is called a *densely defined transformation.* A continuous linear (pseudolinear) transformation can be extended by continuity to the whole space X if A is densely defined. The collection $\mathfrak{A}(X, X')$ ($\bar{\mathfrak{A}}(X, X')$) of all continuous linear (pseudolinear) transformations from X to X', with the common domain X, is a vector subsystem of the system $\mathfrak{A}(\mathfrak{B}, \mathfrak{B}')$ ($\bar{\mathfrak{A}}(\mathfrak{B}, \mathfrak{B}')$) of all linear (pseudolinear) transformations of the vector system $\mathfrak{B}$ of the space X into the vector system $\mathfrak{B}'$ of the space X'.

THEOREM 4.1.3. *The vector system $\mathfrak{A}(X, X')$ ($\bar{\mathfrak{A}}(X, X')$) is a normed space if we define $\|A\|$ for $A \in \mathfrak{A}(X, X')$ ($A \in \mathfrak{A}(X, X')$) by*

$$\|A\| = \gamma_A = \sup_{\|f\|=1} \|Af\| \cdot \tag{9}$$

The space $\mathfrak{A}(X, X')$ ($\overline{\mathfrak{A}}(X, X')$) *is complete if the space* X' *is complete.*

In fact, the definition (9) obviously satisfies the conditions 2) and 3) of the definition of norm (Sec. 3.1.1), and the inequality of 1) is obtained in the following manner: Let A and B belong to $\mathfrak{A}(X, X')$ ($\overline{\mathfrak{A}}(X, X')$). Then

$$\|(A+B)f\| = \|Af+Bf\| \leqslant \|Af\| + \|Bf\| \leqslant \|A\| + \|B\|$$

for $\|f\| = 1$, and hence

$$\|A+B\| \leqslant \|A\| + \|B\|.$$

Now let $\{A_n\}$ be a Cauchy sequence in $\mathfrak{A}(X, X')$ ($\overline{\mathfrak{A}}(X, X')$). Then for any $\varepsilon > 0$, there exists an n_0 such that for $m \geqslant n_0$ and $n \geqslant n_0$,

$$\|A_m - A_n\| < \varepsilon,$$

i.e.

$$\|A_m f - A_n f\| \leqslant \varepsilon \|f\|. \tag{10}$$

Thus $\{A_n f\}$ is a Cauchy sequence in X' for any $f \in X$, and hence it converges strongly to $f' = Af \in X'$ if X' is complete. A transformation A with domain X and range in X' is obviously linear (pseudolinear). Passing to the limit $m \to \infty$ in (10), we obtain for $n \geqslant n_0$,

$$\|Af - A_n f\| \leqslant \varepsilon \|f\|,$$

i.e. the transformation $A - A_n$ is bounded, and

$$\|A - A_n\| \leqslant \varepsilon, \quad n \geqslant n_0. \tag{11}$$

When $A - A_n$ is bounded, the transformation $A = (A - A_n) + A_n$ is also bounded, and thus A belongs to $\mathfrak{A}(X, X')$ ($\overline{\mathfrak{A}}(X, X')$). Finally, the inequality (11) means that $\lim \|A_n - A\| = 0$, i.e. the Cauchy sequence $\{A_n\}$ has a (strong) limit; hence the space $\mathfrak{A}(X, X')$ ($\overline{\mathfrak{A}}(X, X')$) is complete.

1.4. The normed ring of operators $\mathfrak{A}(X)$. In the case when the spaces X and X' coincide and X is complete, the space $\mathfrak{A}(X) = \mathfrak{A}(X, X)$ is a complete normed space of continuous linear operators defined on the entire space X. Operators in $\mathfrak{A}(X)$ are linear operators in a vector system $\mathfrak{B}$ of the space X, i.e. they are elements of the algebra $\mathfrak{A}(\mathfrak{B})$ (Sec. 1.4.2). If the operators A and B are continuous, then their product $C = AB$ ($Cf = A(Bf)$) is also continuous, and thus it belongs to $\mathfrak{A}(X)$. Thus $\mathfrak{A}(X)$ is closed under multiplication, i.e. the normed space $\mathfrak{A}(X)$ is a ring of operators and a subalgebra of the algebra $\mathfrak{A}(\mathfrak{B})$.

The norm of a product in $\mathfrak{A}(X)$ satisfies the inequality

$$\|AB\| \leqslant \|A\| \, \|B\|. \tag{12}$$

Indeed, for $\|f\| = 1$,

$$\|ABf\| = \|A(Bf)\| \leqslant \|A\| \, \|Bf\| \leqslant \|A\| \, \|B\|,$$

and (12) follows.

In addition to the normed ring $\mathfrak{A}(X)$, we can consider the normed space $\overline{\mathfrak{A}}(X) = \overline{\mathfrak{A}}(X, X)$ of pseudolinear operators in X. For example, to $\mathfrak{A}(X)$ belong the pseudolinear operators A for which the mapping $f \to f' = Af$, $f \in X$, is an involutive pseudo-automorphism of the space X. Such operators will be called *mirror operators of the space* X. In particular, a pseudo-involution of a normed $*$-space is a mirror operator.

1.5. Linear and pseudolinear functionals. The set $K_c(K_r)$ of complex (real) numbers can be considered to be a complex (real) normed space if we set $\|\alpha\| = |\alpha|$ for an element α of the complex (real) vector system $K_c(K_r)$. For an arbitrary complex Banach space X, transformations in $\mathfrak{A}(X, K_c)$ and $\overline{\mathfrak{A}}(X, K_c)$ are continuous linear and pseudolinear forms which are defined on the entire space X. Such forms are called, respectively, *linear*, and *pseudolinear functionals* in X. The complete normed spaces $\mathfrak{A}(X, K_c)$ and $\overline{\mathfrak{A}}(X, K_c)$ of these linear and pseudolinear functionals are the *conjugate* and *pseudo-conjugate spaces of the Banach space X*.

REMARK. If X is a real Banach space and if X' coincides with K_r, then the conjugate space $\mathfrak{A}(X, K_r)$ is also a real Banach space.

For fixed $g(\tau) \in X_q^{(\mu)}$, and variable $f(\tau) \in X_p^{(\mu)}$, $\left(\frac{1}{p} + \frac{1}{q} = 1\right)$, the integral

$$W(f, g) = \int_T f(\tau) \bar{g}(\tau) \mu(\underset{\tau}{M}) \tag{13}$$

is a linear functional $l_g(f)$ in $X_p^{(\mu)}$, and for fixed $f(\tau) \in X_p^{(\mu)}$ and variable $g(\tau) \in X_q^{(\mu)}$, it is a pseudolinear functional $l'_f(g)$ in $X_q^{(\mu)}$, moreover

$$\|l_g\| = \|g\|, \qquad \|l'_f\| = \|f\|, \tag{14}$$

where

$$\|g\| = w_q(g), \qquad \|f\| = w_p(f).$$

In fact, $W(f, g)$ is obviously a linear form with respect to f and a pseudo-linear form with respect to g, and by virtue of the inequality (3.2.10),

$$|l_g(f)| = |W(f, g)| \leqslant w_p(f) \, w_q(g) = w_q(g) \|f\|.$$

Thus the linear form $l_q(f)$ is bounded, and hence is a linear functional; moreover

$$\|l_g\| \leqslant w_q(g). \tag{15}$$

Since $(q-1)p = q$, the function $f_1(\tau) = |g(\tau)|^{q-1}\operatorname{sgn} g(\tau)$, where

$$\operatorname{sgn} g(\tau) = \begin{cases} \dfrac{g(\tau)}{|g(\tau)|}, & g(\tau) \neq 0, \\ 0, & g(\tau) = 0, \end{cases}$$

belongs to the space $X_p^{(\mu)}$, and

$$\|f_1\| = w_p(f_1) = [w_q(g)]^{q/p} = [w_q(g)]^{q-1}.$$

Therefore

$$w_q(g) = l_g(f_1) \leqslant \|l_g\|\,\|f_1\| = \|l_g\|(w_q(g))^{q-1}.$$

By dividing this inequality by $(w_q(g))^{q-1}$, we obtain $w_q(g) \leqslant \|l_g\|$, which together with (15), proves the first equation in (14). The second equation in (14) can be reduced to the first by the transition $\overline{l'_f(g)} = \overline{w(f, g)}$, or it can be proved in a similar manner.

Remark. For $f(\tau) \in X^{(\mu)}$ and $g(\tau) \in X_\infty^{(\mu)}$ ($p = 1$, $q = \infty$), the integral (13) exists and the above properties are preserved. In particular, it is not difficult to prove (14).

§ 2. Convergence of Transformations

2.1. Some lemmas concerning the quasinorm (convex functional). A real functional $w(f)$, defined on a set M of a normed space X, is said to be *lower semicontinuous* at a point $f = f_0 \in M$ if for each $\varepsilon > 0$, a $\delta > 0$ can be found such that

$$w(f) > w(f_0) - \varepsilon$$

for all $f \in M$ in the neighborhood

$$\|f - f_0\| < \delta.$$

The condition of lower semicontinuity can also be written in the form

$$\underline{\lim}\, w(f_n) \geqslant w(f), \quad f_n \Rightarrow f, \quad f_n \in M, \quad f \in M.$$

Lemma 4.2.1. *Let $\{w_\tau(f)\}$ be a family of functionals on a set M which are lower semicontinuous at the point $f = f_0$ and let $\sup\limits_\tau w_\tau(f_0) < \infty$. Then the functional*

$$w(f) = \sup_\tau w_\tau(f), \tag{1}$$

defined on the set M' of those elements $f \in M$ for which the upper bound in (1) is finite, is also lower semicontinuous at the point $f = f_0$.

In fact, for a given $\varepsilon > 0$, there exists a τ such that

$$w_\tau(f_0) - \frac{\varepsilon}{2} > w(f_0) - \varepsilon. \tag{2}$$

The inequality $w_\tau(f) > w_\tau(f_0) - \frac{\varepsilon}{2}$ is satisfied for all $f \in M$ (and for $f \in M'$) in some neighborhood $\|f - f_0\| < \delta$. In view of the inequality $w(f) \geqslant w_\tau(f)$, and (2), the inequality $w(f) > w(f_0) - \varepsilon$ holds in the same neighborhood for elements $f \in M'$, i.e. the function $w(f)$ is lower semicontinuous for $f = f_0$.

LEMMA 4.2.2. *A quasinorm $w(f)$ defined in a Banach space X is bounded if it is lower semicontinuous for all $f \in X$.*

Indeed, a quasinorm $w(f)$ which is unbounded on the sphere $\|f\| = 1$ is also unbounded on any sphere $\|f - f_0\| = \delta$. For the sphere $\|f\| = \delta$ this follows from

$$w(f) = w(\delta h) = \delta w(h), \quad \text{where} \quad f = \delta h, \quad \|h\| = 1;$$

For the general case, this follows from the inequality

$$w(f) \geqslant w(g) - w(-f_0),$$

where $g = f - f_0$ since, as has already been shown, the quasinorm $w(g)$ is unbounded for $\|g\| = \delta$. By assuming that the quasinorm $w(f)$ is unbounded, we can construct a sequence of closed spheres K_n: $\|f - f_n\| \leqslant \rho_n$ such that

$$K_{n+1} \subset K_n, \quad \lim \rho_n = 0 \tag{3}$$

and

$$w(f) > n, \quad f \in K_n. \tag{4}$$

The center of the first sphere K_1 is obtained by choosing an element f_1 for which $w(f_1) > 1$. We choose the radius $\rho_1 > 0$ of K_1 so that for $\|f - f_1\| \leqslant \rho_1$, the inequality $w(f) > 1$ is satisfied. This is possible by virtue of the lower semicontinuity of the functional $w(f)$. Let us now assume that the spheres $K_1, K_2, \ldots, K_{n-1}$ have been constructed. Then in the sphere

$$\|f - f_{n-1}\| = \frac{\rho_{n-1}}{2}$$

there exists an element f_n for which $w(f_n) > n$. By the lower semicontinuity of the functional $w(f)$ there exists a sphere

$$K_n: \|f - f_n\| \leqslant \rho_n \leqslant \frac{\rho_{n-1}}{2},$$

for which (4) is satisfied.

Since $K_n \subset K_{n-1}$, and since $\rho_n = \frac{\rho_1}{2^{n-1}}$, the sequence $\{K_n\}$ so constructed satisfies (3) and (4), and for the sequence of centers $\{f_n\}$, we have $\|f_n - f_m\| < \rho_n$

for $n > m$. The sequence $\{f_n\}$ is a Cauchy sequence and therefore converges to some element $f' \in X$. Like the elements $f_n, f_{n+1}, \ldots$, their limit f' belongs to the closed sphere K_n. Therefore f' belongs to the intersection of all the K_n, and hence by (4), $w(f') > n$ for all n. This contradicts the assumption that the functional $w(f)$ is finite.

LEMMA 4.2.3. *If a family $\{w_\tau(f)\}$ of continuous quasinorms $w_\tau(f)$ in a Banach space X satisfies the condition*

$$\sup_\tau \{w_\tau(f)\} < \infty \tag{5}$$

for each $f \in X$, then the functional $w(f) = \sup_\tau \{w_\tau(f)\}$ is a continuous quasinorm in X.

In fact, from

$$w_\tau(f+g) \leqslant w_\tau(f) + w_\tau(g) \leqslant w(f) + w(g)$$

and

$$w_\tau(\alpha f) = |\alpha| w_\tau(f)$$

it follows that

$$w(f+g) \leqslant w(f) + w(g), \qquad w(\alpha f) = |\alpha| w(f),$$

i.e. $w(f)$ is a quasınorm. By virtue of Lemma 4.2.1, it is lower semicontinuous for each $f \in X$. By applying Lemma 4.2.2, we now see that the quasinorm $w(f)$ is bounded, i.e. it is continuous.

REMARK. Lemmas 4.2.2 and 4.2.3 still hold if $w(f)$ in Lemma 4.2.2, and $w_\tau(f)$ and $w(f)$ in Lemma 4.2.3, are convex functionals (Sec. 3.6.2), but are not quasinorms. In fact, the proof of Lemma 4.2.2 is carried over and it implies that the convex functional $w(f)$ is bounded above. The fact that it is bounded below follows from the inequality $w(f) \geqslant -w(-f)$. For convex functionals, Lemma 4.2.3 now follows from a generalization of Lemma 4.2.2.

An important application of Lemma 4.2.3 is the following theorem on the boundedness of the norms of linear or pseudolinear transformations.

THEOREM 4.2.1. *Let X be a Banach space. A family $\{A_\tau\}$ of linear (pseudo-linear) transformations in $\mathfrak{A}(X, X')$ ($\overline{\mathfrak{A}}(X, X')$) is bounded, i.e.*

$$\|A_\tau\| \leqslant \gamma < \infty, \tag{6}$$

if and only if for each fixed $f \in X$, the set $A_\tau f$ is bounded in X', i.e.

$$\|A_\tau f\| \leqslant \gamma_f < \infty. \tag{7}$$

In fact, it follows from (6) that

$$\|A_\tau f\| \leqslant \|A_\tau\| \, \|f\| \leqslant \gamma \|f\|,$$

i.e. in addition to (6), (7) is satisfied with $\gamma_f = \gamma \| f \|$ (uniform boundedness of $A_\tau f$ on the sphere $\| f \| = 1$ or the sphere $\| f \| \leqslant \delta$). Conversely, let (7) hold for each $f \in X$. Then the functional

$$w(f) = \sup \| A_\tau f \| \tag{8}$$

is everywhere finite. Since $A_\tau \in \mathfrak{A}(X, X')$ ($\overline{\mathfrak{A}}(X, X')$), $w_\tau(f) = \| A_\tau f \|$ is a continuous quasinorm in X, and by virtue of Lemma 4.2.3, the functional $w(f)$ defined by (8) is a continuous quasinorm, so that $w(f) \leqslant \gamma < \infty$ for $\| f \| = 1$. Therefore

$$\| A_\tau f \| \leqslant w(f) \leqslant \gamma, \qquad \| f \| = 1,$$

which is equivalent to (6).

2.2. The convergence of transformations on the space X. If a sequence $\{A_n\}$ of transformations from $\mathfrak{A}(X, X')$ converges on a set $M \subset X$, i.e. if for each $f \in M$, the sequence $A_n f$ converges strongly in X', then $\{A_n\}$ also converges on the linear envelope of the set M. Therefore it is sufficient to consider convergence on vector manifolds.

If $\{A_n\}$ converges on a vector manifold $\Omega \subset X$, then the equation

$$A_n f \Rightarrow f' = Af$$

for $f \in \Omega$, defines a linear transformation A from X to X' whose domain is $\Omega_A = \Omega$.

In the case $\Omega = X$, i.e. convergence on the whole space X, we have:

THEOREM 4.2.2. *Let the sequence* $\{A_n\}$ *from* $\mathfrak{A}(X, X')$ ($\overline{\mathfrak{A}}(X, X')$) *converge on the whole Banach space* X. *Then the norms* $\| A_n \|$ *are uniformly bounded and the limit* A *of the sequence* $\{A_n\}$ *belongs to* $\mathfrak{A}(X, X')$ ($\overline{\mathfrak{A}}(X, X')$).

Indeed, if $A_n f \Rightarrow Af$ for any $f \in X$, then for each $f \in X$ the sequence $A_n f$ is bounded in X'. Therefore, according to Theorem 4.2.1, $\| A_n \| \leqslant \gamma < \infty$, and hence for $\| f \| = 1$

$$\| A_n f \| \leqslant \gamma.$$

In the limit we obtain $\| Af \| \leqslant \gamma$, i.e. $\| A \| \leqslant \gamma$, and thus $A \in \mathfrak{A}(X, X')$ ($\overline{\mathfrak{A}}(X, X')$).

Convergence in norm in $\mathfrak{A}(X, X')$ is called *uniform convergence*. This name is justified by the fact that if $\{A_n\}$ converges uniformly to A, then the sequence $\{A_n f\}$ converges for each $f \in X$, since

$$\| A_n f - Af \| \leqslant \| A_n - A \| \, \| f \|,$$

and this convergence is uniform on the sphere $\| f \| = 1$ (or on any sphere $\| f \| \leqslant \delta$).

THEOREM 4.2.3. *Let X and X' be Banach spaces. A sequence $\{A_n\}$ in $\mathfrak{A}(X, X')$ ($\overline{\mathfrak{A}}(X, X')$) converges on all of X if and only if*

1) *the sequence converges on some fundamental set M of the space X;*
2) *the norms $\|A_n\|$ are uniformly bounded.*

The necessity of 1) is obvious, and the necessity of 2) has already been proved (Theorem 4.2.2). We will prove the sufficiency of 1) and 2). Let f be an element in X and let g be an element of the linear envelope $\tilde{M}$ of the fundamental set M. If $\|A_n\| \leqslant \gamma$, then

$$\|A_m - A_n\| \leqslant \|A_m\| + \|A_n\| \leqslant 2\gamma,$$

and

$$\|A_m f - A_n f\| \leqslant \|(A_m - A_n)(f - g)\| + \|A_m g - A_n g\| \leqslant \leqslant 2\gamma\|f - g\| + \|A_m g - A_n g\|. \tag{9}$$

Since $\tilde{M}$ is dense in H, by the suitable choice of a vector g in $\tilde{M}$, $\|f - g\|$ can be made as small as desired, and for fixed g the norm $\|A_m g - A_n g\| \to 0$ as $m \to \infty$ and $n \to \infty$. Therefore, the left-hand side of (9) converges to zero, i.e. $\{A_n f\}$ is a Cauchy sequence in X'. Since X' is complete, $\{A_n f\}$ is strongly convergent in X'.

When $X' = X$, i.e. when the transformations A_n are elements of the normed space $\mathfrak{A}(X)$, the convergence of the sequence $\{A_n\}$ on the entire space X to an operator A is called *strong convergence of the sequence of operators* $\{A_n\}$ *to the operator* A. If the sequence of operators $\{A_n\}$ in $\mathfrak{A}(X)$ is strongly convergent to A, we will write

$$A_n \Rightarrow A.$$

2.3. Weak convergence. We consider the special case $X' = K_c$ (field of complex numbers), when $\mathfrak{A}(X, K_c)$ ($\overline{\mathfrak{A}}(X, K_c)$) is the conjugate (pseudoconjugate) space of linear (pseudolinear) functionals.

If a sequence of linear (pseudolinear) functionals $\{l_n\}$ in the conjugate (pseudoconjugate) space converges on the entire space X to a linear (pseudolinear) functional l, then by definition $\{l_n\}$ *converges weakly to l in the conjugate (pseudoconjugate) space*. Weak convergence of $\{l_n\}$ to l will be written:

$$l_n \to l.$$

Weak convergence in a conjugate (pseudoconjugate) space is called convergence in norm in this space.

In analogy to the concept of weak convergence of linear functionals in a normed space X, i.e. weak convergence of elements of the conjugate space $\mathfrak{A}(X, K_c)$, we introduce the corresponding concept for elements of the fundamental space X, namely: a sequence $\{f_n\}$ in X is by definition weakly convergent to an element $f \in X$ if

$$\lim_{n\to\infty} l(f_n)=l(f) \tag{10}$$

for each functional $l\in\mathfrak{A}(X, K_c)$. We will write $f_n\to f$ if $\{f_n\}$ converges weakly to f.* A Banach space is said to be *complete with respect to weak convergence* if whenever the limit $\lim l(f_n)$ exists for each $l\in\mathfrak{A}(X, K_c)$, there exists an element $f\in X$ to which the sequence $\{f_n\}$ is weakly convergent.

REMARK. If X is a real Banach space, then only the conjugate space $\mathfrak{A}(X, K_r)$ exists (K_r is the field of real numbers). The concepts of weak convergence in $\mathfrak{A}(X, K_r)$ and in X, and also of completeness of X with respect to weak convergence, are preserved.

§ 3. Weak Convergence in a Unitary Space

3.1. The general form of linear and pseudolinear functionals. For fixed g, the scalar product (f, g) of two elements f and g of a unitary space H is a linear functional $l_g f$ (relative to f), and for fixed f it is a pseudolinear functional $l'_f(g)$:

$$l_g(f)=(f, g), \qquad l'_f(g)=(f, g). \tag{1}$$

The functionals $l_g(f)$ and $l'_f(g)$ are respectively elements of the spaces $\mathfrak{A}(H, K_c)$, and $\overline{\mathfrak{A}}(H, K_c)$ (the conjugate and pseudoconjugate spaces, see Sec. 4.1.5).

It follows from the Cauchy-Bunyakovski inequality that

$$|l_g(f)|\leqslant\|f\|\,\|g\|, \qquad \|l'_f(g)\|\leqslant\|f\|\,\|g\|$$

and hence

$$\|l_g\|\leqslant\|g\|, \qquad \|l'_f\|\leqslant\|f\| \tag{2}$$

($\|l_g\|$ and $\|l'_f\|$ are norms in the spaces $\mathfrak{A}(H, K_c)$ and $\overline{\mathfrak{A}}(X, K_c)$, respectively). On the other hand

$$\|g\|^2=(g, g)=l_g(g)\leqslant\|l_g\|\,\|g\|,$$

i.e. $\|g\|\leqslant\|l_g\|$, and similarly $\|f\|\leqslant\|l'_f\|$. Hence

$$\|l_g\|=\|g\|, \qquad \|l'_f\|=\|f\|. \tag{3}$$

---

* By considering the space of linear (or pseudolinear) functionals $\mathfrak{A}(X, K_c)$ to be fundamental, it is possible to consider weak convergence which is defined only by the use of the linear functionals in $\mathfrak{A}(X, K_c)$, i.e. elements of $\mathfrak{A}(\mathfrak{A}(X, K_c), K_c)$. Generally speaking, this convergence will not coincide with the definition given earlier for weak convergence in $\mathfrak{A}(X, K_c)$ when $\mathfrak{A}(X, K_c)$ is a conjugate space.

From (3), and Theorem 4.2.1, we obtain the following criterion for the boundedness of a set in a unitary space H.

THEOREM 4.3.1. *A set $M \subset H$ is bounded if and only if for each $g \in H$*

$$\sup_{f \in M} |(f, g)| = \gamma_g < \infty. \tag{4}$$

In fact, for each element $f \in M$, we consider the pseudolinear functional $l'_f(g) = (f, g)$. Then, in view of (4), the hypothesis of Theorem 4.2.1 is satisfied by the family $A_f g = l'_f(g)$, $f \in M$, and by virtue of (3), $\|A_f\| = \|l'_f\| = \|f\|$. Hence $\|f\| \leqslant \gamma < \infty$, $f \in M$, and the theorem is proved.

Let $\mathfrak{L}$ and $\mathfrak{L}'$, respectively, be the collections of linear and pseudolinear functions $l_g(f)$ and $l'_g(f)$ of the form (1). Consider the mappings of the space H onto $\mathfrak{L}'$ and $\mathfrak{L}$:

$$f \mapsto l'_f, \qquad g \mapsto l_g, \tag{5}$$

given by (1). The first is linear and the second is pseudolinear. Indeed, the equations

$$(f_1 + f_2, g) = (f_1, g) + (f_2, g), \qquad (\alpha f, g) = \alpha (f, g)$$

imply that

$$l'_{f_1+f_2} = l'_{f_1} + l'_{f_2}, \qquad l'_{\alpha f} = \alpha l'_f.$$

In a similar manner, we have

$$l_{g_1+g_2} = l_{g_1} + l_{g_2}, \qquad l_{\alpha g} = \bar{\alpha} l_g.$$

The mappings (5) are one to one: if $l_{g_1} = l_{g_2}$, then $(f, g_1 - g_2) = 0$ for $f \in H$, so that $g_1 - g_2 = 0$ and $g_2 = g_1$. Similarly, it follows from $l'_{f_1} = l'_{f_2}$ that $f_1 = f_2$. Thus the mappings (5) are, respectively, an isomorphism and pseudoisomorphism of the vector system H onto $\mathfrak{L}'$ and $\mathfrak{L}$. Moreover, since by virtue of (3) the mappings (5) are norm preserving, *the mappings* (5) *are, respectively, an isomorphism and pseudoisomorphism of the space onto $\mathfrak{L}'$ and $\mathfrak{L}$.*

Therefore, we can consider $\mathfrak{L}$ and $\mathfrak{L}'$ to be unitary spaces. We have

$$(l_1, l_2) = \overline{(g_1, g_2)}, \qquad (l'_1, l'_2) = (f_1, f_2),$$

for the scalar product in the spaces $\mathfrak{L}$ and $\mathfrak{L}'$, where $l_k = l_{g_k}$ and $l'_k = l'_{f_k}$.

The following theorem shows that all linear and pseudolinear functionals in H are exhausted by the functionals l_g and l'_f from $\mathfrak{L}$ and $\mathfrak{L}'$.

THEOREM 4.3.2. *Every linear functional in H has the form $l = l_g \in \mathfrak{L}$, and every pseudolinear functional in H has the form $l' = l'_f \in \mathfrak{L}'$.*

It is sufficient to give the proof for a linear functional l. First let the space H be finite dimensional, and let $h_1, h_2, \ldots, h_n$ be an orthogonal basis. Then, for any $f \in H$,

$$f = \sum_{k=1}^{n} (f, h_k) h_k,$$

and hence

$$l(f) = \sum_{k=1}^{n} (f, h_k) l(h_k) = (f, g),$$

where

$$g = \sum_{k=1}^{n} \overline{l(h_k)} h_k. \tag{6}$$

Thus, in the special case under consideration, we have obtained the desired representation $l = l_g$ with the explicit expression (6) for the vector g. It follows from (3) and (6) that

$$\| l \|^2 = \| g \|^2 = \sum_{k=1}^{n} | l(h_k) |^2. \tag{7}$$

In the general case of an arbitrary unitary space H, in place of (7) we have the inequality

$$\sum_{k=1}^{n} | (l(h_k) |^2 \leqslant \| l \|^2, \tag{7'}$$

where $h_1, h_2, \ldots, h_n$ is any finite orthonormal system in H. Indeed, the linear hull of the elements $h_1, \ldots, h_n$ is an n-dimensional space H_n, and according to (7) the sum on the left-hand side of (7′) is equal to $\sup |l(f)|^2$ for $\|f\| = 1, f \in H_n$. Since $\|l\| = \sup |l(f)|$ for $\|f\| = 1, f \in H$, the inequality (7′) has been proved. If $h_1, h_2, \ldots$ is an infinite orthonormal system, then by passing to the limit in (7′), we obtain

$$\sum_{k=1}^{\infty} | l(h_k) |^2 \leqslant \| l \|^2, \tag{8}$$

which holds for any countable orthonormal system $\{h_k\}$.

Now let $\{h_\tau\}$ be an orthogonal basis of the space H ($\{h_k\}$, if H is separable). It follows from (7′) (cf. Sec. 3.4.1) that the values of the $l(h_\tau)$ are different from zero only for a finite or countable number of vectors h_τ. By virtue of (8) the series $\sum_\tau |l(h_\tau)|^2$ converges, and the equation

$$g = \sum_{\tau} \overline{l(h_\tau)} h_\tau \tag{9}$$

defines some element. The vector g gives a representation $l = l_g$ of the functional $l(f)$. In fact

$$f = \sum_{\tau} (f, h_\tau) h_\tau,$$

and by the continuity of the functional $l(f)$ and the continuity of the fundamental form (f, g) with respect to g,

$$l(f) = \sum_{\tau} (f, h_\tau) l(h_\tau) = \left(f, \sum_{\tau} \overline{l(h_\tau)} h_\tau\right) = (f, g).$$

Correspondingly, for a pseudolinear functional l' we have $l' = l'_f$ and

$$f = \sum_{\tau} l'(h_\tau)\, h_\tau. \tag{9'}$$

THEOREM 4.3.3. *The mappings* (5) *are, respectively, an isomorphism and a pseudoisomorphism of the space H onto the pseudoconjugate and conjugate space.*

Indeed, by Theorem 4.3.2, $\mathfrak{L} = \mathfrak{A}(H, K_c)$ and $\mathfrak{L}' = \overline{\mathfrak{A}}(H, K_c)$.

The mappings (5) are called *the natural isomorphism and the natural pseudoisomorphism of the space H*.

3.2. Properties of weak convergence. For a unitary space H the limit equality (4.2.10) serves to define the weak convergence $f_n \to f$ of a sequence $\{f_n\}$ to f. By virtue of Theorem 4.3.2 this is written in the form

$$\lim (f_n, g) = (f, g), \qquad g \in H. \tag{10}$$

THEOREM 4.3.4. *If $f_n \to f$, then the norms $\| f_n \|$ are uniformly bounded.*

Indeed, in view of (10), the condition (4) of Theorem 4.3.1 is satisfied for $M = \{f_n\}$, and consequently $\| f_n \| \leqslant \gamma$.

In particulár, it follows from Theorem 4.3.4 that: *if the orthogonal series $\sum_k g_k$ is weakly convergent, then it is strongly convergent.*

Indeed, let $f_n = \sum_{k=1}^{n} g_k$. It is clear that

$$\| f_n \|^2 = \sum_{k=1}^{n} \| g_k \|^2. \tag{11}$$

Then, by Theorem 4.3.4, the weak convergence of the sequence $\{f_n\}$, and (11), imply that the partial sums of the series $\sum_k \| g_k \|^2$ are bounded, and therefore the series is strongly convergent (Lemma 3.4.2).

THEOREM 4.3.5. *The norm $\| f \|$ is lower semicontinuous with respect to weak convergence, i.e. if $f_n \to f$, then*

$$\underline{\lim}_{n \to \infty} \| f_n \| \geqslant \| f \|. \tag{12}$$

The sequence $\{f_n\}$, which is weakly convergent to f, is strongly convergent if and only if

$$\lim_{n \to \infty} \| f_n \| = \| f \|. \tag{13}$$

Indeed, for any $g \in H$ and for $f_n \to f$,

$$|(f, g)| = \lim_{n \to \infty} |(f_n, g)| \leqslant \underline{\lim}_{n \to \infty} \| f_n \| \, \| g \|.$$

and hence $\|f\| \leqslant \underline{\lim}\|f_n\|$, i.e. (12) holds. That (13) is necessary follows from the continuity of the norm with respect to strong convergence, and that it is sufficient follows from the identity

$$\|f-f_n\|^2=[(f_n, f_n)-(f, f)]+[(f, f)-(f, f_n)]+[(f, f)-(f_n, f)].$$

The scalar product (f, g), which is continuous with respect to strong convergence, is not continuous with respect to weak convergence of both arguments f and g. However, the following is true:

LEMMA 4.3.1. *If $f_n \to f$ and $g_n \Rightarrow g$, then*

$$\lim_{n\to\infty}(f_n, g_n)=(f, g). \tag{14}$$

In fact, by Theorem 4.3.4, $\|f_n\| \leqslant \gamma < \infty$, and therefore

$$|(f_n, g_n)-(f, g)| \leqslant |(f_n, g_n)-(f_n, g)|+$$
$$+|(f_n, g)-(f, g)| \leqslant \gamma\|g_n-g\|+|(f_n, g)-(f, g)|,$$

whence (14) follows.

Condition (10) can also be written in the form

$$\lim_{n\to\infty} l'_n(g)=l'(g), \tag{10'}$$

where l'_n and l' are pseudolinear functionals which correspond to the vectors f_n and f in H under the natural isomorphism of the space H (Sec. 3.1). By using this isomorphism, and known properties of weak convergence of pseudolinear functionals, it is possible to prove the following two theorems:

THEOREM 4.3.6. *The space H is complete with respect to weak convergence.*

THEOREM 4.3.7. *A sequence $\{f_n\}$ in H is weakly convergent if and only if the limit $\lim_{n\to\infty}(f_n, g)$ exists for elements g of a fundamental set M and the norms $\|f_n\|$ are uniformly bounded.*

We begin with the proof of Theorem 4.3.6. If the limit

$$\lim_{n\to\infty}(f_n, g)=\lim_{n\to\infty} l'_n(g) \tag{15}$$

exists for each $g \in H$, then according to Theorem 4.2.2 the functional l', which is defined on all of H by (10′), belongs to $\mathfrak{A}(H, K_c)$. Hence $f_n \to f$, where f is the vector in H which corresponds to the pseudolinear functional l' (Theorem 4.3.3) under the natural isomorphism.

We proceed to Theorem 4.3.7. Since the limits (15) exist in the fundamental set M, and since $\|l'_n\|=\|f_n\| \leqslant \gamma$ (cf. (13)), it follows from Theorem 4.2.3 that (10′) holds for each $g \in H$ and determines a pseudolinear functional $l'(g)$. Therefore $f_n \to f$, where the vector $f \in H$ corresponds to the functional l' under the natural isomorphism $f \mapsto l'_f$.

3.3 Weak convergence of transformations. A sequence of transformations $\{A_n\}$ in $\mathfrak{A}(H, H')$ is by definition *weakly convergent* to the transformation A if for each $f \in H$,

$$A_n f \rightharpoonup Af. \tag{16}$$

THEOREM 4.3.8. *The norms* $\|A_n\|$ *of a weakly convergent sequenee* $\{A_n\}$ *in* $\mathfrak{A}(H, H')$ *are bounded and the (weak) limit* A *is a linear transformation in* $\mathfrak{A}(H, H')$.

Indeed, according to Theorem 4.3.4, it follows from (16) that the set $\{A_n f\}$ is bounded for any fixed $f \in H$. Then by Theorem 4.2.1, the sequence $\{\|A_n\|\}$: $\|A_n\| \leqslant \gamma$, $n = 1, 2, \ldots$ is also bounded. Thus, it follows from

$$|(A_n f, g')| \leqslant \|A_n f\| \, \|g'\| \leqslant \|A_n\| \, \|f\| \, \|g'\|$$

that

$$|(Af, g')| \leqslant \gamma \|f\| \, \|g'\|. \tag{17}$$

The linearity of the transformation A is obvious, and by (17), $\|Af\| \leqslant \gamma \|f\|$, i.e. $\|A\| \leqslant \lambda$ and $A \in \mathfrak{A}(H, H')$.

3.4 Closure with respect to weak convergence. Weak compactness. A set $M \in H$ is *closed with respect to weak convergence* if the (weak) limit of any weakly convergent sequence $\{f_n\}$ in M also belongs to M.

If the set M is closed with respect to weak convergence then it is closed (with respect to strong convergence).

Indeed, if $f_n \in M$ and if $f_n \Rightarrow f$, then $f_n \rightharpoonup f$ and hence $f \in M$.

Generally speaking, the converse is not true. For example, the sphere $\|f\| = 1$ in H is closed. However, in the case when H is infinite dimensional, it is not closed with respect to weak convergence. This follows since there exists an orthonormal system $\{h_n\}$ which lies on the sphere $\|f\| = 1$ and which converges weakly to the zero vector. For any orthonormal system $\{h_n\}$ this last fact follows from Bessel's inequality (3.4.7):

$$\sum_n |(g, h_n)|^2 \leqslant \|g\|^2,$$

which shows that for any $g \in H$

$$\lim (h_n, g) = 0,$$

i.e. $h_n \rightharpoonup 0$. This argument is possible only in infinite dimensional spaces, and for such spaces we have the problem of determining conditions under which the closed set M is closed with respect to weak convergence. Here we indicate only two simple examples.

(*a*) *The closed sphere* $\|f - f_0\| \leqslant \delta$, $\delta > 0$, *is closed with respect to weak convergence.*

Indeed, it is sufficient to consider the sphere $\|f\| \leqslant \delta$. The assertion follows from the inequality (12), since if $f_n \rightharpoonup f$ and $\|f_n\| \leqslant \delta$, then

$$\| f \| \leqslant \lim_{n \to \infty} \| f_n \| \leqslant \delta.$$

(*b*) *Every subspace of the space H is closed with respect to weak convergence.*

In fact, let the elements f_n belong to the subspace H' and let $f_n \to f$. Then, for any element g of the orthogonal complement H'' of H, the equation (10) gives $(f, g) = 0$, and hence the limit f belongs to H'.

It follows from (b) that: *If the sequence* $\{f_n\}$ *is weakly convergent to* f, *then there exists a sequence of linear combinations of the set* $\{f_n\}$ *which is strongly convergent to* f.

Indeed, let H' be the *closed* linear hull of the set $\{f_n\}$. Then by virtue of (b), the weak limit f also belongs to H', and therefore a sequence of elements exists (in the linear hull of the set $\{f_n\}$) which weakly converges to f.

A set $M \subset H$ is said to be *weakly compact* in H (in M) if any sequence $\{f_n\}$ in M contains a subsequence which is weakly convergent to the element $f \in H$ (correspondingly to $f \in M$).

For finite dimensional spaces, weak convergence coincides with strong convergence, and hence weak compactness coincides with compactness. Therefore we will consider the following theorem as a generalization of the well-known theorem on compactness of bounded sets in infinite dimensional spaces (the Bolzano-Weierstrass principle). Here we consider the set $M \subset H$ to be *bounded* if it is contained in some sphere.

THEOREM 4.3.9. *A set* $M \subset H$ *is weakly compact in* H *if and only if it is bounded.*

The condition is necessary, since if M is not bounded then a sequence $\{f_n\}$ exists in M with $\lim \| f_n \| = \infty$, and such a sequence does not have a weakly convergent subsequence. The sufficiency of the condition will be proved if it can be shown that any bounded sequence $\{f_n\}$ ($\| f_n \| \leqslant \gamma$) contains a weakly convergent subsequence: We will denote the closed linear hull of the set $\{f_n\}$ by H'. Since

$$|(f_n, f_1)| \leqslant \gamma \| f_1 \|,$$

for all n from our sequence, a subsequence $\{f_n^{(1)}\}$ can be chosen for which the sequence $\{(f_n^{(1)}, f_1)\}$ converges. Similarly, it is possible to choose from $\{f_n^{(1)}\}$ a subsequence $\{f_n^{(2)}\}$ for which the sequence $\{(f_n^{(2)}, f_2)\}$ converges. By continuing this process, we construct an infinite sequence of nested sequences $\{f_n^{(k)}\}$ for which the sequence $(f_1^{(k)}, f_m)$ $(f_2^{(k)}, f_m)$, . . . with $m \leqslant k (k = 1, 2, \ldots)$ converges. Finally, by choosing from each sequence $\{f_n^{(k)}\}$ an element $f_k^{(k)}$, we obtain a diagonal sequence $h_n = f_n^{(n)}$ for which the sequence

$$(h_1, f_k), (h_2, f_k), \ldots$$

$(k = 1, 2, \ldots)$ converges. Since the norms of the elements h_n are bounded, the conditions of Theorem 4.3.7 hold in the space H', and there exists an element $f \in H'$ such that for all $g' \in H'$

$$\lim (h_n, g') = (f, g'). \tag{18}$$

But (18) remains true when g' is replaced by any element $g \in H$. Indeed, let g' be the projection of g on H'. Then $(h_n, g') = (h_n, g)$ and $(f, g') = (f, g)$. Hence, by (18)

$$\lim (h_n, g) = (f, g),$$

for any $g \in H$, i.e. $h_n \to f$.

COROLLARY. *The closed sphere* $\|f - f_0\| \leqslant \delta$ *is weakly compact in itself.* Indeed, this sphere is bounded and closed with respect to weak convergence.

§ 4. Bilinear Functionals and Linear Transformations in Unitary Spaces

4.1. The correspondence between linear transformations and bilinear functionals. Let $W(f, g')$ be a bilinear form (Sec. 3.3.1), where the elements f and g' range independently over vector manifolds $\mathfrak{B}$ and $\mathfrak{B}'$ of unitary spaces H and H' ($\mathfrak{B} \subset H$, $\mathfrak{B}' \subset H'$). A bilinear form $W(f, g')$ is continuous for $f = 0$ and $g' = 0$ if and only if it is bounded (cf. Sec. 4.1.1), i.e. when ($f \in \mathfrak{B}$, $g' \in \mathfrak{B}'$)

$$\sup | W(f, g')) = \gamma_W < \infty, \quad \|f\| = 1, \quad \|g'\| = 1$$

or

$$|W(f, g')| \leqslant \gamma \|f\| \, \|g'\|, \quad \gamma < \infty, \tag{1}$$

where γ_W is the smallest of the numbers γ in (1). From

$$W(f, g') - W(f_n, g'_n) = W(f - f_n, g') + W(f_n, g' - g'_n)$$

by using (1), we obtain the inequality

$$|W(f, g') - W(f_n, g'_n)| \leqslant \gamma \|f - f_n\| \, \|g'\| + \gamma \|f_n\| \, \|g' - g'_n\| \tag{1'}$$

which shows that the bounded bilinear form $W(f, g')$ is continuous.

Continuous (bounded) bilinear forms $W(f, g')$ will also be called *bilinear functionals.* Bilinear functionals of the form $W(f\ g')$ ($W(g', f)$) form a vector subsystem $\mathfrak{W}(\mathfrak{B}, \mathfrak{B}')$ ($\mathfrak{W}(\mathfrak{B}', \mathfrak{B})$) of the system $B(\mathfrak{B}, \mathfrak{B}')$ ($B(\mathfrak{B}', \mathfrak{B})$) which was considered in Sec: 3.3.1.

The inequality (1') shows that a bilinear functional $W(f, g') \in \mathfrak{W}(\mathfrak{B}, \mathfrak{B}')$ can be extended by continuity to a bilinear functional in $\mathfrak{W}(\overline{\mathfrak{B}}, \overline{\mathfrak{B}}')$, where the closures $\mathfrak{B}$ and $\mathfrak{B}'$ can be taken to be $\mathfrak{B} = H$ and $\mathfrak{B}' = H'$. Thus it is sufficient to consider bilinear functionals in $\mathfrak{W}(H, H')$.

We will compare the bilinear form

$$W_A(f, g') = (Af, g'), \quad f \in H, \quad g' \in H' \tag{2}$$

to the linear transformation $A \in \mathfrak{A}(H, H')$. By applying the Cauchy-Bunyakovski inequality to (2) for $\|g'\| = 1$, we obtain the estimate $|W_A(f, g')| \leqslant \|Af\|$, while for $\|f\| = 1$ and $\|g'\| = 1$, we obtain

$$|W_A(f, g')| \leqslant \|A\|. \tag{3}$$

Thus the bilinear form $W_A(f, g')$ is bounded (continuous) and is a bilinear functional in $\mathfrak{W}(H, H')$.

In a similar manner, a bilinear functional in $\mathfrak{W}(H', H)$ corresponds to a transformation $A' \in \mathfrak{A}(H', H)$;

$$W_{A'}(g', f) = (A'g', f),$$

where the first argument g' ranges over the space H', and the second argument f ranges over the space H. Now

$$|W_{A'}(g', f)| \leqslant \|A'\|. \tag{3'}$$

THEOREM 4.4.1. *To each bilinear functional* $W(f, g') \in \mathfrak{W}(H, H')$, *there corresponds a linear transformation* $A \in \mathfrak{A}(H, H')$ *such that* $W(f, g') = W_A(f, g')$. *The transformation* A *is unique, and*

$$\|A\| = \gamma_W. \tag{4}$$

Indeed, for fixed f the form $W(f, g')$ is, by (1), a pseudolinear functional with respect to g'. Therefore, according to Theorem 4.3.2,

$$W(f, g') = (f', g'), \tag{5}$$

where the vector $f' \in H'$ is uniquely defined by the element $f : f' = Af$. The transformation A is defined on the entire space H. It is linear since the functional $W(f, g')$ is linear in the first argument and since the representation (5) is unique. By letting $g' = f' = Af$ in (5), and by using (1) with $\gamma = \gamma_W$, we have

$$\|Af\|^2 = (f', g') = W(f, g') \leqslant \gamma_W \|f\| \|g'\| = \gamma_W \|f\| \|Af\|$$

and hence $\|Af\| \leqslant \gamma_W \|f\|$. Thus, $A \in \mathfrak{A}(H, H')$ and

$$\|A\| \leqslant \gamma_W. \tag{4'}$$

Since by (5)

$$W(f, g') = (Af, g') = W_A(f, g'),$$

the inequality (3) gives $\gamma_W \leqslant \|A\|$, and when compared with (4′) it proves (4).

The one-to-one correspondence between bilinear functionals and transformations in $\mathfrak{A}(H, H')$ is obviously an isomorphism between the vector systems $\mathfrak{W}(H, H')$ and $\mathfrak{A}(H, H')$. If we set $\|W\| = \gamma_W$ for a form $W \in \mathfrak{W}(H, H')$, then in view of (4), *the vector system* $\mathfrak{W}(H, H')$ *is converted into a normed space which is isomorphic to the space* $\mathfrak{A}(H, H')$.

The weak convergence $A_n \to A$ of the sequence $\{A_n\}$ in $\mathfrak{A}(H, H')$ is written explicitly in the form (cf. (4.3.16))

$$\lim_{n\to\infty} (A_n f, g') = (Af, g') \tag{6}$$

for all $f \in H$ and $g' \in H'$. Thus, the weak convergence of a sequence $\{A_n\}$ to A is equivalent to the convergence of the corresponding sequence of bilinear functionals $W_n(f, g') = (A_n f, g')$ to $W(f, g') = (Af, g')$. The connection between the linear transformations and bilinear functionals allows Theorem 4.3.8 to be formulated in the following form:

If a sequence $W_n(f, g)$ *of bilinear functionals is everywhere convergent, then the norms* $\|W_n\|$ *are uniformly bounded and the limit* $W(f, g') =$ $\lim W_n(f, g')$ *is again a bilinear functional.*

4.2. The adjoint transformation. By generalizing the concept of $*$-operation for bilinear forms on a vector system (Sec. 3.3.1), we define a *pseudoisomorphism between the spaces* $\mathfrak{W}(H, H')$ and $\mathfrak{W}(H', H)$. We let

$$W^*(g', f) = \overline{W(f, g')}. \tag{7}$$

If $W(f, g') \in \mathfrak{W}(H, H')$, *then* $W^*(g', f)$ *is also a bilinear functional, i.e.* $W^*(g', f) \in \mathfrak{W}(H', H)$, *and*

$$\|W^*\| = \|W\|. \tag{8}$$

Indeed, $W^*(g', f)$ is a linear form in the first argument g', and it is pseudolinear in the second argument f. For $\|f\| = \|g'\| = 1$,

$$\gamma_{W^*} = \sup |W^*(g', f)| = \gamma_W < \infty.$$

Equation (7) is equivalent to

$$W(f, g') = \overline{W^*(g', f)}, \tag{7′}$$

and by using (7) it is easy to verify that

$$(W_1 + W_2)^* = W_1^* + W_2^*, \quad (\alpha W)^* = \bar{\alpha} W^*. \tag{9}$$

In addition, it follows from (7) and (7′) that

$$(W^*)^* = W, \tag{9′}$$

i.e. the mapping $W' \mapsto (W')^*$ ($W' \in \mathfrak{W}(H', H)$) is the inverse of the mapping $W \mapsto W^*$. Thus, it follows from (9), (9′), and (8) that the *mappings* $W \mapsto W^*$ *and* $W' \mapsto (W')^*$ *are mutually inverse pseudoisomorphisms of the normed spaces* $\mathfrak{W}(H, H')$ *and* $\mathfrak{W}(H', H)$.

The bilinear functionals $W \in \mathfrak{W}(H, H')$ and $W^* \in \mathfrak{W}(H', H)$ are said to be *conjugate*.

In view of the isomorphism between the normed spaces $\mathfrak{W}(H, H')$ and $\mathfrak{A}(H, H')$ which was established in Theorem 4.4.1, the $*$-operation is transferred from the space $\mathfrak{W}(H, H')$ into the space $\mathfrak{A}(H, H')$. Namely, a transformation $A^* \in \mathfrak{A}(H', H)$ is by definition *adjoint* to a transformation $A \in \mathfrak{A}(H, H')$ if the corresponding bilinear functionals $W_A(f, g') = (Af, g')$ and $W_{A^*}(g', f) = (A^*g', f)$ are mutually conjugate. By virtue of (7), or equivalently (7′), the condition that two transformations A and A^* be adjoint can be written

$$(Af, g') = (f, A^*g'), \tag{10}$$

moreover, in view of (8)

$$\|A^*\| = \|A\|. \tag{11}$$

The mapping $A \mapsto A^*$ is a pseudoisomorphism of the space $\mathfrak{A}(H, H')$ onto $\mathfrak{A}(H', H)$, i.e. in addition to (11), the following equations also hold (cf. (9) and (9′)):

$$(A_1 + A_2)^* = A_1^* + A_2^*, \quad (\alpha A)^* = \bar{\alpha} A^*, \quad (A^*)^* = A. \tag{12}$$

The mapping $A' \mapsto (A')^*$ ($A' \in \mathfrak{A}(H', H)$) is a (inverse to $A \mapsto A^*$) pseudoisomorphism of the space $\mathfrak{A}(H', H)$ onto $\mathfrak{A}(H, H')$.

We use the mapping $A \mapsto A^*$ to prove the following theorem.

THEOREM 4.4.2. *A transformation* $A \in \mathfrak{A}(H, H')$ *is continuous with respect to weak convergence,* i.e. $f_n \rightharpoonup f$ *in* H *implies that* $Af_n \rightharpoonup Af$ *in* H'.

Indeed, $(Af_n, g') = (f_n, A^*g')$ for any $g' \in H'$, and hence

$$\lim_{n\to\infty} (Af_n, g') = (f, A^*g') = (Af, g').$$

COROLLARY. *For a bilinear functional* $W(f, g')$ ($f \in H$, $g' \in H'$)

$$\lim W(f_n, g_n') = W(f, g'),$$

if one of the sequences $\{f_n\}$ *or* $\{g_n'\}$ *converges strongly and the other converges weakly to* f *or* g, *respectively.*

In fact, $W(f, g') = (Af, g')$, where $A \in \mathfrak{A}(H, H')$, and our assertion follows from Lemma 4.3.1 since, by virtue of Theorem 4.4.2, one of the sequences $\{Af_n\}$ or $\{g_n'\}$ converges strongly and the other converges weakly.

4.3. Hermitian operators. An important case occurs when $H' = H$ and

$\mathfrak{A}(H, H') = \mathfrak{A}(H', H) = \mathfrak{A}(H)$. The condition that two operators A and A^* be adjoint can then be written

$$(Af, g) = (f, A^*g), \qquad f, g \in H. \tag{10'}$$

The mapping $A \mapsto A^*$ is then an involutive pseudoautomorphism of the normed space $\mathfrak{A}(H)$, i.e. it is a pseudoinvolution of this space. The Hermitian elements with respect to this pseudoinvolution, i.e. the operators $A \in \mathfrak{A}(H)$ which satisfy the condition

$$A^* = A, \tag{13}$$

will be called *Hermitian* or *self-adjoint* operators.

Condition (13) is equivalent to the condition that the bilinear form $W_A(f, g) = (Af, g)$ be self-adjoint in the vector system H, i.e.

$$(Af, g) = (f, Ag), \qquad f, g \in H. \tag{14}$$

Therefore an operator A is Hermitian if and only if the quadratic form

$$Q(f) = Q_A(f) = (Af, f) \tag{15}$$

is real for any $f \in H$ (Theorem 3.3.1).

A quadratic form $Q(f)$ which is defined on a vector manifold $\mathfrak{B}$ of the space H and is continuous in $\mathfrak{B}$ will be called a *quadratic functional*. Quadratic functionals are bounded quadratic forms, i.e. they can be characterized by the condition

$$\sup_{\|f\|=1} |Q(f)| = \gamma_Q < \infty, \qquad f \in \mathfrak{B}.$$

Indeed, the boundedness of the quadratic form $Q(f)$ readily follows from its continuity for $f = 0$ (Sec. 4.1.1), and the boundedness of $Q(f)$, by virtue of (3.3.4), implies the boundedness of the corresponding bilinear form:

$$|W(f, g)| \leqslant \gamma_Q(\|h_1\|^2 + \|h_2\|^2 + \|h_3\|^2 + \|h_4\|^2) = \gamma_Q(\|f\|^2 + \|g\|^2);$$

hence

$$\sup |W(f, g)| \leqslant 2\gamma_Q, \qquad \|f\| = 1, \qquad \|g\| = 1. \tag{16}$$

Therefore the bilinear form $W(f, g)$ is continuous, and thus the quadratic form $Q(f) = W(f, f)$ is also continuous for all $f \in \mathfrak{B}$, i.e. $Q(f)$ is a quadratic functional generated by the bilinear functional $W(f, g)$. Thus, by virtue of Theorem 4.4.1, a one-to-one correspondence is established between operators in $\mathfrak{A}(H)$ and all quadratic functionals $Q(f)$. Namely, the quadratic functional (15) corresponds to the operator $A \in \mathfrak{A}(H)$. For the system $\mathfrak{A}_r(H)$ of Hermitian operators whose elements correspond to the real functionals $Q(f)$, this correspondence is preferred to the correspondence between $\mathfrak{A}(H)$ and $\mathfrak{W}(H) = \mathfrak{W}(H, H)$, which was considered previously.

Equation (6), which is the condition for weak convergence in $\mathfrak{A}(H)$ $(g' = g \in H)$, can be replaced by

$$\lim (A_n f, f) = (Af, f), \qquad f \in H,$$

i.e. the convergence of the corresponding quadratic functionals. As for bilinear functionals, *the limit of an everywhere convergent sequence of quadratic functionals is also a quadratic functional.*

§ 5. Supplementary Information

5.1. General lemmas on convex functionals. A subset M of a normed space X is said to be *nowhere dense* if in each open set there can be found an open subset which does not contain points of the set M. A set M is by definition a *set of the first category* in X if it can be represented as a *countable* sum of nowhere dense sets. A set M is said to be a *set of the second category* in X if it is not a set of the first category. Any subset of a set of the first category is also a set of the first category, and hence a set which contains a set of the second category is a set of the second category. Thus, only in the ease when the space X is itself a set of the second category do these concepts establish a distinction between sets. In fact, the following proposition shows that this occurs if X is a Banach space.

If X is a Banach space and M is a set of the first category in X, then $X \setminus M$ is dense in X.

Indeed, let $M = \bigcup\limits_k M_k$, where each of the sets M_k is nowhere dense in X, and let K_1 be an arbitrary sphere. Then, just as in the proof of Lemma 4.2.2, we can construct a sequence of closed spheres $\{K_n\}$ which has the following properties:

$$K_n \subset K_{n-1} \setminus M_n, \quad \rho_n \leqslant \frac{\rho_{n-1}}{2}, \qquad n = 2, 3, \ldots$$

Such a sequence $\{K_n\}$ exists, since the sets M_n are nowhere dense. From the inequality $\rho_n \leqslant \frac{\rho_1}{2^{n-1}}$, we conclude that the centers of the spheres K_n converge to some point $f \in X$; moreover, $f \in X \setminus M$ since

$$\bigcap_n K_n \subset \bigcap_n (X \setminus M_n) = X \setminus \bigcup_n M_n = X \setminus M.$$

Thus the arbitrarily chosen sphere K_1 contains a point of the set $X \setminus M$, i.e. $X \setminus M$ is dense in X.

The following simple lemma is a generalization of Lemma 4.2.2 (Sec. 4.2.3) to the case when the quasinorms or convex functionals are not defined on the whole space.

LEMMA 4.5.1. *If a functional $w(f)$ is lower semicontinuous on a set M, then for any real α the set $M(w(f)\leqslant\alpha)$ is closed in M*.*

Indeed, if

$$f_n \in M(w(f)\leqslant\alpha), \quad f_n \Rightarrow f_0$$

the inequality

$$w(f_0)\leqslant \varliminf_{n\to\infty} w(f_n)\leqslant\alpha;$$

holds, and hence $f_0 \in M\{w(f)\leqslant\alpha\}$.

LEMMA 4.5.2. *Let a non-negative convex functional $w(f)$ be defined on a vector manifold $\mathfrak{B}$ of a Banach space X and let $\mathfrak{B}$ be a set of the second category in X. Then the functional $w(f)$ is bounded if it is semicontinuous on $\mathfrak{B}$.*

In fact, we have

$$\mathfrak{B}=\bigcup_n \mathfrak{B}(w(f)\leqslant n)=\bigcup_n M_n,$$

where at least one M_k is not nowhere dense in X, i.e. there exists a closed sphere K_1: $\|f-f_1\|\leqslant\rho_1$ such that the intersection $K_1 \cap M_k$ is dense in K_1, and thus it is dense in $K_1 \cap \mathfrak{B}$. But the set M_n is closed in $\mathfrak{B}$ (Lemma 4.5.1), and hence $K_1 \cap M_k$ coincides with $K_1 \cap \mathfrak{B}$. Thus, $w(f)\leqslant k$ in $K_1 \cap \mathfrak{B}$, and the functional $w(f)$, which is bounded in the sphere $K_1 \cap \mathfrak{B}$, is also bounded in any sphere.

REMARK. If the vector manifold $\mathfrak{B}$ is a set of the second category in X, then it is dense in X since there exists a subset of the set $\mathfrak{B}$ which is dense in some sphere, and the same is true for any sphere. Therefore the functional $w(f)$ in Lemma 4.5.2 can be extended by continuity to the entire space X.

Lemma 4.2.3 can be generalized in a manner similar to that of Lemma 4.2.2; namely, it is sufficient to assume that the condition (4.2.5) is satisfied for elements of a vector manifold $\mathfrak{B}$ which is of the second category in X. Using these generalizations and Lemma 4.2.3, we can sharpen Theorems 4.2.1, 4.3.1 and 4.2.2. In Theorem 4.2.1 (or Theorem 4.3.1), it is sufficient to assume that the condition (4.2.7) (or (4.3.4)) holds for elements of some set (or of the vector manifold $\mathfrak{B}$) of the second category in X; Theorem 4.2.2 can be generalized in a similar manner.

5.2. Another proof of Theorem 4.3.2. Equation (4.3.6) can be used, in a somewhat different form than previously given, to construct the vector g in the representation (4.3.1), i.e. to prove Theorem 4.3.2. We can assume that $l \neq 0$ since otherwise it is sufficient to let $g = 0$.

* As is easily seen, the converse is also true: if for any real α, the set $M(w(f)\leqslant\alpha)$ is closed in M, then the functional $w(f)$ is semicontinuous on M.

Since the functional $l(f)$ is continuous, the vector manifold H_0 of elements g which satisfy the equation

$$l(g)=0, \tag{1}$$

is closed and does not coincide with H. The orthogonal complement H_1 of the subspace H_0 is one dimensional. Indeed, let h_1 be a fixed normalized vector, and let f_1 be an arbitrary vector in H_1.

Then the vector

$$g=f_1-\frac{l(f_1)}{l(h_1)}h_1,$$

which belongs to H_1, satisfies (1), and hence $g \in H_0$. Therefore $g=0$ and $f_1 = \dfrac{l(f_1)}{l(h_1)}h_1$.

In the decomposition $f=f_0+f_1$ of a vector $f \in H$, if the elements f_0 and f_1 belong to H_0 and H_1, respectively, then $l(f_0)=0$ and

$$l(f)=l(f_1),$$

so that it is sufficient to consider the functional on the one dimensional space H_1. The vector

$$g=\overline{l(h_1)}\,h_1, \tag{2}$$

which, according to (4.3.6), satisfies the first of the equations in (4.3.1) for elements $f=f_1 \in H_1$, gives the desired representation $l=l_g$ in all of H: if $f=f_0+f_1$, $f_0 \in H_0$, and $f_1 \in H_1$, then

$$l(f)=l(f_1)=(f_1, g)=(f_0+f_1, g)=(f, g).$$

Equation (2) is the special case of (4.3.9) when the basis H is composed of the orthogonal bases of the subspaces H_0 and H_1. The derivation of the representation $l(f)=(f, g)$, carried out in Sec. 4.3.1, has the advantage that it gives an explicit expression for the elements g for an *arbitrary* choice of the orthogonal basis in H.

CHAPTER 5

Bounded Operators, Defined on the Whole Space

§ 1. A Normed $*$-algebra of Operators

1.1. Convergence in $\mathfrak{A}(H)$. An algebra $\mathfrak{A}$ is said to be *normed* if a norm is defined in the vector system $\mathfrak{A}$ which satisfies the condition

$$\|AB\| \leqslant \|A\| \, \|B\|, \tag{1}$$

where A and B are any elements of $\mathfrak{A}$.

A normed algebra $\mathfrak{A}$ is said to be *complete* if the normed space $\mathfrak{A}$ is complete. A complete normed algebra will also be called a *Gel'fand algebra*. An example of such an algebra is the normed ring of operators $\mathfrak{A}(H)$.*

The continuity of multiplication with respect to convergence in norm in the space $\mathfrak{A}$ follows from (1):

$$\lim \|A_n B_n - AB\| = 0, \tag{2}$$

if $\lim \|A_n - A\| = 0$ and $\lim \|B_n - B\| = 0$.

Indeed,

$$A_n B_n - AB = A_n (B_n - B) + (A_n - A) B \tag{3}$$

and by the properties of the norm

$$\|A_n B_n - AB\| \leqslant \|A_n\| \, \|B_n - B\| + \|A_n - A\| \, \|B\|,$$

which proves (2), since the norms $\|A_n\|$ are bounded.

An algebra $\mathfrak{A}$ is called a *normed algebra with pseudoinvolution* or *a normed $*$-algebra* if $\mathfrak{A}$ is both a normed algebra and an algebra with pseudoinvolution (Sec. 1.5.2). The given pseudoinvolution $A \to A^*$ ($A \in \mathfrak{A}$) satisfies the condition

$$\|A^*\| = \|A\|,$$

i.e. it is a pseudoinvolution of the normed space $\mathfrak{A}$.

* We recall that the space H is always assumed to be complete.

The pseudoinvolution $A \to A^*$ (transition from the operator A to its adjoint A^*) has been defined in the normed space $\mathfrak{A}(H)$. We show that *$A \to A^*$ is also a pseudoinvolution of the algebra $\mathfrak{A}(H)$, i.e. that*

$$(AB)^* = B^* A^*. \tag{4}$$

Indeed, we have

$$(ABf, g) = (Bf, A^*g) = (f, B^*A^*g)$$

for arbitrary f and g in H, whence (4) follows by the definition of the adjoint operator.

In order to convince ourselves that $\mathfrak{A}(H)$ is a normed $*$-algebra, we recall the equation $\|A^*\| = \|A\|$, which has been proved for $A \in \mathfrak{A}(H)$ in Sec. 4.4.2.

The normed ring of operators $\mathfrak{A}(H)$ is non-commutative, i.e. in general $AB \neq BA$ $(A, B \in \mathfrak{A}(H))$. In the normed ring of operators $\mathfrak{A}(H)$, in addition to convergence in norm (which is called uniform convergence here), we can also consider strong and weak convergence (Sec. 4.2.2 and 4.3.3).

Multiplication in $\mathfrak{A}(H)$ is continuous with respect to strong convergence, i.e. $A_n \Rightarrow A$ and $B_n \Rightarrow B$ imply that

$$A_n B_n \Rightarrow AB. \tag{5}$$

In fact, for any $f \in H$ (see (3)),

$$A_n B_n f - ABf = A_n (B_n - B) f + (A_n - A) Bf$$

and moreover, $\|A_n\| \leqslant \gamma$ (see Theorem 4.2.3). Therefore

$$\|A_n B_n f - ABf\| \leqslant \gamma \|B_n f - Bf\| + \|A_n Bf - ABf\| \tag{5'}$$

and $A_n B_n f - ABf \Rightarrow 0$ since both terms on the right-hand side of (5′) tend to zero.

The $*$-operation in $\mathfrak{A}(H)$ is continuous with respect to weak convergence in $\mathfrak{A}(H)$, i.e. $A_n \to A$ implies that

$$A_n^* \to A^*. \tag{6}$$

Indeed, $A_n \to A$ means that

$$\lim (A_n f, g) = (Af, g)$$

for all f and g in H, or

$$\lim (f, A_n^* g) = (f, A^* g),$$

and this is (6).

The product AB is continuous in each of its variables with respect to weak convergence in $\mathfrak{A}(H)$, i.e.

$$\left.\begin{aligned} &A_n B \to AB, \quad \text{if} \quad A_n \to A, \\ \text{and} \quad &AB_n \to AB, \quad \text{if} \quad B_n \to B. \end{aligned}\right\} \tag{7}$$

Indeed, for any f, g in H,

$$\lim (A_n Bf, g) = \lim (Bf, A_n^* g) = (Bf, A^* g) = (ABf, g).$$

A similar equation proves the second part of (7).

From this property of continuity, we have (see Sec. 1.5.2 and 1.5.3):

THEOREM 5.1.1. *The commutator $\mathfrak{C}(\mathfrak{M})$ and the *-commutator $\mathfrak{C}^*(\mathfrak{M})$ are closed with respect to weak (and this means strong, and also uniform) convergence.*

In fact, let $B \in \mathfrak{M}$, $A_n B = BA_n$, and $A_n \to A$. Then $AB = BA$, and in addition, if $A_n B^* = B^* A_n$, then $AB^* = B^* A$.

Since the *-operation is not continuous with respect to weak convergence in $\mathfrak{A}(H)^*$, it is natural to introduce the concept of **-convergence*

$$A_n \overset{*}{\Rightarrow} A,$$

which by definition holds if and only if both

$$A_n \Rightarrow A, \qquad A_n^* \Rightarrow A^*.$$

*The *-operation is continuous with respect to *-convergence; multiplication is also continuous in $\mathfrak{A}(H)$.*

In fact, if $A_n \overset{*}{\Rightarrow} A$ and $B_n \overset{*}{\Rightarrow} B$, then in addition to (5),

$$(A_n B_n)^* \Rightarrow (AB)^*,$$

holds since this equation is equivalent to $B_n^* A_n^* = B^* A^*$.

1.2. Power series. Although we assume in this section that elements of the algebra are operators of the normed ring $\mathfrak{A}(H)$, all remarks obviously hold for elements of an arbitrary Gel'fand algebra.

We will consider power series of the form

$$\sum_{k=0}^{\infty} \alpha_k A^k = \alpha_0 E + \alpha_1 A + \alpha_2 A^2 + \ldots, \tag{8}$$

where the α_k are complex numbers and $A \in \mathfrak{A}(H)$. If

$$\sum_{k=0}^{\infty} |\alpha_k| \, \|A^k\| < \infty, \tag{9}$$

then the series (8) converges uniformly (in norm in the space $\mathfrak{A}(H)$). In fact,

$$\Big\| \sum_{k=m}^{n} \alpha_k A^k \Big\| \leqslant \sum_{k=m}^{n} |\alpha_k| \, \|A^k\|,$$

---

* The truth of this assertion can be seen from the following example. Let $e_1, e_2, \ldots$ be a basis in H and let the operator A_n be defined by the formula

$$A_n \left(\sum_{k=1}^{\infty} x_k e_k \right) = \sum_{k=1}^{\infty} x_{n+k} e_k.$$

Then $A_n^* \left(\sum_{k=1}^{\infty} y_k e_k \right) = \sum_{k=1}^{\infty} y_k e_{k+n}$. Obviously, $A_n \Rightarrow 0$ and $A_n^* \not\Rightarrow 0$.

and by virtue of (9), the partial sums of the series (8) form a Cauchy sequence which converges to an operator in $\mathfrak{A}(H)$, since $\mathfrak{A}(H)$ is complete.

We will compare the ordinary power series

$$\varphi(z) = \sum_{k=0}^{\infty} \alpha_k z^k, \tag{10}$$

with the series (8). Let $\mathfrak{A}_\varphi$ be the collection of all $A \in \mathfrak{A}(H)$ which satisfy (9). Then

$$\varphi(A) = \sum_{k=0}^{\infty} \alpha_k A^k \tag{11}$$

defines an operator $\varphi(A)$ in $\mathfrak{A}(H)$ for each $A \in \mathfrak{A}_\varphi$. Let $\mathfrak{F}_A$ be the set of power series $\varphi(z)$ which satisfy (9) for fixed A in $\mathfrak{A}(H)$. The conditions $\varphi(z) \in \mathfrak{F}_A$ and $A \in \mathfrak{A}_\varphi$ give the same relation between the power series $\varphi(z)$ and the operator A. The set $\mathfrak{F}_A$ is obviously a vector system (under the usual definitions of the sum of power series and their multiplication by a constant), and

$$\|\varphi\| = \sum_{k=0}^{\infty} |\alpha_k| \, \|A^k\| \tag{12}$$

defines a norm in the system $\mathfrak{F}_A$ which converts $\mathfrak{F}_A$ into a (complete) normed space, if $A^k \neq 0$ for all k*.

Indeed,

$$\|\alpha\varphi\| = \sum_{k=0}^{\infty} |\alpha| \, |\alpha_k| \, \|A^k\| = |\alpha| \, \|\varphi\|,$$

$$\|\varphi + \psi\| \leqslant \sum_{k=0}^{\infty} (|\alpha_k| + |\beta_k|) \|A^k\| \leqslant \|\varphi\| + \|\psi\|,$$

where

$$\psi(z) = \sum_{k=0}^{\infty} \beta_k z^k \tag{10'}$$

is a power series in $\mathfrak{F}_A$. It is easy to verify that $\mathfrak{F}_A$ is complete.

We now define

$$\varphi(z)\psi(z) = \sum_{n=0}^{\infty} \gamma_n z^n, \qquad \gamma_n = \sum_{k=0}^{n} \alpha_k \beta_{n-k},$$

(formally) to be the product of the power series (10) and (10′).

Whenever $\varphi(z)$ and $\psi(z)$ belong to $\mathfrak{F}_A$, their product $\varphi(z)\,\psi(z)$ also belongs to $\mathfrak{F}_A$, and

$$\|\varphi\psi\| \leqslant \|\varphi\| \, \|\psi\|.$$

---

* If k's exist for which $A^k = 0$, and if $k_0 = n+1$ is the least of these, then it is necessary to consider the polynomials $\varphi(z) = \sum_{k=0}^{n} \alpha_k z^k$ instead of the series (10).

This follows since

$$\|\varphi\psi\| = \sum_n \gamma_n \|A^n\| \leqslant \sum_{n,k} |\alpha_k|\, |\beta_{n-k}|\, \|A^k\|\, \|A^{n-k}\| = \|\varphi\|\, \|\psi\|.$$

Thus $\mathfrak{F}_A$ is a normed algebra, and to each element $\varphi(z)$ in $\mathfrak{F}_A$ there corresponds an operator $\varphi(A) \in \mathfrak{A}(H)$ which is defined by (11). This correspondence will be denoted by $\backsim$, and we will write

$$\varphi(z) \backsim \varphi(A). \tag{13}$$

According to the definition (12) of the norm $\|\varphi\|$,

$$\|\varphi(A)\| \leqslant \|\varphi\|.$$

Since the correspondence is obviously linear, i.e. since

$$\alpha\varphi(z) + \beta\psi(z) \backsim \alpha\varphi(A) + \beta\psi(A), \tag{14}$$

we have

$$\|\varphi_n(A) - \varphi(A)\| \leqslant \|\varphi_n - \varphi\|,$$

and the sequence $\{\varphi_n(A)\}$ converges uniformly to $\varphi(A)$ in $\mathfrak{A}(H)$ if $\varphi_n \Rightarrow \varphi$ in the normed space $\mathfrak{F}_A$.

In addition to the linearity property (14), the correspondence (13) also has the *multiplicative* property

$$\varphi(z)\psi(z) \backsim \varphi(A)\psi(A), \tag{15}$$

i.e. it follows from

$$\vartheta(z) = \varphi(z)\psi(z), \quad \varphi(z) \backsim \varphi(A), \quad \psi(z) \backsim \psi(A)$$

that

$$\vartheta(A) = \varphi(A)\psi(A). \tag{15'}$$

It is obvious that (15′) holds if one of the functions $\varphi(z)$ or $\psi(z)$ is a polynominal. In the general case we let

$$\varphi_n(z) = \sum_{k=0}^{n} \alpha_k z^k, \qquad \vartheta_n(z) = \varphi_n(z)\psi(z).$$

Then

$$\|\vartheta_n - \vartheta\| = \|(\varphi_n - \varphi)\psi\| \leqslant \|\varphi_n - \varphi\|\, \|\psi\|$$

and since $\lim\limits_{n\to\infty} \|\varphi_n - \varphi\| = 0$, we have

$$\lim \|\vartheta_n - \vartheta\| = 0. \tag{16}$$

But it follows from (16) that the sequence $\vartheta_n(A)$ is uniformly convergent to $\vartheta(A)$, since $\vartheta_n(A) = \varphi_n(A)\psi(A)$ gives the desired relation (15′) in the limit.

Let us assume now that the radius of convergence ρ of the series (10) is positive. As is known from analysis (the Cauchy criterion), (9) holds if

$$K = \overline{\lim_{n\to\infty}} (|\alpha_n|\, \|A^n\|)^{1/n} < 1.$$

We show below that the limit $\lim_{n\to\infty} \|A^n\|^{1/n}$ exists. Therefore

$$\overline{\lim_{n\to\infty}}\,(|\alpha_n|\,\|A^n\|)^{1/n} = \overline{\lim_{n\to\infty}}|\alpha_n|^{1/n} \lim_{n\to\infty}\|A^n\|^{1/n} = \frac{1}{\rho} \lim_{n\to\infty}\|A^n\|^{1/n}.$$

Hence, the inequality $K<1$ is equivalent to the condition

$$\omega(A) < \rho, \tag{17}$$

where

$$\omega(A) = \lim_{n\to\infty} \|A^n\|^{1/n}. \tag{18}$$

The following lemma is necessary to prove the existence of the limit (18).

LEMMA 5.1.1. *If a (real) sequence $\{c_n\}$ satisfies the condition*

$$c_{m+n} \leqslant c_m + c_n$$

and if $\gamma = \inf_n \frac{c_n}{n}$, then

$$\lim_{n\to\infty} \frac{c_n}{n} = \gamma. \tag{19}$$

In fact, for $n = mk+r$, $0 \leqslant r < m$, we have $c_n \leqslant c_{mk} + c_r \leqslant kc_m + c_r$, and hence $\left(\frac{mk}{n} = 1 - \frac{r}{n}\right)$,

$$\gamma \leqslant \frac{c_n}{n} \leqslant \left(1 - \frac{r}{n}\right)\frac{c_m}{m} + \frac{c_r}{n},$$

If m is chosen such that $\frac{c_m}{m} < \gamma+\varepsilon$, $\varepsilon > 0$, then

$$\gamma \leqslant \frac{c_n}{n} < \left(1 - \frac{r}{n}\right)(\gamma + \varepsilon) + \frac{c_r}{n}.$$

Whence, (19) follows since $\varepsilon > 0$ was arbitrary.

The existence of the limit (18) can now be proved without difficulty. Indeed, we let $c_n = ln\|A^n\|$. Then

$$c_{n+m} = \ln\|A^n \cdot A^m\| \leqslant \ln\|A^n\| + \ln\|A^m\| = c_n + c_m$$

and hence

$$\gamma = \lim \frac{1}{n} \ln\|A^n\| = \ln \omega(A),$$

i.e.

$$0 \leqslant \omega(A) = \lim_{n\to\infty} (\|A^n\|)^{\frac{1}{n}} = e^{\gamma} < \infty.$$

Since $\|A^2\| \leqslant \|A\|^2$, $\|A^k\| \leqslant \|A^{k-1}\|\,\|A\|$, and thus

$$\|A^k\| \leqslant \|A\|^k. \tag{20}$$

Then $(\|A^n\|)^{\frac{1}{n}} \leqslant \|A\|$ and

$$\omega(A) \leqslant \|A\|. \tag{21}$$

If the radius of convergence ρ of the series (10) is positive, then the series (10) represents an analytic function $\varphi(z)$ in the circle $|z| < \rho$, and we can consider $\varphi(A)$ to be an operator function of the operator A with domain $\mathfrak{A}_\varphi$ and range in $\mathfrak{A}(H)$. The set $\mathfrak{A}_\varphi$ contains all operators $A \in \mathfrak{A}(H)$ which satisfy (17). In view of (21), the operators A also satisfy the inequality

$$\|A\| < \rho.$$

If the series (10) is absolutely convergent on the boundary $|z| = \rho$ of its circle of convergence, then the region $\|A\| \leqslant \rho$ also belongs to $\mathfrak{A}_\varphi$.

In fact, by virtue of (20),

$$\sum_k |\alpha_k| \, \|A^k\| \leqslant \sum_k |\alpha_k| \, \|A\|^k.$$

An operator $A \in \mathfrak{A}(H)$ is said to be *regular* if

$$\omega(A) = \|A\|, \tag{22}$$

and is said to be *singular* if $\omega(A) < \|A\|$.

An operator A is regular if and only if

$$\|A^n\| = \|A\|^n. \tag{23}$$

Indeed,

$$\omega(A) = \inf_k (\|A\|^k)^{\frac{1}{k}} \leqslant \|A^n\|^{\frac{1}{n}},$$

and for a regular operator it follows from (22) that

$$\|A\| \leqslant \|A^n\|^{\frac{1}{n}} \leqslant \|A\|,$$

so that $\|A^n\|^{\frac{1}{n}} = \|A\|$, i.e. (23) holds. Conversely, if (23) holds for all n, then

$$\omega(A) = \lim (\|A^n\|)^{\frac{1}{n}} = \|A\|.$$

Singular operators A which satisfy the condition $\omega(A) = 0$ are said to be *generalized nilpotent elements of the algebra* $\mathfrak{A}(H)$ (these exist by virtue of the generalization of the previously noted case when $A^n = 0$ for some natural number n). Condition (17) is fulfilled for a generalized nilpotent operator A if $\rho > 0$, and hence in that case $\mathfrak{F}_A$ contains all power series with positive radii of convergence.

Examples. 1) Since $(1-z)\varphi(z) = 1$ for

$$\varphi(z) = \frac{1}{1-z} = \sum_{k=0}^{\infty} z^k,$$

the multiplicative property of the correspondence (13) implies that

$$(E - A)\,\varphi(A) = \varphi(A)\,(E - A) = E,$$

if $\varphi \in \mathfrak{F}_A$ or if $A \in \mathfrak{A}_\varphi$, i.e. for all $A \in \mathfrak{A}_\varphi$, the element $E - A$ of the ring $\mathfrak{A}(H)$ is regular, and

$$(E-A)^{-1}=\varphi(A)=\sum_{k=0}^{\infty} A^k. \tag{24}$$

In particular, the set $\mathfrak{A}_\varphi$ contains all operators A in $\mathfrak{A}(H)$ which satisfy the condition $\|A\|<1$, or even $\omega(A)<1$, and (24) holds for these operators. Whence it follows that the *set of regular elements of the ring* $\mathfrak{A}(H)$ *is open in the normed space* $\mathfrak{A}(H)$, i.e. whenever A is regular, the operators B of some neighborhood

$$\|B-A\|<\varepsilon \tag{25}$$

of the operator A are also regular.

Indeed, let A be a regular element of $\mathfrak{A}(H)$ and let $A-B=C$. Then

$$B=A-C=(A-C)A^{-1}A=(E-CA^{-1})A,$$

and if $\|C\|<\frac{1}{\|A^{-1}\|}$, then

$$\|CA^{-1}\|\leqslant\|C\|\|A^{-1}\|<1,$$

i.e. $E-CA^{-1}$ is a regular element. In addition, the operator B, which is the product of two regular elements (Sec. 1.5.2), is itself regular. Therefore all operators B of the neighborhood (25) are regular for $\varepsilon=\frac{1}{\|A^{-1}\|}$.

2) Let $k>1$ be an integer, and let*

$$\varphi(z)=(1-z)^{\frac{1}{k}}=\sum_{n=0}^{\infty}(-1)^n\binom{\frac{1}{k}}{n}z^n=1-\sum_{n=1}^{\infty}\left|\binom{\frac{1}{k}}{n}\right|z^n. \tag{26}$$

Since $(\varphi(z))^k=1-z$ for $A\in\mathfrak{A}_\varphi$, we have

$$(\varphi(A))^k=E-A \tag{27}$$

and hence ($\varphi(A)=(E-A)^{1/k}$),

$$(E-A)^{\frac{1}{k}}=E-\sum_{n=1}^{\infty}\left|\binom{\frac{1}{k}}{n}\right|A^n. \tag{28}$$

The series (26) converges absolutely on the boundary of its circle of convergence and

$$\sum_{n=1}^{\infty}\left|\binom{\frac{1}{k}}{n}\right|=1. \tag{29}$$

Therefore $\mathfrak{A}_\varphi$ contains all operators A in $\mathfrak{A}(H)$ for which $\omega(A)\leqslant 1$ ($\rho=1$) or

---

* $\binom{\alpha}{n}=\begin{cases}1 \text{ for } n=0\\ \alpha(\alpha-1)\ldots(\alpha-n+1)/n! \text{ for } n\geqslant 1.\end{cases}$

$$\|A\| \leqslant 1,$$

and (28) holds for these operators.

Similarly, in the case of a regular operator A, it is possible to consider power series of the form

$$\varphi(z) = \sum_{k=-\infty}^{\infty} \alpha_k z^k$$

and corresponding operators ($A^{-n} = (A^{-1})^n$, $n > 0$)

$$\varphi(A) = \sum_{k=-\infty}^{\infty}{}' \alpha_k A^k,$$

which are constructed for operators A of the set $\mathfrak{A}_\varphi$, where $\mathfrak{A}_\varphi$ is now defined by the condition

$$\sum_{k=-\infty}^{\infty} |\alpha_k| \, \|A^k\| < \infty.$$

Another type of power series is represented by the series

$$\sum_{k=0}^{\infty} A_k z^k, \qquad \sum_{k=-\infty}^{\infty} A_k z^k$$

in powers of a complex variable z, where the coefficients A_k are operators in $\mathfrak{A}(H)$. We consider the first of these series in detail:

$$\Phi(z) = \sum_{k=0}^{\infty} A_k z^k. \tag{30}$$

Its radius of convergence is

$$\rho = \frac{1}{\overline{\lim\limits_{n\to\infty}} \|A_n\|^{\frac{1}{n}}}, \tag{31}$$

i.e. the series (30) converges absolutely for all z such that $|z| = r < \rho$

$$\sum_{k=1}^{\infty} \|A_k\| r^k < \infty, \qquad r < \rho, \tag{32}$$

and diverges for $|z| = r > \rho$. Indeed, the number ρ given in (31) is the radius of convergence of the power series (32), and hence this series converges for $r < \rho$. If the series (30) converges for $z = z_0$, then $\lim\limits_{k\to\infty} \|A_k\| \, |z_0|^k = 0$, and (32) holds for any $r < |z_0|$. Thus $\rho \geqslant |z_0|$, and hence the series (30) diverges for $|z| = r > \rho$.

Under the assumption $\rho > 0$, the series (30) converges in the circle $|z| < \rho$ and defines a function $\Phi(z)$ in this circle with values in the normed ring $\mathfrak{A}(H)$. If (32) is also satisfied for $r = \rho$, then the series (30) converges uniformly in the closed circle $|z| \leqslant \rho$, which we will then consider to be the domain of the function $\Phi(z)$.

To each operator $A \in \mathfrak{A}(H)$, we associate a series

$$\Phi_A(z) = \sum_{k=0}^{\infty} A^k z^k. \tag{33}$$

The radius of convergence of this series is

$$\frac{1}{\rho} = \omega(A),$$

and hence it converges for

$$|z| < \frac{1}{\omega(A)} \tag{34}$$

and for these values of z (see 24))

$$\Phi_A(z) = (E - zA)^{-1}. \tag{33'}$$

The series (33) *converges for all values of* z, *i.e. it is an entire function if and only if* $\omega(A) = 0$, *i.e. when* A *is a generalized nilpotent element of the algebra* $\mathfrak{A}(H)$.

Operations on series of the form (30)—addition, multiplication (in a definite order), differentiation, and integration—are performed in the same way as on ordinary power series. To generalize the proofs of corresponding propositions from analysis to the present type of series, it is only necessary to replace the absolute values of the coefficients of the power series (30) by the norms $\|A_k\|$. In particular, if z_0 is a value of z from the circle of convergence $|z| < \rho$, then by substituting $z = (z - z_0) + z_0$, the series (30) can be transformed into the series

$$\sum_{k=0}^{\infty} B_k (z - z_0)^k, \tag{35}$$

which converges in the circle $|z - z_0| < \rho - |z_0|$ to the values of the function $\Phi(z)$.

We will say that a function $\Psi(z)$ whose domain is a connected open set of the complex plane, and whose range belongs to the algebra $\mathfrak{A}(H)$, is *analytic* at the point $z = z_0$ if $\Psi(z)$ can be represented by a convergent series of the form (35) in some neighborhood $|z - z_0| < \rho'$, $\rho' > 0$. The transition from the series (30) to the series (35) is an example of an elementary step in the process of the *continuation* of an analytic function with values in $\mathfrak{A}(H)$.

§ 2. Hermitian Operators in $\mathfrak{A}(H)$

2.1. Upper and lower bounds of an operator. Hermitian operators of a normed $*$-algebra (of a normed ring) are Hermitian elements of the $*$-system $\mathfrak{A}(H)$ and they form a *real* vector system $\mathfrak{A}_r(H)$. $\mathfrak{A}_r(H)$ is *closed with respect* *to both weak and strong convergence of operators*: if $\{A_n\}$ is a sequence of Hermitian operators, and if $A_n \to A$, then

$$(Af, g) = \lim (A_n f, g) = \lim (f, A_n f) = (f, Ag)$$

and A is a Hermitian operator.

In particular, for A Hermitian, the operators $\varphi(A)$ defined by (5.1.11) are also Hermitian if the coefficients of the power series are real, since the partial sums of this series are Hermitian (Sec. 1.5.3).

Since the operator A being Hermitian is characterized by the fact that the corresponding quadratic functional $Q_A(f) = (Af, f)$ (Sec. 4.4.3 and 3.3.1) is real, it is possible to form the numbers

$$m = m(A) = \inf_{\|f\|=1} (Af, f), \qquad m' = m'(A) = \sup_{\|f\|=1} (Af, f), \tag{1}$$

which are called, respectively, the *lower* and *upper bounds of the Hermitian operator A* or of *the form* $Q_A(f)$.

THEOREM 5.2.1. *If A is a Hermitian operator, then*

$$\|A\| = \sup_{\|f\|=1} |(Af, f)| = \sup\{|m(A)|, |m'(A)|\}. \tag{2}$$

Let $A \in \mathfrak{A}(H)$ and let $\gamma = \sup\limits_{\|f\|=1} |(Af, f)|$. Then γ does not exceed sup $|(Af, g)|$ for $\|f\| = 1$ and $\|g\| = 1$, i.e.

$$\gamma \leqslant \|A\| \tag{3}$$

for arbitrary A in $\mathfrak{A}(H)$. If the operator A is also Hermitian, then in identity (3.3.4), for

$$W(f, g) = (Af, g), \qquad Q(f) = (Af, f),$$

the first two terms on the right are real and the last two terms are imaginary. Therefore

$$|\mathrm{Re}\,(Af, g)| \leqslant \gamma \|h_1\|^2 + \gamma \|h_2\|^2 = \gamma \tag{4}$$

for $\|f\| = 1$ and $\|g\| = 1$, since $\|h_1\|^2 + \|h_2\|^2 = \dfrac{\|f\|^2 + \|g\|^2}{2} = 1$. Now let (for fixed f and g) $\alpha = \dfrac{(Af, g)}{|(Af, g)|}$ for $(Af, g) \neq 0$, and let $\alpha = 1$ otherwise; furthermore, let $g' = \alpha g$. Then $\|g'\| = |\alpha|\, \|g\| = \|g\|$, and

$$(Af, g') = |(Af, g)|.$$

Hence, by virtue of (4),

$$\|A\| = \sup |(Af, g)| = \sup (Af, g') = \sup \mathrm{Re}\,(Af, g') \leqslant \gamma,$$

for $\|f\| = \|g\| = 1$, and combined with (3) this proves (2).

REMARK. For any $A \in \mathfrak{A}(H)$, the equation

$$\sup_{\|f\|=1} |(Af, f)| \leqslant \|A\| \leqslant 2 \sup_{\|f\|=1} |(Af, f)|$$

holds. Indeed, (3) holds for any $A \in \mathfrak{A}(H)$, and for an estimate of $\|A\|$ from above, we have the inequality (4.4.16) in which $W(f, g) = (Af, g)$ and $\gamma_Q = \gamma$.

2.2. Positive operators. If the form $Q_A(f) = (Af, f)$ is positive, then the operator A in $\mathfrak{A}(H)$ is said to be *positive*. A positive operator A is a non-zero Hermitian operator with a lower bound $m(A) \geqslant 0$.

A linear combination of positive operators with positive coefficients is a positive operator. The limit (uniform, strong or weak) of a sequence $\{A_n\}$ of positive operators is a positive operator or is equal to zero.

Indeed, if $A_n \to A$, then

$$\lim (A_n f, f) = (Af, f)$$

and $(A_n f, f) \geqslant 0$ implies that $(Af, f) \geqslant 0$.

If $A \in \mathfrak{A}(H)$ and if $A \neq 0$, then the operators A^*A and AA^* are positive.

In fact,

$$(A^*Af, f) = (Af, Af) = \|Af\|^2 \geqslant 0,$$

and

$$AA^* = (A^*)^* A^* \qquad (A^* \in \mathfrak{A}(H),\ A^* \neq 0).$$

In particular, if A is a non-zero Hermitian operator, then A^2 is a positive operator. As will be shown below, any positive operator in $\mathfrak{A}(H)$ can be represented *as the square of a positive operator.*

Let A be a positive operator in $\mathfrak{A}(H)$. Then:

1) The operator A^n is positive or zero. In fact, for $n = 2p+1$

$$(A^n f, f) = (AA^p f, A^p f) \geqslant 0.$$

For $n = 2p$, we have $A^n = (A^p)^2$, and since A^p is a Hermitian operator, the operator A^n is positive if it is not equal to zero.

2) The inverse operator A^{-1} is positive. Indeed, if $g = Af$, then

$$(A^{-1}g, g) = (f, Af) = (Af, f) \tag{5}$$

and hence $(A^{-1}g, g) \geqslant 0$.

We will write $A > 0$ for elements A of the (real) vector system $\mathfrak{A}_r(H)$ of Hermitian operators in $\mathfrak{A}(H)$, if the operator A is positive; correspondingly, we will write (for $B \in \mathfrak{A}_r(H)$) $B > A$ if $B - A > 0$, and $B \geqslant A$ if $B > A$ or $B = A$. In other words, $B \geqslant A$ if for all $f \in H$,

$$(Bf, f) \geqslant (Af, f).$$

In particular, by the definition of the upper and lower bound of an operator A,

$$m(A) E \leqslant A \leqslant m'(A) E. \tag{6}$$

An operator A is said to be *non-negative* if $A \geqslant 0$, i.e. if $m(A) \geqslant 0$.

The relation $\geqslant$ *is a proper order relation in* $\mathfrak{A}_r(H)$ *which converts* $\mathfrak{A}_r(H)$ *into a vector order.*

Indeed, $\alpha A > 0$ and $A+B > 0$ if $A > 0$, $B > 0$ and $\alpha > 0$, so that by remarks in Sec. 1.7.1, $\mathfrak{A}_r(H)$, with the given order relation $\geqslant$, is an ordered vector system. Since, by virtue of (6) and (2), $\|A\|E$ is a non-negative majorant for any A in $\mathfrak{A}_r(H)$, $\mathfrak{A}_r(H)$ is a vector order (Sec. 1.7.2).

Equation (2) connects the norm with the order relation in $\mathfrak{A}_r(H)$. We have:

LEMMA 5.2.1. *A non-negative operator A satisfies the inequality* $A \leqslant \gamma E$ *if and only if* $\|A\| \leqslant \gamma$.

In fact, for non-negative A, (2) becomes

$$\|A\| = m'(A),$$

and the condition $A \leqslant \gamma E$ is equivalent to the inequality $m'(A) \leqslant \gamma$.

A vector $*$-system $\mathfrak{A}(H)$ can thus be considered to be a vector $*$-order.

A $*$-algebra A, with a given (proper) order relation in the set $\mathfrak{A}_r$ of its Hermitian elements, is said to be an ordered $*$-algebra if:

1) $\mathfrak{A}$ is a vector $*$-order,

2) $A^*A \geqslant 0$ for any $A \in \mathfrak{A}$,

3) $AB \geqslant 0$ when $A \geqslant 0$ and $B \geqslant 0$, and the elements A and B in $\mathfrak{A}$ commute.

Then the fact that $\mathfrak{A}(H)$ is an ordered $*$-algebra gives:

THEOREM 5.2.2. *The product of two non-negative operators in* $\mathfrak{A}(H)$ *which commute is a non-negative operator.*

For the proof of this theorem it will be necessary to prove.

THEOREM 5.2.3. *Any positive operator* $A \in \mathfrak{A}(H)$ *can be uniquely represented in the form*

$$A = C^2, \tag{7}$$

where C is a positive operator in $\mathfrak{A}(H)$.

The proof of Theorem 5.2.3 is based on the following lemma.

LEMMA 5.2.2. *If* $A \geqslant 0$, *and if* $(Af, f) = 0$, *then*

$$Af = 0. \tag{8}$$

In fact, for any complex α and any vector $h \in H$, we have (see (2.4.2))

$$(A(f+\alpha h), (f+\alpha h)) = (Af, f) + \bar{\alpha}(Af, h) + \alpha(Ah, f) + \alpha\bar{\alpha}(h, h)$$

and since $A \geqslant 0$ and $(Af, f) = 0$,

$$\bar{\alpha}(Af, h) + \alpha(Ah, f) + |\alpha|^2 \|h\|^2 \geqslant 0.$$

Whence, since α was arbitrary,

$$(Af, h) = 0,$$

which proves (8) since h is any element of the space H.

We turn now to the proof of Theorem 5.2.3. For the construction of the operator C which satisfies (7), without loss of generality it is possible to consider the operator A to be normed by the condition $\|A\| = 1$, and not to be equal to E. For $\gamma = 1$, it follows from Lemma 5.2.1 that whenever $A > 0$, the inequality $E - A > 0$ holds, whence it follows that

$$\|E - A\| \leqslant 1. \tag{9}$$

We consider the power series (Sec. 5.1. 2)

$$\varphi(z) = (1 - z)^{\frac{1}{2}} = 1 - \sum_{n=1}^{\infty} \left| \binom{1/2}{n} \right| z^n,$$

where by virtue of (5.1.29), $\sum_{n=1}^{\infty} \left|\binom{1/2}{n}\right| = 1$. Referring to Sec. 5.1.2, we form the operator

$$C = \varphi(E - A) = E - \sum \left| \binom{1/2}{n} \right| (E - A)^n. \tag{10}$$

By virtue of (5.1.27), $C^2 = E - (E - A) = A$.

To prove that $C > 0$, it is sufficient to show that the operator

$$D = \sum_{n=1}^{\infty} \left| \binom{1/2}{n} \right| (E - A)^n$$

is positive and that $\|D\| \leqslant 1$, since according to Lemma 5.2.1, the operator $C = E - D$ is also positive. Since whenever the operator $C - A$ is positive, the operator $(E - A)^n$ is also positive, the first assertion is obvious, and in view of (9)

$$\|(E - A)^n\| \leqslant \|E - A\|^n \leqslant 1,$$

and hence

$$\|D\| \leqslant \sum_{n=1}^{\infty} \left| \binom{1/2}{n} \right| = 1.$$

Now in addition to (7), let $A = C_1^2$, where C_1 is also a positive operator. Then $C_1^2 = C^2$, and if we let $f = (C_1 - C)h$ for arbitrary $h \in H$, then $(C_1 - C$ is a Hermitian operator)

$$\|f\|^2 = (f, (C_1 - C)h) = ((C_1 - C)f, h). \tag{11}$$

But $Cf = C_1 f = 0$. Indeed, $C_1 A = AC_1 = C_1^3$, and the operator C_1, which commutes with A, commutes with $E - A$ and with any power $(E - A)^n$. This means it commutes with the operator C defined by (10) (cf., Theorem 5.1.1). Therefore

$$(C+C_1)(C_1-C)=C_1^2-C^2=0$$

and

$$(Cf,\, f)+(C_1f,\, f)=((C+C_1)f,\, f)=((C_1^2-C^2)h,\, f)=0.$$

Whence, since $C \geqslant 0$ and $C_1 \geqslant 0$, it follows that

$$(Cf,\, f)=0, \qquad (C_1f,\, f)=0,$$

By Lemma 5.2.2, and since Cf = 0, it follows that $C_1 f = 0$. Hence (11) implies that $f = 0$, or that $C_1 h =$ Ch for any $h \in H$, i.e. $C_1 = C$.

REMARK. The argument which we used to prove that C_1 and C commute still holds if C_1 is replaced by any operator B in $\mathfrak{A}(H)$ which commutes with A. Thus, *an operator* $B \in \mathfrak{A}(H)$, *which commutates with a positive operator* $A \in \mathfrak{A}(H)$, *also commutes with the operator* $C = \sqrt{A}$.

Proof of Theorem 5.2.2. Let the operators $A \geqslant 0$ and $B \geqslant 0$ in $\mathfrak{A}(H)$ commute, and let $A = C^2$, where $C \geqslant 0$. Then the operator B commutes with C (see the last remark), and

$$(ABf,\, f)=(C^2Bf,\, f)=(CBCf,\, f)=(BCf,\, Cf) \geqslant 0,$$

since $B \geqslant 0$, i.e. $AB \geqslant 0$.

In $\mathfrak{A}(H)$, as in any ordered *-algebra, it is in some cases possible to multiply inequalities between Hermitian elements. For example, if $B \geqslant A$ for two operators A and B in $\mathfrak{A}_r(H)$, and if the operator $C \geqslant 0$ commutes with A and B, or perhaps only with their difference $B-A$, then

$$BC \geqslant AC, \qquad CB \geqslant CA. \tag{12}$$

Indeed, the non-negative operator $B-A$ commutes with C, $C \geqslant 0$, and hence

$$BC-AC=(B-A)C \geqslant 0, \qquad CB-CA=C(B-A) \geqslant 0.$$

Whence it follows that:

If the operators A, B, C, and D in $\mathfrak{A}_r(H)$ *mutually commute, if*

$$B \geqslant A, \qquad D \geqslant C \tag{13}$$

and if $A \geqslant 0$, $D \geqslant 0$ (*or even* $B \geqslant 0$, $C \geqslant 0$), *then*

$$BD \geqslant AC. \tag{14}$$

In fact, by multiplying the first of the inequalities in (13) by D and the second by A, it follows from (12) that

$$BD \geqslant AD, \qquad AD \geqslant AC,$$

whence (14) follows.

In particular, by multiplying the inequality $A \leqslant \|A\|E$ by A, we obtain*

$$A^2 \leqslant \|A\| A \tag{15}$$

for $A \geqslant 0$.

It follows from (15) that:

LEMMA 5.2.3. *If a sequence $\{A_n\}$ of non-negative operators in $\mathfrak{A}(H)$ is weakly convergent to zero, then it converges strongly.*

Indeed, it follows from (15) that

$$\| A_n f \|^2 = (A_n^2 f, f) \leqslant \| A_n \| (A_n f, f)$$

for any $f \in H$, and since $\{A_n\}$ converges weakly to zero, the norms $\|A_n\|$ are bounded and lim $(A_n f, f) = 0$. Therefore $A_n f \Rightarrow 0$, i.e. $A_n \Rightarrow 0$.

THEOREM 5.2.4. *If*

$$A_n \leqslant A_{n+1}, \quad A_n \leqslant C \in \mathfrak{A}_r(H),$$

for a sequence $\{A_n\}$ in $\mathfrak{A}(H)$, then $\{A_n\}$ is strongly convergent in $\mathfrak{A}(H)$.

In fact, for any $f \in H$

$$(A_n f, f) \leqslant (A_{n+1} f, f) \leqslant (Cf, f),$$

so that quadratic functionals converge everywhere, i.e. the sequence $\{A_n\}$ converges weakly to some operator $A \leqslant C$. Thus $A - A_n$ converges weakly to zero, and since $A - A_n \geqslant 0$, it follows from Lemma 5.2.3 that $A - A_n \Rightarrow 0$, i.e. $A_n \Rightarrow A$.

§ 3. Projection Operators

3.1. Properties of projection operators. The transformation P, which assigns to an element $f \in H$ its unique orothogonal projection $f' = Pf$ on a subspace H', is linear, i.e.

---

* Inequality (15) can be obtained in a more elementary manner. It is obvious that it is sufficient to prove it in the special case when $\| A \| = 1$. Under this condition, (15) follows from

$$A - A^2 = A(E - A)A + (E - A)A(E - A),$$

where both terms on the right-hand side are non-negative whenever A and $E - A$ are non-negative, since

$$A(E - A)Af, f) = ((E - A)Af, Af) \geqslant 0$$

and

$$((E - A)A(E - A)f, f) = (A(E - A)f, (E - A)f) \geqslant 0.$$

$$P(f+g)=Pf+Pg, \quad P(\alpha f)=\alpha Pf.$$

Indeed, let $f'=Pf$, $g'=Pg$, $f''=f-f'$, and $g''=g-g'$. Then f'' and g'' belong to the orthogonal complement H'' of the subspace H', and in the equations

$$f+g=(f'+g')+(f''+g''), \quad \alpha f=\alpha f'+\alpha f''$$

the elements $f'+g'$ and $\alpha f'$ belong to H', and the elements $f''+g''$ and $\alpha f''$ belong to the subspace H''. Therefore

$$P(f+g)=f'+g'=Pf+Pg, \quad P(\alpha f)=\alpha f'=\alpha Pf.$$

A linear operator P in H, with $\Omega_P=H$, is called a *projection operator*, or in other words a *projector of the subspace H'*.

If $f\in H'$ and $g\in H''$, then

$$Pf=f, \quad Pg=0, \tag{1}$$

and for $H'=H$ and $H''=H$, these equations show that the identity operator $E\,(Ef=f, f\in H)$ and the zero operator (which associates to each vector f in H the value (0)) are the projection operators of the subspaces H and (0). Since

$$f''=f-f'=f-Pf=(E-P)f,$$

the operator $E-P$ is the projection operator of the orthogonal complement H'' of H', and the representation

$$f=Pf+(E-P)f$$

is the decomposition of the vector f with respect to the orthogonal pair $[H', H'']$.

Thus

$$\|f\|^2=\|Pf\|^2+\|(E-P)f\|^2, \tag{2}$$

and hence

$$\|Pf\|\leqslant\|f\|, \tag{3}$$

so that the projection operator P is bounded, i.e. $P\in\mathfrak{A}(H)$. Here

$$\|P\|=1$$

in the normed space $\mathfrak{A}(H)$ if $H'\neq 0$, since in this case non-zero vectors $f\in H'$ exist such that $\|Pf\|=\|f\|$.

It also follows from (2) that the *equation*

$$\|Pf\|=\|f\| \tag{4}$$

holds if and only if $(E-P)f=0$, *i.e. when* $Pf=f$ *or* $f\in H'$. Therefore, for $f\neq 0$, the equation (4) is equivalent to the first of the equations in (1).

A non-zero projection operator P is a positive operator in $\mathfrak{A}(H)$.

Indeed, if P is the projection operator of the subspace H', and if $f = f' + f''$ is the decomposition of a vector $f \in H$ with respect to the orthogonal pair $[H', H'']$, then for $f' = Pf$,

$$(Pf, f) = (Pf, f' + f'') = (Pf, f') + (Pf, f'') = (Pf, Pf)$$

and the quadratic form

$$Q(f) = (Pf, f) = \| Pf \|^2 \tag{5}$$

which is generated by the bilinear form $W(f, g) = (Pf, g)$ is non-negative.

THEOREM 5.3.1. *An operator $P \in \mathfrak{A}(H)$ is a projection operator of some subspace H' if and only if it is Hermitian and satisfies the condition*

$$P^2 = P. \tag{6}$$

In fact, a projection operator P is Hermitian, and since $P^2 f = PPf = Pf$, it satisfies (6), i.e. the condition is necessary. Conversely, let P be a Hermitian operator which satisfies (6). We define H' to be the set $\Omega_P^{-1} = PH$ of all elements of the form Ph. H' is a subspace, and hence H' coincides with the subspace of elements f which satisfy the equation

$$Pf = f. \tag{7}$$

Indeed, if $f = Ph$, then

$$Pf = P^2 h = Ph = f.$$

We show that P is the projection operator of the subspace H'. Since

$$f = Pf + (E - P) f = f' + f''$$

for any vector $f \in H$, and since $f' = Pf \in H'$, it is only necessary to prove that the vectors f' and f'' are orthogonal. Since the operator P is Hermitian, it follows that

$$(f', f'') = (Pf, (E - P) f) = (f, P(E - P) f) = 0,$$

since $P(E-P)f = (P-P^2)f = 0$.

Two projection operators P_1 and P_2 are said to be *orthogonal* if $P_1 P_2 = 0$. This is equivalent to $P_2 P_1 = 0$, since $P_1 P_2 = 0$ implies that

$$P_2 P_1 = P_2^* P_1^* = (P_1 P_2)^* = 0,$$

and conversely, it follows from $P_2 P_1 = 0$ that $P_1 P_2 = 0$.

In particular, the projection operators P and $E-P$ are orthogonal: $P(E-P) = P - P^2 = 0$.

Two subspaces H_1 and H_2 are orthogonal if and only if their projection operators $P_1 = P_{H_1}$ and $P_2 = P_{H_2}$ are orthogonal.

In fact, if H_1 and H_2 are orthogonal, then for each $f \in H$, the vector $P_2 f \in H_2$ is orthogonal to H_1, and therefore $P_1 P_2 f = 0$, i.e. $P_1 P_2 = 0$. Conversely, if $P_1 P_2 = 0$, then

$$(f_1, f_2) = (P_1 f_1, P_2 f_2) = (f_1, P_1 P_2 f_2) = (f_1, 0) = 0$$

for any pair of elements $f_1 \in H_1$ and $f_2 \in H_2$, i.e. H_1 and H_2 are orthogonal.

THEOREM 5.3.2. *The sum $P_1 + P_2$ of the projection operators $P_1 = P_{H_1}$ and $P_2 = P_{H_2}$ is a projection operator if and only if P_1 and P_2 are orthogonal. If this condition is satisfied, then*

$$P_1 + P_2 = P_{H_1} + P_{H_2} = P_{H'},$$

where $H' = H_1 \oplus H_2$.

Indeed, if $P_1 P_2 = P_2 P_1 = 0$, then

$$(P_1 + P_2)^2 = P_1^2 + P_1 P_2 + P_2 P_1 + P_2^2 = P_1 + P_2.$$

Conversely, let $P_1 + P_2$ be a projection operator, i.e. let $(P_1 + P_2)^2 = P_1 + P_2$. Then

$$P_1 P_2 + P_2 P_1 = 0,$$

and multiplying this equation on the left by P_1:

$$P_1 P_2 + P_1 P_2 P_1 = 0, \tag{8}$$

and then multiplying Equation (8) on the right by P_1, we obtain $P_1 P_2 P_1 = 0$, which with (8) gives $P_1 P_2 = 0$, so that P_1 and P_2 are orthogonal. Now let $P_1 = P_{H_1}$ and $P_2 = P_{H_2}$ be orthogonal, and let $P = P_1 + P_2$. Then for any $h \in H$,

$$Ph = P_1 h + P_2 h \in H_1 \oplus H_2 = H',$$

If $f = f_1 + f_2$, where $f_1 \in H_1$ and $f_2 \in H_2$, and if f is any element in H', then $f_1 = P_1 f$, $f_2 = P_2 f$, and $f = P_1 f + P_2 f = Pf$. Thus, the domain of P coincides with H', and $P = P_1 + P_2$ is the projection operator of the subspace H'.

THEOREM 5.3.3. *The product of the projection operators $P_1 = P_{H_1}$ and $P_2 = P_{H_2}$ is a projection operator if and only if P_1 and P_2 commute. If this condition is satisfied, then $P = P_1 P_2$ is the projection operator of the intersection H' of the subspaces H_1 and H_2.*

In fact, the condition of commutativity is already necessary in order that the product $P_1 P_2$ be Hermitian. This condition is also sufficient, since if $P_1 P_2 = P_2 P_1$, then $P = P_1 P_2$ is a Hermitian operator (Theorem 1.5.2), and

$$P^2 = (P_1 P_2)^2 = P_1^2 P_2^2 = P_1 P_2 = P.$$

Furthermore, it follows from the equation $Ph = P_1P_2h = P_2P_1h$ that each vector f, for which $Pf = f$, belongs both to H_1 and to H_2, and that any element f which belongs to the intersection H' satisfies this equation. Hence, P is the projection operator of the subspace H'.

THEOREM 5.3.4. *The condition*

$$P_2P_1 = P_1 \tag{9}$$

is necessary and sufficient for the difference $P_2 - P_1$ of the projection operators $P_2 = P_{H_2}$ and $P_1 = P_{H_1}$ be a projection operator. If this condition is satisfied, then H_1 is a subspace of the space H_2, and $P_2 - P_1$ is the projection operator of the orthogonal complement H' of H_1 in H_2. ($H_2 = H_1 \oplus H'$.)

Indeed, it is obvious that whenever the operator $P_2 - P_1$ is a projection operator, the operator

$$E - (P_2 - P_1) = (E - P_2) + P_1 \tag{10}$$

is also a projection operator, and conversely. As we have seen, the right-hand side of (19) is a projection operator if and only if $(E - P_2)P_1 = 0$, i.e. if $P_1 = P_2P_1$. We now assume that (9) holds. Then $(P_2 - P_1)P_1 = 0$, so that P_1 and $P_2 - P_1$ are orthogonal, and since $P_2 = P_1 + (P_2 - P_1)$ and $H_2 = H_1 \oplus H'$, the operator $P_2 - P_1$ is the projection operator of the subspace H'.

3.2. The lattice of projection operators and orthogonal series of projection operators. The order relation $\geqslant$ in the system of Hermitian operators $\mathfrak{A}_r(H)$ induces an order relation in the set Π of projection operators in $\mathfrak{A}_r(H)$, and converts Π into an ordered set.

THEOREM 5.3.5. *The relation*

$$P_2 \geqslant P_1 \tag{11}$$

of subordination for projection operators P_1 and P_2 in Π is equivalent to the equation $P_2P_1 = P_1$.

Indeed, if $P_2P_1 = P_1$, then $P_2 - P_1$ is a projection operator, and hence $P_2 - P_1 \geqslant 0$ and (11) is satisfied. Conversely, for any $f \in H$, (11) implies that $(P_2f, f) \geqslant (P_1f, f)$ which is equivalent (by (5)) to the inequality

$$\|P_2f\| \geqslant \|P_1f\|. \tag{12}$$

For the vector $f = P_1h \in H_1$, (12) implies

$$\|P_2f\| \geqslant \|P_1f\| = \|f\|.$$

But since $\|P_2f\| \leqslant \|f\|$, we have $\|P_2f\| = \|f\|$ or $P_2f = f$. By substituting $f = P_1h$, we obtain $P_2P_1h = P_1h$ for any $h \in H$, i.e. $P_2P_1 = P_1$.

COROLLARY 1. *A projection operator P' which is orthogonal to a projection* *operator P is also orthogonal to any projection operator P_1 which is subordinate to P.*

Indeed, $P_1 = PP_1$ and

$$P'P_1 = P'PP_1 = 0P_1 = 0.$$

COROLLARY 2. *A projection operator $P_1 = P_{H_1}$ is subordinate to a projection operator $P_2 = P_{H_2}$ if and only if H_1 is a subspace of H_2.*

This corollary follows from Theorems 5.3.5 and 5.3.4.

According to Corollary 2, the one-to-one correspondence between subspaces $H' \in H$ and their projection operators $P = P_{H'}$, is an isomorphism between the ordered set Π and the lattice of subspaces of the space H. Since this lattice is bounded and complete (see the remark in Sec. 3.5), the following theorem holds:

THEOREM 5.3.6. *The set Π is a complete bounded lattice. If two projection operators P_1 and P_2 are orthogonal, then*

$$\sup \{P_1, P_2\} = P_1 + P_2, \tag{13}$$

and if they commute, then

$$\sup \{P_1, P_2\} = P_1 + P_2 - P_1P_2, \quad \inf \{P_1, P_2\} = P_1P_2. \tag{14}$$

In fact, (13) and the second of the equations in (14) are already known (P_1+P_2 is the projection operator of the sum, and P_1P_2 the projection operator of the intersection, of the corresponding subspaces). Furthermore,

$$\sup \{P_1, P_2\} = \sup \{P_1, P_2 - P_1P_2\},$$

since the left-hand side is the projection operator of the sum of the subspaces H_1 and H_2, whose projection operators are the operators P_1 and P_2. The right-hand side is the projection operator of the sum of the subspace H_1 and of the orthogonal complement of the intersection $H' = H_1 \cap H_2$ in H_2; moreover, these sums obviously coincide. However, the projection operator $P_2 - P_1P_2 = (E - P_1)P_2$ is orthogonal to P_1, and hence by (13)

$$\sup \{P_1, P_2\} = \sup \{P_1, P_2 - P_1P_2\} = P_1 + P_2 - P_1P_2.$$

Equation (13) immediately generalizes to a finite number of mutually orthogonal projection operators $P_1, P_2, \ldots, P_n$:

$$\sup \{P_1, P_2, \ldots, P_n\} = \sum_{k=1}^{n} P_k. \tag{13'}$$

Equation (13′) is also true for a countable sequence $\{P_k\}$ of mutually orthogonal projection operators. Namely, in that case

$$\sup_k \{P_k\} = \sum_{k=1}^{\infty} P_k, \tag{15}$$

where the *series on the right-hand side is strongly convergent in* $\mathfrak{A}(H)$. In fact, the operator

$$P'_n = \sum_{k=1}^{n} P_k = \sup\{P_1, P_2, \ldots, P_n\} \tag{16}$$

is the projection operator of the subspace

$$H'_n = \sum_{k=1}^{n} \oplus H_k,$$

where H_k is the subspace whose projection operator is P_k; moreover,

$$P'_n \leqslant P_{n+1}, \qquad P'_n \leqslant E$$

for all n. By virtue of Theorem 5.2.4, the sequence $\{P'_n\}$ converges strongly to a Hermitian operator P' which is also a projection operator, since under passage to the limit, the equation $(P'_n)^2 = P'_n$ still holds (continuity of multiplication with respect to strong convergence in $\mathfrak{A}(H)$). We will prove that the projection operator $P' = \sum_{k=1}^{\infty} P_k$ is equal to $\sup\{P_1, P_2, \ldots, P_n, \ldots\}$. In fact, $P'_n \leqslant P'$, and if $P'_n \leqslant C \in \mathfrak{A}(H)$, then

$$(P'f, f) = \lim_{n\to\infty} (P'_n f, f) \leqslant (Cf, f),$$

so that $P' \leqslant C$. Thus

$$P' = \sum_{k=1}^{\infty} P_k = \sup_n \{P'_n\}, \tag{15'}$$

and (15) follows from (16).

By virtue of (15), the operator $P' = \sum_{k=1}^{\infty} P_k$ is the projection operator of the subspace

$$H' = \sum_{k=1}^{\infty} \oplus H_k.$$

Remark. If, in the sense of strong (or weak) convergence,

$$P' = \sum_{k=1}^{\infty} P_k,$$

where the operator P' and all the operators P_k are projection operators, then the *projection operators* P_k *are mutually orthogonal.* Indeed

$$P' = P_n + \sum_{k\neq n} P_k = P_n + P''_n.$$

Now $P_n \leqslant P'$, and hence the operator $P''_n = P' - P_n$ is a projection operator which is orthogonal to P_n. But then (Corollary 1 of Theorem 5.3.5) P_n is orthogonal to any $P_k \leqslant P''_n$, i.e. the projection operators P_k are mutually orthogonal.

For a *non-denumerable* collection $\{P_\tau\}$ of mutually orthogonal projection operators, the series $\sum_\tau P_\tau$ can also be given a definite meaning. For each $f \in H$ only a finite or countable number of the elements $P_\tau f$ are different from zero, and

$$\sum_\tau \|P_\tau f\|^2 \leqslant \|f\|^2. \tag{17}$$

We now define the sum

$$P' = \sum_\tau P_\tau,$$

by letting

$$P'f = \sum_\tau P_\tau f, \tag{18}$$

for each $f \in H$, where the series on the right-hand side converges strongly by (17). P' is a projection operator, since

$$(P'f, f) = \overline{(P'f, f)}, \qquad (P')^2 f = P'f. \tag{19}$$

(For given f, the series (18) consists of not more than a countable number of terms, and passage to the limit shows that (19) holds.)

By comparing (18) with the definition of orthogonal sum of subspaces given in Sec. 3.5.1 ($f_\tau = P_\tau f$), we see that P' is a projection operator of the subspace

$$H' = \sum_\tau \oplus H_\tau,$$

where each subspace H_τ has P_τ as its projection operator. Since H' is the smallest subspace which contains all of the H_τ, by the isomorphism between Π and the lattice of the subspaces, we have (in the generalization (15))

$$\sup \{P_\tau\} = \sum_\tau P_\tau.$$

3.3. The aperture of subspaces. Let P_1 and P_2 be the projection operators of the subspaces H_1 and H_2. The number

$$D(H_1, H_2) = \|P_1 - P_2\|$$

is said to be the *aperture of the subspaces* H_1 and H_2.

The operators $E-P_1$ and $E-P_2$ are the projection operators of the orthogonal complements H'_1 and H'_2 of the subspaces H_1 and H_2, and since $(E-P_1)-(E-P_2) = P_2-P_1$, it follows that

$$D(H_1, H_2) = D(H'_1, H'_2). \tag{20}$$

By virtue the identity

$$P_1 - P_2 = (E - P_2) P_1 - P_2 (E - P_1)$$

we have the decomposition

$$(P_1 - P_2) h = f_1 - f_2, \tag{21}$$
$$f_1 = (E - P_2) P_1 h, \qquad f_2 = P_2 (E - P_1) h,$$

for any $f \in H$, where the elements f_1 and f_2 are orthogonal since f_2 belongs to the subspace H_2 and f_1 belongs to its orthogonal complement H_2'. Therefore

$$\| (P_1 - P_2) h \|^2 = \| f_1 \|^2 + \| f_2 \|^2, \tag{22}$$

and hence

$$D(H_1, H_2) = \sup_{h \in H} \frac{\sqrt{\| f_1 \|^2 + \| f_2 \|^2}}{\| h \|}. \tag{23}$$

From (21) we have

$$\| f_1 \| \leqslant \| P_1 h \|, \qquad \| f_2 \| \leqslant \| (E - P_1) h \|,$$

which implies by (22) that

$$\| (P_1 - P_2) h \|^2 \leqslant \| P_1 h \|^2 + \| (E - P_1) h \|^2 = \| h \|^2$$

and hence

$$D(H_1, H_2) \leqslant 1.$$

If $H_1 \supset H_2$, $H_1 \neq H_2$, then obviously $D(H_1, H_2) = 1$. If H_1 and H_2 are two lines in the plane, then $D(H_1, H_2) = \sin \alpha$, where α is the angle between the lines $\left(0 \leqslant \alpha \leqslant \frac{\pi}{2}\right)$.

LEMMA 5.3.1. *If the decompositions of some non-zero vector $h \in H$ with respect to the orthogonal pairs $[H_1, H_1']$ and $[H_2', H_2]$ coincide, then $D(H_1, H_2) = 1$.*

In fact, under the hypothesis,

$$P_1 h = (E - P_2) h = h_1, \qquad (E - P_1) h = P_2 h = h_2,$$
$$f_1 = (E - P_2) P_1 h = h_1, \qquad f_2 = P_2 (E - P_1) h = h_2,$$

so that

$$\| (P_1 - P_2) h \|^2 = \| h_1 \|^2 + \| h_2 \|^2 = \| h \|^2$$

for $h \neq 0$, and hence $\| P_1 - P_2 \| = 1$.

In particular, the condition of Lemma 5.3.1 is satisfied *if H_1 contains a non-zero vector h which is orthogonal to H_2*. Indeed, in this case $h = h + 0$ is a decomposition with respect to both of the orthogonal pairs $[H_1, H_1']$ and $[H_2', H_2]$.

THEOREM 5.3.7. *The dimensionalities of the subspaces H_1 and H_2 are identical if the aperture $D(H_1, H_2)$ is less than one.*

In fact, let the dimensionalities of H_1 and H_2 be unequal, say where the dimensionality of H_1 is greater than the dimensionality of H_2. Let $\mathfrak{B} = P_1H_2$ be the projection of H_2 on H_1. Since a fundamental set in H_2 is projected into a fundamental set of a subspace of $\overline{\mathfrak{B}}$, the dimensionality of $\overline{\mathfrak{B}}$ does not exceed the dimensionality of H_2, and hence it is less than the dimensionality of H_1. Therefore, a non-zero vector h can be found in H_1 which is orthogonal to $\mathfrak{B}$ and which is also orthogonal to H_2: for any $g \in H_2$,

$$(h, g) = (P_1h, g) = (h, P_1g) = 0,$$

since $P_1g \in \mathfrak{B}$. Now, by applying the special case of Lemma 5.3.1 to the subspaces H_1 and H_2, we obtain $D(H_1, H_2) = 1$, which contradicts the hypothesis of the theorem.

An upper bound for the aperture $D(H_1, H_2)$ is given in the following lemma.

LEMMA 5.3.2. *If*

$$\|g_1 - P_2g_1\| \leqslant \alpha\|g_1\|, \quad \|g_2 - P_1g_2\| \leqslant \alpha\|g_2\|, \tag{24}$$

for the projection operators P_1 and P_2 of the subspaces H_1 and H_2, where g_1 and g_2 are any elements in H_1 and H_2, respectively, then

$$D(H_1, H_2) \leqslant \alpha. \tag{25}$$

In fact, by letting $g_1 = P_1h$ for any $h \in H$, we have

$$f_1 = (E - P_2)P_1h = g_1 - P_2g_1$$

for an element f_1 in the decomposition (21), and hence by the first of the inequalities in (24),

$$\|f_1\| \leqslant \alpha\|P_1h\|. \tag{26}$$

We obtain

$$\|f_2\|^2 = (P_2(E - P_1)h, f_2) = \\ = ((E - P_1)h, f_2) = ((E - P_1)h, (E - P_1)f_2). \tag{27}$$

for a vector $f_2 = P_2(E-P_1)h$ in the same decomposition ($P_2f_2 = f_2$). By letting $g_2 = f_2 \in H_2$, we obtain the following estimate for the norm of the vector $(E-P_1)f_2 = g_2-P_1g_2$ from the second inequality in (24):

$$\|(E - P_1)f_2\| \leqslant \alpha\|f_2\|. \tag{28}$$

It now follows from (27) and (28) that

$$\|f_2\|^2 \leqslant \|(E - P_1)h\| \, \|(E - P_1)f_2\| \leqslant \alpha\|f_2\| \, \|(E - P_1)h\|.$$

By dividing by $\|f_2\|$, we finally obtain

$$\|f_2\| \leqslant \alpha \|(E - P_1) h\|. \tag{29}$$

By using the estimates (26) and (29) for the elements f_1 and f_2 in (22), we find

$$\|(P_1 - P_2) h\|^2 \leqslant \alpha^2 \|h\|^2$$

for any $h \in H$, or $\|(P_1 - P_2)h\| \leqslant \alpha \|h\|$, i.e. the inequality (25) follows.

REMARK. We let

$$\alpha_1 = \sup_{g_1 \in H_1} \frac{\|g_1 - P_2 g_1\|}{\|g_1\|}; \quad \alpha_2 = \sup_{g_2 \in H_2} \frac{\|g_2 - P_1 g_2\|}{\|g_2\|}.$$

Then it is possible to let $\alpha = \sup\{\alpha_1, \alpha_2\}$ in (24), hence

$$D(H_1, H_2) \leqslant \sup\{\alpha_1, \alpha_2\}. \tag{30}$$

On the other hand, restricting the vector h in (23) to range through elements g_1 in H_1, we find that $D(H_1, H_2) \geqslant \alpha_1$, and in view of the equality of H_1 and H_2, we also find that $D(H_1, H_2) \geqslant \alpha_2$. Thus $D(H_1, H_2) \geqslant \sup\{\alpha_1, \alpha_2\}$, which with (30) gives

$$D(H_1, H_2) = \sup\{\alpha_1, \alpha_2\}. \tag{31}$$

Equation (31) also serves as a *definition of the aperture*, and since

$$\alpha_1 = \sup_{\|g_1\|=1} d(g_1, H_2), \ g_1 \in H_1; \ \alpha_2 = \sup_{\|g_2\|=1} d(g_2, H_1), \ g_2 \in H_2,$$

where $d(f; M)$ is the distance of f from M, this definition can be extended to an arbitrary Banach space.

The concept of aperture can also be introduced for vector manifolds. Namely, for two vector manifolds $\mathfrak{B}_1$ and $\mathfrak{B}_2$ with closures $\overline{\mathfrak{B}}_1$ and $\overline{\mathfrak{B}}_2$, we have

$$D(\mathfrak{B}_1, \mathfrak{B}_2) = D(\overline{\mathfrak{B}}_1, \overline{\mathfrak{B}}_2).$$

§ 4. Some Classes of Linear Operators

4.1. Normal operators. *For an arbitrary operator A of the normed ring $\mathfrak{A}(H)$,*

$$\|A^*A\| = \|AA^*\| = \|A\|^2. \tag{1}$$

In fact (since $\|A^*\| = \|A\|$),

$$\|A^*A\| \leqslant \|A^*\| \, \|A\| = \|A\|^2. \tag{2}$$

At the same time,

$$\|Af\|^2 = (Af, Af) = (A^*Af, f) \leqslant \|A^*Af\| \, \|f\| \leqslant \|A^*A\| \, \|f\|^2,$$

i.e.

$$\|Af\| \leqslant \sqrt{\|A^*A\|}\,\|f\|,$$

which is equivalent to $\|A\| \leqslant \sqrt{\|A^*A\|}$, or to $\|A\|^2 \leqslant \|A^*A\|$. The last inequality together with (2) proves (1), since

$$\|AA^*\| = \|(A^*A)^*\| = \|A^*A\|.$$

If the operator A is Hermitian, then $A^* = A$, and hence

$$\|A^2\| = \|A\|^2. \tag{3}$$

An operator $A \in \mathfrak{A}(H)$ is said to be *normal* if it $*$-commutes with itself, i.e. if

$$A^*A = AA^*. \tag{4}$$

In particular, all Hermitian operators are normal. Equation (4) is equivalent to the condition

$$\|Af\| = \|A^*f\|, \qquad f \in H, \tag{5}$$

since

$$\|Af\|^2 = (A^*Af, f), \qquad \|A^*f\|^2 = (AA^*f, f). \tag{6}$$

Equation (3), *which is true for Hermitian operators, is also true for normal operators A*. In fact, by virtue of (5),

$$\|A^2f\| = \|AAf\| = \|A^*Af\|,$$

and hence

$$\|A^2\| = \|A^*A\|,$$

which, with (1) gives (3).

REMARK. Normal operators A are regular operators, i.e. $\omega(A) = \|A\|$ (Sec. 5.1.2). Indeed, since A^n is a normal operator, it follows from (3) that

$$\|A^{2^m}\| = \|A^{2^{m-1}}\|^2 = \dots = \|A\|^{2^m}$$

and hence

$$\omega(A) = \lim_{n\to\infty} (\|A^n\|)^{\frac{1}{n}} = \lim_{m\to\infty} \left(\|A^{2^m}\|^{\frac{1}{2^m}}\right) = \|A\|.$$

Thus, for normal operators,

$$\|A^n\| = \|A\|^n \tag{3'}$$

for any natural n.

4.2. Unitary operators. An operator $U \in \mathfrak{A}(H)$ is said to be *unitary* if the mapping

$$f \to f' = Uf$$

is an automorphism of the unitary space H. The definition of a unitary operator U is equivalent to the two requirements:

1) For all $f \in H$, the operator U in $\mathfrak{A}(H)$ satisfies the isometry condition

$$\|Uf\| = \|f\|; \tag{7}$$

2) the inverse operator U^{-1} is contained in $\mathfrak{A}(H)$.
It follows from (7) that $\|U\| = 1$, and that

$$(Uf, Ug) = (f, g) \tag{8}$$

is equivalent to (7) for any f and g in H, and, in turn, that (8) is equivalent to

$$U^*U = E. \tag{9}$$

Indeed,

$$(f, g) = (Uf, Ug) = (f, U^*Ug)$$

and $U^*Ug = g$, since $f \in H$ was arbitrary.

By virtue of (9), and since $U^{-1} \in \mathfrak{A}(H)$, it follows that

$$U^* = U^{-1}. \tag{10}$$

This equation also serves to define the fact that an operator U is unitary: it follows from (10) that $U^{-1} = U^* \in \mathfrak{A}(H)$, and furthermore that (9) holds, and thus (8) and (7) follow. Moreover, it follows from (10) that

$$UU^* = E, \tag{11}$$

but (9) and (11) are equivalent to (10), i.e. to the definition of unitary operator.

A unitary operator U is normal, since $U^*U = UU^* = E$. Since the product of unitary operators is again a unitary operator, unitary transformations form a group with respect to multiplication—*the rotation group in the geometry of a unitary space.*

4.3. Semi-unitary operators. The possibility that $U^{-1} \notin \mathfrak{A}(H)$ leads to a generalization of the concept of unitary operator. An operator $U \in \mathfrak{A}(H)$ is said to be *semi-unitary* if (7) holds for all $f \in H$. We also have $\|U\| = 1$ for a semi-unitary operator U. The product of two semi-unitary operators U and V is semi-unitary, since

$$\|UVf\| = \|U(Vf)\| = \|Vf\| = \|f\|;$$

in particular, whenever U is semi-unitary, the operators U^n, $n \geqslant 0$ are also semi-unitary.

Whenever (7) holds for a semi-unitary operator U, the equivalent equation (8) holds, and hence (9) also holds.

Thus, from (9), which can serve as the definition of a semi-unitary operator U, it follows that the operator U^* is a left inverse to U in the $*$-algebra $\mathfrak{A}(H)$. If a semi-unitary operator U is not unitary, then $U^* \neq U^{-1}$, and the operator

U^{-1}, whose domain is $\Omega_{U^{-1}} = UH = H_1$, does not belong to $\mathfrak{A}(H)$, i.e. $H_1 \neq H$.

For any vector $g = Uf \in H_1 \in \Omega_{U^{-1}}$, in view of (9), we have ($f = U^{-1}g$, $U^*Uf = U^*g$)

$$U^*g = U^{-1}g, \tag{12}$$

i.e.

$$U^* \supset U^{-1}. \tag{12'}$$

If $g \in H_0$, where $H = H_0 \oplus H_1$, then

$$U^*g = 0. \tag{13}$$

Indeed, $Uf \in H_1$ for any $f \in H$, and

$$(f, U^*g) = (Uf, g) = 0,$$

so that the element U^*g, which is orthogonal to H, is equal to zero. Both (12) and (13) can be combined by introducing the projection operator P_1 of the subspace H_1. Namely,

$$U^*g = U^{-1}P_1g$$

for any $g \in H$, or

$$U^* = U^{-1}P_1. \tag{14}$$

If U is a unitary operator, then $H_1 = H$, $P_1 = E$, and (14) reduces to (10).

Example. Let $\sum_{k=0}^{\infty} |a_k|^2 < \infty$ and let

$$f(z) = \sum_{k=0}^{\infty} a_k z^k$$

be a function which is analytic in the circle $|z| < 1$. We define the scalar product

$$(f, g) = \sum_{k=0}^{\infty} a_k \bar{b}_k,$$

where

$$f(z) = \sum_{k=0}^{\infty} a_k z^k, \qquad g(z) = \sum_{k=0}^{\infty} b_k z^k,$$

in the vector system $\mathfrak{F}$ of all such functions. This converts $\mathfrak{F}$ into a unitary space H which is isomorphic to $H(N_0)$, where N_0 is the set of natural numbers.

The transformation

$$Uf(z) = \tilde{f}(z) = zf(z) \tag{15}$$

defines a semi-unitary operator U in H. Indeed,

$$\tilde{f}(z) = \sum_{k=0}^{\infty} a_k z^{k+1} = \sum_{k=0}^{\infty} \tilde{a}_k z^k,$$

where $\tilde{a}_0 = 0$, $\tilde{a}_k = a_{k-1}$, $k \geqslant 0$, and hence

$$\| Uf \|^2 = \sum_{k=0}^{\infty} | \tilde{a}_k |^2 = \sum_{k=0}^{\infty} | a_k |^2 = \| f \|^2.$$

However the operator U is not unitary since $H_1 = UH$ does not coincide with H, namely: H_1 consists of those $f(z) \in H$ for which $a_0 = 0$, and thus H_0 consists of constants. Therefore

$$P_1 f = \sum_{k=1}^{\infty} a_k z^k$$

and hence

$$U^* f(z) = U^{-1} P_1 f = z^{-1} \sum_{k=1}^{\infty} a_k z^k,$$

or

$$U^* f = \sum_{k=0}^{\infty} a_{k+1} z^k.$$

Instead of the functions $f(z)$, their boundary values on the circle κ: $z = e^{i\theta}$ of radius one can be considered:

$$f(e^{i\theta}) = \lim_{r \to 1-0} f(re^{i\theta}).$$

Boundary values exist for almost all $z = e^{i\theta} \in \kappa$ with respect to Lebesgue measure $\mu(M)$, which we normalize by the condition $\mu(\kappa) = 1$.

We form the space $H^{(\mu)}$ (Sec. 3.2.2), in which the sequence $\{e^{ik\vartheta}\}$ ($k = 0, 1, 2, \ldots$) is an orthonormal system, and hence is an orthogonal basis of some space $H_+^{(\mu)} \subset H^{(\mu)}$. For any function $f(e^{i\vartheta}) \in H_+^{(\mu)}$ which is defined almost everywhere on κ,

$$f(e^{i\vartheta}) = \sum_{k=0}^{\infty} a_k e^{ik\vartheta},$$

$$a_k = \frac{1}{2\pi} \int_{\varkappa} f(e^{i\vartheta}) e^{-ik\vartheta} d\vartheta, \quad \| f \|^2 = \sum_{k=0}^{\infty} | a_k |^2,$$

so that $f(z) = f(re^{i\vartheta}) \to f(e^{i\vartheta})$ is an isomorphism of the space H onto $H_+^{(\mu)}$. Under this isomorphism, to an operator U in H there corresponds an isomorphic operator in $H_+^{(\mu)}$, which we also denote by U. For this last operator, all equations which were introduced above hold for the operator U in H if we replace z by $e^{i\vartheta}$. In particular, the definition (15) of the operator U is preserved:

$$\tilde{f}(e^{i\vartheta}) = Uf(e^{i\vartheta}) = e^{i\vartheta} f(e^{i\vartheta}). \tag{15'}$$

As is easily seen, in the space $H^{(\mu)}$ with orthogonal basis $e^{ik\vartheta}$ (k integer, $-\infty < k < \infty$), the operator U given by (15′) is unitary, and

$$U^* f = U^{-1} f = e^{-i\vartheta} f(e^{i\vartheta}).$$

§ 5. The Vector *-order $\mathfrak{A}(H)$

We show here that a vector *-order $\mathfrak{A}(H)$ has all the properties which we assumed to be satisfied in the theory of the integral, for extensions of linear operations, and for measures (Chapter 2). Moreover, some connections between order convergence in $\mathfrak{A}(H)$, and strong and uniform convergence in a normed ring $\mathfrak{A}(H)$, will be indicated.

THEOREM 5.5.1. *A vector *-order* $\mathfrak{A}(H)$ *is complete and has a sufficient number of completely linear positive functionals, i.e. it satisfies the condition* (1) *of Sec.* 1.9.2.

In fact, any set $\mathfrak{N}$ in $\mathfrak{A}_r(H)$ which is bounded below has a lower bound $\inf \mathfrak{N}$ if it is directed below (the completeness of $\mathfrak{A}_r(H)$). Let C be a minorant of the set $\mathfrak{N}$. Then for each $A \in \mathfrak{N}$ and for any $f \in H$, we have $(Cf, f) \leqslant (Af, f)$, and hence by letting

$$Q(f) = \inf_{A \in \mathfrak{N}} (Af, f),$$

for any $A \in \mathfrak{N}$, we can write

$$(Cf, f) \leqslant Q(f) \leqslant (Af, f). \tag{1}$$

If $Q(f) = (Bf, f)$, where $B \in \mathfrak{A}_\tau(H)$, then in view of (1), B is the greatest minorant of the set $\mathfrak{N}$, i.e. $B = \inf \mathfrak{N}$.

We show that $Q(f)$ is a quadratic functional. With this aim, and guided by (3.3.4), we form the expression

$$W(f, g) = Q(h_1) - Q(h_2) + iQ(h_3) - iQ(h_4), \tag{2}$$

in which the elements h_k are given by (cf., (3.4.3))

$$h_1 = \frac{f+g}{2}, \quad h_2 = \frac{f-g}{2}, \quad h_3 = \frac{f+ig}{2}, \quad h_4 = \frac{f-ig}{2}, \tag{3}$$

and we will show that $W(f, g)$ is a bilinear form. We prove first that for any finite number of pairs $[f_k, g_k]$, $k = 1, 2, \ldots, r$ of elements in H, there exists a sequence of operators $A_n \in \mathfrak{N}$ for which

$$W(f_k, g_k) = \lim_{n\to\infty} (A_n f_k, g_k). \tag{4}$$

In fact, for each f_k there exists a sequence $\{A_n^{(k)}\}$ in $\mathfrak{N}$ such that

$$\lim \left(A_n^{(k)} f_k, f_k\right) = Q(f_k). \tag{5}$$

For any n, an operator $A_n \in \mathfrak{N}$ can be found for which $A_n \leqslant A_n^{(k)}$ ($k = 1, 2, \ldots, r$), since the set $\mathfrak{N}$ is directed below. Therefore

$$Q(f_k) \leqslant (A_n f_k, f_k) \leqslant (A_n^{(k)} f_k, f_k),$$

and hence by (5)

$$\lim (A_n f_k, f_k) = Q(f_k). \tag{6}$$

for each k. Thus, our assertion is proved in the case when $g_k = f_k$. The general case is reduced to the special case if (6) is applied to the $4r$ elements $h_1^{(k)}$, $h_2^{(k)}$, $h_3^{(k)}$, and $h_4^{(k)}$, which are obtained by substituting the pairs $[f_k, g_k]$ instead of $[f, g]$ in (3), and then by using (2).

Now let $f = \alpha f' + \beta f''$, and let $\{A_n\}$ be a sequence in $\mathfrak{N}$ which satisfies

$$W(f', g) = \lim (A_n f, g), \quad W(f'', g) = \lim (A_n f'', g),$$
$$W(f, g) = \lim (A_n f, g).$$

for the three pairs $[f', g]$, $[f'', g]$, and $[f, g]$. Then

$$W(f, g) = \alpha \lim (A_n f', g) + \beta \lim (A_n f'', g) = \\ = \alpha W(f', g) + \beta W(f'', g).$$

Similarily, it is proved that

$$W(f, g) = \bar{\alpha} W(f, g') + \bar{\beta} W(f, g''),$$

if $g = \alpha g' + \beta g''$. Thus, $W(f, g)$ is a bilinear form and $Q(f) = W(f, f)$ is a quadratic form in H. By virtue of (1), $Q(f)$ is also bounded, i.e. it is a quadratic functional and can be represented in the form $Q(f) = (Bf, f)$, where the operator B belongs to $\mathfrak{A}_r(H)$.

By the same token, we note that for a set $\mathfrak{N} \subset \mathfrak{A}_r(H)$ which is directed below and bounded below, we have proved

$$\inf_{A \in \mathfrak{N}} (Af, f) = (Bf, f), \tag{7}$$

where $B = \inf \mathfrak{N}$.

In the vector order $\mathfrak{A}_r(H)$, each vector $f \in H$ determines a linear form

$$p(A) = (Af, f), \tag{8}$$

(A ranges through $\mathfrak{A}_r(H)$) which is positive for $f \neq 0$, since by definition $(Af, f) \geqslant 0$, if $A \geqslant 0$.

$p(A)$ is a completely linear functional (Sec. 1.9.1). Indeed, let $A(\tau)$ be a non-increasing function on a directed above set T with values in $\mathfrak{A}_r(H)$, and let $\inf A(\tau) = 0$. Then by (7)

$$\inf_{\tau} \{p(A(\tau))\} = \inf_{\tau} (A(\tau) f, f) = 0.$$

Finally, the set of functionals $p(A)$ of the form (8) is sufficient by the very definition of non-negative operator.

For the study of interrelations between convergence in a complete vector $*$-order $\mathfrak{A}(H)$ and in a normed ring $\mathfrak{A}(H)$, we confine ourselves to the following three theorems.

THEOREM 5.5.2. *A monotone sequence* $\{A_n\}$ *in* $\mathfrak{A}(H)$ *has an order limit if and only if it is strongly convergent.*

Indeed, for monotone sequences, order convergence coincides with weak convergence, and the theorem follows from Theorem 5.2.4.

THEOREM 5.5.3. *If a sequence* $\{A_n\}$ *in* $\mathfrak{A}(H)$ *has an order limit* A, *then it* $*$*-converges to* A, *i.e.* $A_n \overset{*}{\Rightarrow} A$.

In fact, let B_n be an ordered non-negative infinitely small sequence in $\mathfrak{A}(H)$, and let $\{C_n\}$ be a monotone infinitely small majorant of $\{B_n\}$. Then

$$0 \leqslant (B_n f, f) \leqslant (C_n f, f)$$

for any $f \in H$. Since $\inf\{C_n\} = \lim C_n = 0$, it follows from (7) that

$$0 \leqslant \overline{\lim}\,(B_n f, f) \leqslant \lim\,(C_n f, f) = 0.$$

Thus, $\lim\,(B_n f, f) = 0$ for any $f \in H$, i.e. $\{B_n\}$ converges weakly to 0, and by Lemma 5.2.3,

$$B_n \Rightarrow 0. \tag{9}$$

Equation (9) holds for any infinitely small sequence, since any infinitely small sequence can be represented in the form (cf. (1.8.1))

$$B_n = B_n^{(1)} - B_n^{(2)} + iB_n^{(3)} - iB_n^{(4)},$$

where the $B_n^{(k)}$ are non-negative infinitely small sequences, and hence according to (9), $B_n^{(k)} \Rightarrow 0$. Therefore

$$B_n^* = B_n^{(1)} - B_n^{(2)} - iB_n^{(3)} + iB_n^{(4)} \Rightarrow 0,$$

which, with (9), gives $B_n \overset{*}{\Rightarrow} 0$. For $B_n = A_n - A$, this is the assertion of Theorem 5.5.3.

THEOREM 5.5.4. *If a sequence* $\{A_n\}$ *in* $\mathfrak{A}(H)$ *has a uniform order limit* A, *then it converges uniformly in the normed ring* $\mathfrak{A}(H)$, *i.e.* $\lim \|A_n - A\| = 0$.

Indeed, we let $B_n = A_n - A$ and let $\{\lambda_n C\}$, $\lim \lambda_n = 0$, be an absolute majorant for the uniformly infinitely small sequence $\{B_n\}$. Then $\{\lambda_n \|C\| E\}$ is an absolute majorant for $\{B_n\}$ (since $C \leqslant \|C\| E$), and hence for $\|f\| = 1$,

$$-\lambda_n \|C\| \leqslant (B_n' f, f) \leqslant \lambda_n \|C\|, \quad -\lambda_n \|C\| \leqslant (B_n'' f, f) \leqslant \lambda_n \|C\|,$$

where B_n' and B_n'' are the Hermitian components in the representation $B_n = B_n' + iB_n''$. Therefore

$$\|B_n'\| = \sup_{\|f\|=1} |(B_n' f, f)| \leqslant \lambda_n \|C\|$$

and similarly

$$\|B_n''\| \leqslant \lambda_n \|C\|.$$

Hence, $\lim \|B_n'\| = \lim \|B_n''\| = 0$, and

$$\overline{\lim}\,\|B_n\| \leqslant \lim \|B_n'\| + \lim \|B_n''\| = 0.$$

§ 6. Completely Continuous Operators

6.1. An ideal of completely continuous operators. If an algebra (∗-algebra) A is normed, then we include the following requirement in the definition of an ideal (∗-ideal) in Sec. 1.5.4:

An ideal $\mathfrak{a}$ *is a closed subspace of the normed space* $\mathfrak{A}$.

If the space H is infinite dimensional, a normed ∗-algebra $\mathfrak{A}(H)$ has a non-trivial two-sided ∗-ideal of completely continuous operators whose definition will be given below.

For the bilinear functional (Af, g) which corresponds to an operator $A \in \mathfrak{A}(H)$,

$$\lim (Af_n, g_n) = (Af, g), \tag{1}$$

if one of the sequences $\{f_n\}$ or $\{g_n\}$ converges strongly and the other converges weakly to f or g, respectively (see the corollary of Theorem 4.4.2). An operator $A \in \mathfrak{A}(H)$, and the corresponding functionals (Af, g) and $Q(f) = (Af, f)$, are said to be *completely continuous* if (1) holds when both sequences $\{f_n\}$ and $\{g_n\}$ are weakly convergent, i.e. when

$$f_n \to f, \quad g_n \to g. \tag{2}$$

Conditions (1–2) are equivalent to

$$Af_n \Rightarrow Af, \quad \text{if} \quad f_n \to f. \tag{3}$$

Indeed, any operator A which satisfies (3) obviously satisfies the conditions (1–2) (see Lemma 4.3.1). Conversely, if the conditions (1–2) are satisfied, then in view of

$$\| Af_n - Af \|^2 = (Af_n, Af_n - Af) - (Af, Af_n - Af),$$

(3) is satisfied; in fact, if $f_n \to f$, then

$$Af_n \to Af, \quad g_n = Af_n - Af \to 0.$$

Thus, there exists two equivalent definitions of a completely continuous operator A:

1) if $f_n \to f$ and $g_n \to g$, then $\lim (Af_n, g_n) = (Af, g)$;

2) if $f_n \to f$, then $Af_n \Rightarrow Af$.

We note yet another definition:

3) an operator A is completely continuous if $f' = Af$ maps every bounded set into a set which is compact in H.

Definition 3) is equivalent to the two preceding ones. In fact, we assume that A is completely continuous in the sense of 2). We show that A then satisfies 3). Let M be a bounded set in H and let $g_n = Af_n, f_n \in M$, be an infinite sequence of elements from its image AM. From the bounded sequence $\{f_n\}$, it is possible to select (Theorem 4.3.9) a weakly convergent subsequence

$\{f_{n_k}\}$, and according to 2), the sequence $g_{n_k} = Af_{n_k}$ converges strongly. Hence the image AM is compact in H.

Conversely, let A be completely continuous in the sense of 3) and let $f_n \to f$. Then the sequence $\{f_n\}$ is bounded, the set $\{Af_n\}$ is compact in H, and

$$Af_n \to Af. \tag{4}$$

It is easy to see that

$$Af_n \Rightarrow Af. \tag{4'}$$

We assume the contrary. Then a subsequence Af_{n_k} exists such that

$$\| Af_{n_k} - Af \| \geqslant \delta > 0 \quad (k = 1, 2, \ldots).$$

By the compactness, a strongly convergent subsequence

$$g_{kj} = Af_{n_{kj}} \Rightarrow g_0,$$

can be extracted from the sequence $g_k = Af_{n_k}$; moreover, it is obvious that $g_0 \neq Af$. It is clear that $Af_{n_{kj}} \to g_0$ and $Af_{n_{kj}} \to Af$ (since $Af_n \to Af$). Since the limit $g_0 = Af$ is unique, we arrive at a contradiction which proves (4′).

THEOREM 5.6.1. *Completely continuous operators form a two-sided $*$-ideal in the normed $*$-algebra $\mathfrak{A}(H)$.*

For the proof of the theorem we successively verify the conditions of the definition of a $*$-ideal for the family $\mathfrak{a}$ of completely continuous operators in $\mathfrak{A}(H)$.

(a) *Whenever A and B belong to $\mathfrak{a}$, the operator $\alpha A + \beta B$ also belongs to $\mathfrak{a}$, i.e. $\mathfrak{a}$ is a vector subsystem of $\mathfrak{A}(H)$.* Indeed, let $f_n \to f$; then

$$Af_n \Rightarrow Af, \quad Bf_n \Rightarrow Bf$$

and hence

$$(\alpha A + \beta B) f_n \Rightarrow \alpha Af + \beta Bf = (\alpha A + \beta B) f.$$

(b) *If $A \in \mathfrak{a}$ and $B \in \mathfrak{A}(H)$, then the operators AB and BA are completely continuous.* Indeed, $f_n \to f$ implies that

$$Af_n \Rightarrow Af, \qquad Bf_n \to Bf$$

and

$$BAf_n \Rightarrow BAf, \quad ABf_n \Rightarrow ABf.$$

(c) *If A is completely continuous, A^* is also completely continuous.* This follows immediately from 1), since

$$(Af, g) = (f, A^*g).$$

Finally, we show that the vector manifold $\mathfrak{a}$ is closed in the space $\mathfrak{A}(H)$. In fact, let $A_m \in \mathfrak{a}$,

$$\lim \| A_m - A \| = 0 \tag{5}$$

and let $f_n \to f$. Then

$$\|Af_n - Af\| \leqslant \|Af_n - A_m f_n\| + \|A_m f_n - A_m f\| + \\ + \|A_m f - Af\| \leqslant \|A_m f_n - A_m f\| + \|A - A_m\|(\|f_n\| + \|f\|).$$

In view of (5), and since $\|f_n\| \leqslant \gamma$, the second term in the last sum can be made as small as desired for sufficiently large m, and for each fixed m the first term of the sum tends to zero for $n \to \infty$ since the operator A_m is completely continuous. Hence $A_m f_n \Rightarrow A_m f$, and therefore $\lim \|Af_n - Af\| = 0$, i.e. $Af_n \Rightarrow Af$ and $A \in \mathfrak{a}$ (the definition 2)). Hence, the theorem is proved.

The operator E (the identity of the algebra $\mathfrak{A}(H)$) is completely continuous if and only if the space H is finite dimensional. This is true since in an infinite dimensional space H there exists an orthonormal system $f_1, f_2, \ldots, f_n, \ldots$, and it is obvious that $f_n \to 0$, but $Ef_n = f_n \not\Rightarrow 0$ since $\|f_n\| = 1$. Whence it follows (Theorem 5.6.1) that in an infinite dimensional space, a completely continuous operator has neither a left nor a right inverse and thus does not have an inverse $A^{-1} \in \mathfrak{A}(H)$. If the space H is finite dimensional, then all operators in $\mathfrak{A}(H)$ are completely continuous, and the two sided $*$-ideal of completely continuous operators coincides with the whole algebra $\mathfrak{A}(H)$, i.e. it is trivial.

The following theorem gives a criterion for complete continuity of an operator.

THEOREM 5.6.2. *An operator A is completely continuous if and only if the operator $(A^*A)^k$ is completely continuous for some natural number k.*

The necessity of the condition is obvious. We first prove it is sufficient for $k = 1$. Let the operator A^*A be completely continuous and let $f_n \to f$. Then $A^*A(f_n - f) \Rightarrow 0$ (the definition 2)), and hence (see (5.4.6)),

$$\lim \|Af_n - Af\|^2 = \lim (A^*A(f_n - f), f_n - f) = 0,$$

i.e., $Af_n \Rightarrow Af$ and A is a completely continuous operator. Applying the above remarks to A^*A instead of to A, we see that whenever $(A^*A)^2$ is completely continuous, the operators A^*A and A are completely continuous. Similarly, we obtain the theorem for $k = 2^m (m > 1)$. In the general case let $k < 2^m$. Then whenever $(A^*A)^k$ is completely continuous, the operator $(A^*A)2^m$ is also completely continuous, and hence so is A.

6.2. Operators with a finite unitary norm. We assume that a non-negative operator A satisfies the condition

$$\sum_k (Ah_{\tau_k}, h_{\tau_k}) \leqslant \gamma < \infty, \tag{6}$$

with respect to a fixed orthogonal basis $\{h_\tau\}$ of the space H, where $\{h_{\tau_k}\}$ is any *finite* subset of the basis $\{h_\tau\}$, and where the constant γ depends only on A. It follows from (6) that $(Ah_\tau, h_\tau) \geqslant \varepsilon > 0$ only for a finite number of elements, and hence that (Ah_τ, h_τ) is different from zero for no more than a countable set of vectors h_τ. Therefore it is possible to form the series $\sum_\tau (Ah_\tau, h_\tau)$, and

$$\varphi(A) = \sum_\tau (Ah_\tau, h_\tau) \leqslant \gamma. \tag{6'}$$

We will show that a non-negative operator A also has the property (6) with respect to an arbitrary orthogonal basis $\{g_\sigma\}$, and that the value of $\varphi(A)$ does not depend on the choice of the basis $\{h_\tau\}$. If $A = C^2$, where $C \geqslant 0$ (Theorem 5.2.3), then

$$\sum_\tau (Ah_\tau, h_\tau) = \sum_\tau \|Ch_\tau\|^2 = \sum_\tau \sum_\sigma |(Ch_\tau, g_\sigma)|^2 \tag{7}$$

and the set of pairs σ, τ for which $(g_\sigma, Ch_\tau) = (Ch_\tau, g_\sigma) \neq 0$, is finite or countable. Therefore, it follows from

$$(Ag_\sigma, g_\sigma) = \|Cg_\sigma\|^2 = \sum_\tau |(Cg_\sigma, h_\tau)|^2$$

that $(Ag_\sigma, g_\sigma) \neq 0$ only when σ belongs to a finite or countable set, and it is possible to form the sum

$$\sum_\sigma (Ag_\sigma, g_\sigma) = \sum_\sigma \sum_\tau |(Cg_\sigma, h_\tau)|^2 = \sum_\tau \sum_\sigma |(Ch_\tau, g_\sigma)|^2.$$

By comparing it with (7), we obtain

$$\sum_\sigma (Ag_\sigma, g_\sigma) = \sum_\tau (Ah_\tau, h_\tau),$$

which proves our assertion with respect to the basis $\{g_\sigma\}$, and proves that $\varphi(A)$ is independent of the choice of the basis.

The number $\varphi(A)$, which thus depends only on the operator A, is by definition the *trace* of the non-negative operator A, and non-negative operators which satisfy (6) are said to be *operators with a finite trace.*

We have

$$\varphi(A) = \sum_\tau (Ah_\tau, h_\tau) = \sum_{\tau, \sigma} |(Ch_\tau, h_\sigma)|^2, \tag{8}$$

where $C \geqslant 0$ and $C^2 = A$. Non-negative operators A with a finite trace form a semivector system in which $\varphi(A)$ is a linear form:

$$\varphi(A+B) = \varphi(A) + \varphi(B), \qquad \varphi(\alpha A) = \alpha\varphi(A), \qquad \alpha \geqslant 0.$$

If the operators $A \geqslant 0$ and $B \geqslant 0$ commute, then the operator AB has a finite trace when at least one of the operators A or B has this property. Indeed, $AB \leqslant \|A\| B$ and $AB \leqslant \|B\| A$, so that (6) is satisfied for AB if it is satisfied for A or B.

In particular, whenever a non-negative operator A has a finite trace, all of the operators $A^n (n \geqslant 1)$ have a finite trace.

We consider the collection $\mathfrak{A}_u(H)$ of operators A in $\mathfrak{A}(H)$ for which the (non-negative) operator A^*A has a finite trace. For an operator $A \in \mathfrak{A}_u(H)$,

$$\varphi(A^*A) = \sum_\tau \|Ah_\tau\|^2 = \sum_{\tau, \sigma} |(Ah_\tau, h_\sigma)|^2, \tag{9}$$

where $\{h_\tau\}$ is an orthogonal basis of the space H.

THEOREM 5.6.3. *An operator A is completely continuous if it belongs to $\mathfrak{A}_u(H)$.*

In fact, let $A \in \mathfrak{A}_u(H)$, $f_n \to f$, and let

$$f = \sum_\tau \alpha_\tau h_\tau,$$
$$f_n = \sum_\tau \alpha_\tau^{(n)} h_\tau$$

be orthogonal decompositions of f and f_n with respect to $\{h_\tau\}$. Then $\|f_n\| \leqslant \gamma < \infty$,

$$|\alpha_\tau| = |(f, h_\tau)| \leqslant \|f\|, \quad |\alpha_\tau^{(n)}| \leqslant \gamma, \quad \lim_{n\to\infty} \alpha_\tau^{(n)} = \alpha_\tau, \tag{10}$$

and

$$\|Af_n - Af\|^2 = \sum_\tau |\alpha_\tau^{(n)} - \alpha_\tau|^2 \|Ah_\tau\|^2 =$$
$$= \sum{}' |\alpha_\tau^{(n)} - \alpha_\tau|^2 \|Ah_\tau\|^2 + \sum{}'' |\alpha_\tau^{(n)} - \alpha_\tau|^2 \|Ah_\tau\|^2, \tag{11}$$

where the sum $\sum'$ extends over a finite number of indices $\tau \in T'$ and $\sum''$ extends over the remaining indices. Since, by virtue of (9), the latter sum does not exceed $2(\gamma^2 + \|f\|^2) \sum'' \|Ah_\tau\|^2$, for a suitable choice of T', it can be made as small as desired. For such fixed T', the first sum on the right-hand side of (11) tends to zero by the limit equation in (10), so that $\lim \|Af_n - Af\| = 0$. Thus, it follows from $f_n \to f$ that $Af_n \Rightarrow Af$, and the operator $A \in \mathfrak{A}_u(H)$ is completely continuous.

REMARK. If A is a non-negative operator with a finite trace, then $C = \sqrt{A} \in \mathfrak{A}_u(H)$, and hence the operator $C = \sqrt{A}$ is completely continuous. Thus the operator $A = C^2$ is also completely continuous.

THEOREM 5.6.4. *$\mathfrak{A}_u(H)$ is a two-sided $*$-ideal of the $*$-algebra $\mathfrak{A}(H)$*, and*

$$\|A\|_u = \sqrt{\varphi(A^*A)}, \quad A \in \mathfrak{A}_u(H), \tag{12}$$

defines a unitary norm in the $*$-algebra $\mathfrak{A}_u(H)$ which converts the latter into a complete normed $*$-algebra.

PROOF. 1) *$\mathfrak{A}_u(H)$ is a vector subsystem of $\mathfrak{A}(H)$ in which $\|A\|_u$ is the norm.* Indeed, whenever $\mathfrak{A}_u(H)$ contains an operator A, it also contains the operator αA. Moreover, $\|\alpha A\|_u = |\alpha| \|A\|_u$. For operators A and B in $\mathfrak{A}_u(H)$,

$$\Big(\sum_{\tau,\sigma} |((A+B)h_\tau, h_\sigma)|^2\Big)^{1/2} \leqslant \Big(\sum_{\tau,\sigma} |(Ah_\tau, h_\sigma)|^2\Big)^{1/2} + \Big(\sum_{\tau,\sigma} |(Bh_\tau, h_\sigma)|^2\Big)^{1/2}$$

* We note that $\mathfrak{A}_u(H)$ *is not an ideal of the normed $*$-algebra* $\mathfrak{A}(H)$ since the set $\mathfrak{A}_u(H)$ is obviously not closed in the uniform norm in $\mathfrak{A}(H)$ (see Sec. 5.6.1).

(Cauchy-Bunyakovski inequality) so that (see (9)) whenever A and B belong to $\mathfrak{A}_u(H)$, the sum $A+B$ also belongs to $\mathfrak{A}_u(H)$, and (compare (12))

$$\|A+B\|_u \leqslant \|A\|_u + \|B\|_u. \tag{13}$$

Finally, if $\|A\|_u = 0$, then $Ah_\tau = 0$ for all h_τ, i.e. $A = 0$.

2) *The norm* $\|A\|_u$ *is unitary*. For operators A and B in $\mathfrak{A}_u(H)$, we let

$$W(A, B) = \sum_\tau (Ah_\tau, Bh_\tau), \tag{14}$$

where the inequalities

$$|(Ah_\tau, Bh_\tau)| \leqslant \|Ah_\tau\| \, \|Bh_\tau\|,$$

$$\sum_\tau \|Ah_\tau\| \|Bh_\tau\| \leqslant \left(\sum_\tau \|Ah_\tau\|^2\right)^{1/2} \left(\sum_\tau \|Bh_\tau\|^2\right)^{1/2},$$

imply that the series on the right-hand side converges absolutely, and

$$|W(A, B)| \leqslant \|A\|_u \|B_u\|. \tag{15}$$

$W(A, B)$ is a bilinear form in the (complex) vector system $\mathfrak{A}_u(H)$, and the unique quadratic form $W(A,A) = \sum_\tau \|Ah_\tau\|^2 = \phi(A^*A) = (\|A\|_u)^2$ generated by it is independent of the choice of the basis $\{h_\tau\}$.

3) *The transformation* $A \to A^*$ *is a pseudoinvolution of the normed space* $\mathfrak{A}_u(H)$. In fact

$$\begin{aligned} W(A, B) = \sum_\tau \sum_\sigma (Ah_\tau, h_\sigma)(h_\sigma, Bh_\tau) = \\ = \sum_\tau \sum_\sigma (B^*h_\sigma, h_\tau)(h_\tau, A^*h_\sigma), \end{aligned} \tag{16}$$

moreover, the series converges absolutely, since

$$\begin{aligned} \sum_{\tau, \sigma} |(Ah_\tau, h_\sigma)| \, |(h_\sigma, Bh_\tau)| \leqslant \\ \leqslant \left(\sum_{\tau, \sigma} |(Ah_\tau, h_\sigma)|^2\right)^{1/2} \left(\sum_{\tau, \sigma} |h_\sigma, Bh_\tau|^2\right)^{1/2}. \end{aligned}$$

Therefore it is possible to change the order of summation in (16), and hence

$$W(B^*, A^*) = W(A, B). \tag{17}$$

Letting $B = A$, we have

$$\varphi(AA^*) = W(A^*, A^*) = W(A, A) = \varphi(A^*A).$$

Thus, whenever $A \in \mathfrak{A}_u(H)$, the operator A^* also belongs to $\mathfrak{A}_u(H)$, and

$$\|A^*\|_u = \|A\|_u. \tag{18}$$

4) $\mathfrak{A}_u(H)$ *is complete with respect to the norm* $\|A\|_u$, *i.e. it is a unitary space with the scalar product*

$$(A, B) = W(A, B). \tag{19}$$

In fact, to each operator $A \in \mathfrak{A}_u(H)$ there corresponds a function

$$a(\sigma, \tau) = (Ah_\tau, h_\sigma),$$

which is different from zero for a finite or countable set of pairs of values σ and τ, and it satisfies the condition

$$\sum_{\sigma, \tau} |a(\sigma, \tau)|^2 < \infty. \tag{20}$$

Then

$$x'_\sigma = \sum_\tau a(\sigma, \tau) x_\tau, \tag{21}$$

where

$$x_\tau = (f, h_\tau), \quad x'_\sigma = (Af, h_\sigma). \tag{22}$$

To verify this it is sufficient to substitute the decomposition $f = \sum_\tau x_\tau h_\tau$ into the second equation in (22). Conversely, if $a(\sigma, \tau)$ is a function defined on the set $T \times T$, where T is the set of indices τ, and it satisfies (20), then (21) and (22) define an operator A in $\mathfrak{A}(H)$ $(Af = \sum_\sigma x'_\sigma h_\sigma)$, since

$$\sum_\sigma |x'_\sigma|^2 \leqslant \sum_\sigma \left(\sum_\tau |a(\sigma, \tau)|^2 \sum_\tau |x_\tau|^2\right) = \sum_\tau |x_\tau|^2 \sum_{\sigma, \tau} |a(\sigma, \tau)|^2 < \infty,$$

i.e.

$$\|Af\| \leqslant \|f\| \left(\sum_{\sigma, \tau} |a(\sigma, \tau)|^2\right)^{1/2}$$

and hence

$$\|A\| \leqslant \left(\sum_{\sigma, \tau} |a(\sigma, \tau)|^2\right)^{1/2}. \tag{23}$$

The operator A belongs to $\mathfrak{A}_u(H)$ since

$$\sum_\tau |(Ah_\tau, h_\tau)|^2 = \sum_{\sigma, \tau} |a(\sigma, \tau)|^2 < \infty. \tag{20'}$$

We will consider the functions $a(\sigma, \tau)$ to be elements of the unitary space $H(T \times T)$ (see Sec. 3.3.3 and 3.2.2) with the norm

$$\|a(\sigma, \tau)\| = \left(\sum_{\sigma, \tau} |a(\sigma, \tau)|^2\right)^{1/2}. \tag{24}$$

The one-to-one correspondence $A \to a(\sigma, \tau)$ thus established between elements of the normed spaces $\mathfrak{A}_u(H)$ and $H(T \times T)$ is obviously an isomorphism of the vector systems $\mathfrak{A}_u(H)$ and $H(T \times T)$. By (20′) and (24), it is also an isomorphism of the corresponding spaces. Since the (unitary) space $H(T \times T)$ is complete, the space $\mathfrak{A}_u(H)$, which is isomorphic to it, is also complete and has the unitary norm $\|A\|_u$.

5) *If $A \in \mathfrak{A}(H)$ and $B \in \mathfrak{A}_u(H)$, then the operators AB and BA belong to $\mathfrak{A}_u(H)$, and*

$$\|AB\|_u \leqslant \|A\| \, \|B\|_u, \quad \|BA\|_u \leqslant \|A\| \, \|B\|_u. \tag{25}$$

Indeed, since $A^*A \leqslant \|A^*A\| E$,

$$(ABh_\tau, ABh_\tau) = \sum_\tau (A^*ABh_\tau, Bh_\tau) \leqslant \| A^*A \| \sum_\tau (Bh_\tau, Bh_\tau)$$

and hence $AB \in \mathfrak{A}_u(H)$. Since $\|A^*A\| = \|A\|^2$ (see (5.5.1)), the first inequality of (25) holds. The second inequality is obtained by the application of the first inequality to A^*B^*; namely, by (18), and since $(BA)^* = A^*B^*$, we find

$$\| BA \|_u = \| A^*B^* \|_u \leqslant \| A^* \| \, \| B^* \|_u = \| A \| \, \| B \|_u.$$

6) $\mathfrak{A}_u(H)$ *is a complete normed $*$-algebra.* From 1), 3) and 4), $\mathfrak{A}_u(H)$ is a $*$-algebra which is complete with respect to the norm (12). It remains to prove that this norm satisfies

$$\| AB \|_u \leqslant \| A \|_u \| B \|_u. \tag{26}$$

But this follows from the first equation in (25) if not only B, but also A, belongs to $\mathfrak{A}_u(H)$, since in this case it is possible to use (23) which, in view of (24) and since $\|a(\sigma, \tau)\| = \|A\|_u$, can be written in the form

$$\| A \| \leqslant \| A \|_u. \tag{23'}$$

Assertions 5) and 6) together with 2) completely prove Theorem 5.6.4.

We will say that the norm $\|A\|_u$ is the *unitary norm of the operator A*, and that operators in $\mathfrak{A}_u(H)$ are *operators with finite unitary norms.*

REMARK. The fact that $\mathfrak{A}_u(H)$ is both a normed $*$-algebra and a unitary space adds a special completeness to the results of the theory of operators with finite unitary norm.

In particular, if we consider a functional equation of the form

$$f - zAf = g$$

for an operator A in $\mathfrak{A}_u(H)$, then in order to study the convergence in the normed algebra $\mathfrak{A}_u(H)$ of the power series $\sum_{k=1}^{\infty} A^k z^k$, which is related to the operator $(E - zA)^{-1}$ (see (5.1.33–33')), we must replace $\omega(A)$ by

$$\omega_u(A) = \lim_{n \to \infty} (\| A^n \|_u)^{\frac{1}{n}} = \lim_{n \to \infty} (\varphi((A^*)^n A^n))^{\frac{1}{2n}}. \tag{27}$$

The radius of convergence $\rho' = \dfrac{1}{\omega_u(A)}$ of the above series with respect to the unitary norm in $\mathfrak{A}_u(H)$ does not exceed the radius of convergence $\rho = \dfrac{1}{\omega_u(A)}$ in $\mathfrak{A}(H)$, since $\|A^n\| \leqslant \|A^n\|_u$ (compare (23')). If the operator $A \in \mathfrak{A}_u(H)$ is Hermitian, then (27) becomes

$$\omega_u(A) = \lim_{n \to \infty} (\varphi(A^{2n}))^{\frac{1}{2n}}. \tag{28}$$

In the case of a finite dimensional space H, all operators in $\mathfrak{A}(H)$ have a finite unitary norm, and for the study of continuous linear transformations, it is possible to start with $\mathfrak{A}_u(H)$ instead of $\mathfrak{A}(H)$. If the space H has countable dimensionality, then the representation (21)–(22) of an operator $A \in \mathfrak{A}_u(H)$ with respect to an orthogonal basis $\{h_n\}$ takes the form $(a_{mn} = a(m, n))$

$$x'_m = \sum_{n=1}^{\infty} a_{mn} x_n, \tag{29}$$

where

$$x_n = (f, h_n), \quad x'_m = (Af, h_m), \quad a_{mn} = (Ah_n, h_m), \tag{30}$$

and the matrix $[a_{mn}]$ satisfies the condition

$$\sum_{m,n} |a_{mn}|^2 < \infty. \tag{31}$$

Conversely, if a given matrix $[a_{mn}]$ satisfies (31), then (29) and (30) define an operator $A \in \mathfrak{A}_u(H)$. The continuous analog of such a matrix operator is the integral operator which, in the classical case of the space $H = L_2(\Delta)$, has the form

$$\tilde{f}(t) = Af(t) = \int_\Delta a(t, s) f(s)\, ds, \tag{32}$$

where $f(t)$ and $g(t)$ belong to $L_2(\Delta)$, $\Delta = (a, b)$. Moreover, (31) is now replaced by

$$\int_\Delta \int_\Delta |a(t, s)|^2\, dt\, ds < \infty, \tag{31'}$$

i.e. the function $a(t, s)$ must be Lebesgue integrable together with $|a(t, s)|^2$ on the square $\Delta \times \Delta$. The operator A defined by (32) (with the condition (31')) belongs to $\mathfrak{A}_u(L_2(\Delta))$, and hence it is completely continuous. This can be shown by using a matrix representation of the integral operator A with respect to an orthogonal basis $\{h_n(t)\}$ in $L_2(\Delta)$, namely

$$a_{mn} = \int_\Delta \int_\Delta a(t, s) h_n(t) \bar{h}_m(s)\, dt\, ds,$$

$$\sum_{m,n} |a_{mn}|^2 = \int_\Delta \int_\Delta |a(t, s)|^2\, dt\, ds < \infty.$$

It is easy to show that the operator which is adjoint to the integral operator (32) is given by

$$\tilde{g}(t) = A^* g(t) = \int_\Delta \overline{a(s, t)}\, g(s)\, ds, \tag{32'}$$

and hence the operator A is Hermitian if and only if

$$\overline{a(s, t)} = a(t, s) \tag{33}$$

almost everywhere in the square $\Delta\times\Delta$. In particular, for real $a(s, t)$, the condition (33) means that the kernel $a(s, t)$ is symmetric: $a(s, t) = a(t, s)$.

6.3. The trace. For the product AB of two operators A and B in $A_u(H)$, the series

$$\sum_{\tau}(ABh_{\tau}, h_{\tau}) = \sum_{\tau}(Bh_{\tau}, A^*h_{\tau}) = (B, A^*),$$

where $\{h_{\tau}\}$ is an orthogonal basis of H (see (14) and (19)), converges absolutely and its sum is independent of the choice of the basis $\{h_{\tau}\}$. *The trace* $\varphi(AB)$ *of the operator* AB ($A\in\mathfrak{A}_u(H)$, $B\in\mathfrak{A}_u(H)$) is defined by.*

$$\varphi(AB) = \sum_{\tau}(ABh_{\tau}, h_{\tau}) = (B, A^*). \tag{34}$$

Then (see (14))

$$(A, B) = \varphi(B^*A) \tag{35}$$

and by virtue of (15), (19), and $\|A^*\|_u = \|A_u\|$,

$$|\varphi(AB)| \leqslant \|A\|_u \|B\|_u. \tag{36}$$

The trace $\varphi(AB)$ is a linear functional in the unitary space $A_u(H)$ with respect to each of the arguments A and B, and moreover, it has the following properties:

$$\varphi(BA) = \varphi(AB); \quad \varphi(A^*B^*) = \overline{\varphi(AB)}. \tag{37}$$

Indeed, by virtue of (34), (19), and (17)

$$\varphi(AB) = (B, A^*) = (A, B^*) = \varphi(BA),$$

$$\varphi(A^*B^*) = \sum_{\tau}(A^*B^*h_{\tau}, h_{\tau}) = \sum_{\tau}(h_{\tau}, BAh_{\tau}) = \overline{\varphi(BA)}.$$

We now consider the linear hull $\mathfrak{A}_s(H)$ of the set of all products AB (A and B in $\mathfrak{A}_u(H)$). For the operator

$$C = \sum_{k=1}^{n} c_k A_k B_k \quad (A_k \in \mathfrak{A}_u(H), \ B_k \in \mathfrak{A}_u(H)) \tag{38}$$

from $\mathfrak{A}_s(H)$, we define the *trace* $\varphi(C)$ by

$$\varphi(C) = \sum_{k=1}^{n} c_k \varphi(A_k B_k). \tag{39}$$

Since

$$\varphi(C) = \sum_{k=1}^{n} c_k \sum_{\tau}(A_k B_k h_{\tau}, h_{\tau}) = \sum_{\tau}(Ch_{\tau}, h_{\tau}),$$

the definition (39) is correct, since it is independent of the choice of the

* The definition given in Sec. 6.2 for the trace of a non-negative operator which satisfies (6) obviously agrees with this more general definition of trace, since $C \geqslant 0$ and $\varphi(C) < +\infty$ imply that $C = \sqrt{C}\cdot\sqrt{C}$, where $\sqrt{C}\in\mathfrak{A}_u(H)$ and (34) coincides with (6′).

representation (38) of the operator $C \in \mathfrak{A}_s(H)$. For the trace $\varphi(C)$, we have

$$\varphi(C_1 + C_2) = \varphi(C_1) + \varphi(C_2), \quad \varphi(\alpha C) = \alpha\varphi(C),$$
$$\varphi(C^*) = \overline{\varphi(C)}. \tag{40}$$

The first two of these are obvious, and with respect to the third, we note that whenever the vector system $\mathfrak{A}_s(H)$ contains the operator C which is given by (38), it also contains the operator $C^* = \sum_{k=1}^{n} \bar{c}_k B_k^* A_k^*$ whose trace is

$$\varphi(C^*) = \sum_{k=1}^{n} \overline{c_k} \varphi(B_k^* A_k^*) = \overline{\sum_k c_k \varphi(A_k B_k)} = \overline{\varphi(C)}.$$

$\mathfrak{A}_s(H)$ is a two-sided $*$-ideal of the $*$-algebra $\mathfrak{A}(H)$.

Indeed, $\mathfrak{A}_s(H)$ is a vector $*$-subsystem of $\mathfrak{A}(H)$, and if $A \in \mathfrak{A}(H)$, then

$$AC = \sum_k c_k (AA_k) B_k \in \mathfrak{A}_s(H),$$
$$CA = \sum_k c_k A_k (B_k A) \in \mathfrak{A}_s(H),$$

since

$$AA_k \in \mathfrak{A}_u(H) \text{ and } B_k A \in \mathfrak{A}_u(H).$$

Thus, for any $A \in \mathfrak{A}(H)$ and $C \in \mathfrak{A}_s(H)$, it is possible to form the traces $\varphi(AC)$ and $\varphi(CA)$ which are oviously linear forms with respect to each of the arguments A and C. Moreover, they satisfy the equations

$$\varphi(CA) = \varphi(AC), \quad \varphi(A^* C^*) = \overline{\varphi(AC)}. \tag{41}$$

In fact, for the operators A_k and B_k in the representation (38) of the operator C, we have

$$\varphi(A_k B_k A) = \varphi(B_k A A_k), \quad \varphi(A A_k B_k) = \varphi(B_k A A_k),$$

by virtue of (37). Whence the first of the equations in (41) follows, and the second equation follows from (40) since $(CA)^* = A^* C^*$.

§ 7. Matrix Representation of Linear Operators

The derivation of the representation (5.6.21) remains valid for any operator $A \in \mathfrak{A}(H)$, since only the continuity of the operator A and of the scalar product (f, g) was used in this derivation. This representation, which thus holds for any $A \in \mathfrak{A}(H)$, will be considered in this section for the case of a countable dimensional space H (with respect to the orthogonal basis $\{h_n\}$). According to (5.6.29–30),

$$x'_m = \sum_{n=1}^{\infty} a_{mn} x_n, \tag{1}$$

$$x_n = (f, h_n), \quad x'_m = (Af, h_m), \quad a_{mn} = (Ah_n, h_m). \tag{1'}$$

Hence, we have

$$f=\sum_{n=1}^{\infty} x_n h_n, \quad Af=\sum_{m=1}^{\infty} x'_m h_m. \tag{2}$$

The matrix $[a_{mn}]$ is by definition the *matrix of the operator* $A \in \mathfrak{A}(H)$ with respect to the basis $\{h_n\}$. The matrices $[a_{mn}+b_{mn}]$ and $[\alpha a_{mn}]$, where $[a_{mn}]$ and $[b_{mn}]$ are the matrices of the operators A and B, correspond (for a fixed basis) to the operators $A+B$ and αA, respectively. Since

$$(A^* h_n,\ h_m)=(h_n,\ Ah_m)=\overline{(Ah_m,\ h_n)}$$

the adjoint matrix

$$[a^*_{mn}]=[\bar{a}_{nm}],$$

corresponds to the adjoint operator A^*, and hence a Hermitian matrix $[a_{mn}]$ corresponds to a Hermitian operator:

$$a_{nm}=\bar{a}_{mn}.$$

Finally, the matrix $[c_{mn}]$, where

$$c_{mn}=\sum_{k=1}^{\infty} a_{mk} b_{kn}$$

corresponds to the operator $C = AB$. Indeed,

$$c_{mn}=(ABh_n,\ h_m)=(Bh_n,\ A^*h_m)=$$

$$=\sum_{k=1}^{\infty}(Bh_n,\ h_k)\overline{(A^*h_m,\ h_k)}=\sum_{k=1}^{\infty} a_{mk} b_{kn}.$$

The bilinear functional (Af, g) can now be represented in the form

$$(Af,\ g)=\sum_{m=1}^{\infty}\sum_{n=1}^{\infty} a_{mn} x_n \bar{y}_m, \tag{3}$$

where $y_m = (g, h_m)$. In fact,

$$(Af,\ g)=\sum_{m=1}^{\infty}(Af,\ h_m)\overline{(g,\ h_m)}=\sum_{m=1}^{\infty} x'_m \bar{y}_m.$$

In the sum (3), it is possible to interchange the order of summation on m and n, since

$$(Af,\ g)=(f,\ A^*g)=\overline{(A^*g,\ f)}=\sum_{n=1}^{\infty}\sum_{m=1}^{\infty} a_{mn} x_n \bar{y}_m. \tag{3'}$$

The matrix $[a_{mn}]$, which we associate with the operator A, satisfies the condition (5.6.31)

$$\sum_{m,\ n=1}^{\infty} |a_{mn}|^2 < +\infty$$

if and only if A has a finite unitary norm. In the general case, *the matrix* $[a_{mn}]$ *is a matrix of the operator* A *in* $\mathfrak{A}(H)$ *whose norm* $\|A\|$ *does not exceed*

γ *if and only if for any complex values of the variables* $x_1, x_2, \ldots, x_N; y_1, y_2, \ldots, y_M$,

$$\left|\sum_{m=1}^{M}\sum_{n=1}^{N} a_{mn}x_n\bar{y}_m\right| \leqslant \gamma\left(\sum_{n=1}^{N}|x_n|^2\right)^{1/2}\left(\sum_{m=1}^{M}|y_m|^2\right)^{1/2}. \tag{4}$$

In fact, (4) coincides with

$$|(Af_N, g_M)| \leqslant \gamma\|f_N\|\,\|g_M\|,$$

where

$$f_N = \sum_{n=1}^{N} x_n h_n, \qquad g_M = \sum_{m=1}^{M} y_m h_m,$$

$$\|f_N\|^2 = \sum_{n=1}^{N}|x_n|^2, \quad \|g_M\|^2 = \sum_{m=1}^{M}|y_m|^2,$$

if (3), with $f = f_N$ and $g = g_M$, is used for (Af_N, g_M). Thus, the condition (4) is necessary. We now assume that (4) is satisfied. By letting

$$M = N, \qquad y_m = \sum_{k=1}^{N} a_{mk}x_k,$$

we substitute these values of y_m in (4). Dividing both sides of (4) by $\left(\sum_{m=1}^{N}\left|\sum_{n=1}^{N} a_{mn}x_n\right|^2\right)^{1/2}$, we obtain

$$\left(\sum_{m=1}^{N}\left|\sum_{n=1}^{N} a_{mn}x_n\right|^2\right)^{1/2} \leqslant \gamma\left(\sum_{n=1}^{N}|x_n|^2\right)^{1/2}. \tag{5}$$

On the other hand, by the convergence of the series $\sum_{n=1}^{\infty}|x_n|^2$ $(= \|f\|^2)$, it follows from (5) that

$$\lim \sum_{n=p}^{q} a_{mn}x_n = 0 \qquad (p, q \to \infty).$$

Therefore the series $\sum_{n=1}^{\infty} a_{mn}x_n$ converges, and it follows from (5) that

$$\left(\sum_{m=1}^{\infty}\left|\sum_{n=1}^{\infty} a_{mn}x_n\right|^2\right)^{1/2} \leqslant \gamma\left(\sum_{n=1}^{\infty}|x_n|^2\right)^{1/2}$$

or that $\|Af\| \leqslant \gamma\|f\|$, only if (2) and (1) hold, i.e. if an operator $A \in \mathfrak{A}(H)$ with $\|A\| \leqslant \gamma$ corresponds to the matrix $[a_{mn}]$.

REMARK. It follows from $f_N \Rightarrow f$ and $g_M \Rightarrow g$ that $\lim (Af_N, g_M) = (Af, g)$ for $M \to \infty$, $N \to \infty$, and hence, in addition to (3) and (3′), we have the representation

$$(Af, g) = \lim_{\substack{M\to\infty \\ N\to\infty}} \sum_{m=1}^{M}\sum_{n=1}^{N} a_{mn}x_n\bar{y}_m. \tag{3″}$$

A matrix $[a_{mn}]$ which satisfies (4) ($\gamma < \infty$), is said to be *bounded*.

For a fixed orthogonal basis $\{h_n\}$ of the space H, the mapping

$$f \to \{x_n\} = \{(f, h_n)\}$$

is an isomorphism of the space H onto the space $H(N_0)$, where N_0 is the set of natural numbers (see Sec. 3.4.4, Theorem 3.4.6), and the sequence $e_n = \{\delta_{kn}\}$, $\delta_{kn} = 0$, $n \neq k$; $\delta_{nn} = 1$, corresponds to an element h_k of the basis. Therefore, for a given *bounded* matrix $[a_{mn}]$, (1) immediately defines an operator A in the unitary space of sequences $H = H(N_0)$, and the operator A belongs to $\mathfrak{A}(H)$. Such an operator will be called a *matrix operator*. It coincides with its matrix representation with respect to the basis $\{e_n\}$.

A matrix representation of the operator A in $\mathfrak{A}(H)$ (H countable dimensional) can be considered to be a matrix operator in the space $H(N_0)$ which is isomorphic to A. Under the transition to matrix operators, operations in the $*$-algebra $\mathfrak{A}(H)$ (vector operations, multiplication, $*$-operation) are written, as we have seen, in simple matrix form, and hence the theory of operators in $\mathfrak{A}(H)$ can be presented in purely matrix form. Hence matrix calculus for matrices of finite order which correspond to a finite dimensional space H has been generalized.